Plant Pigments

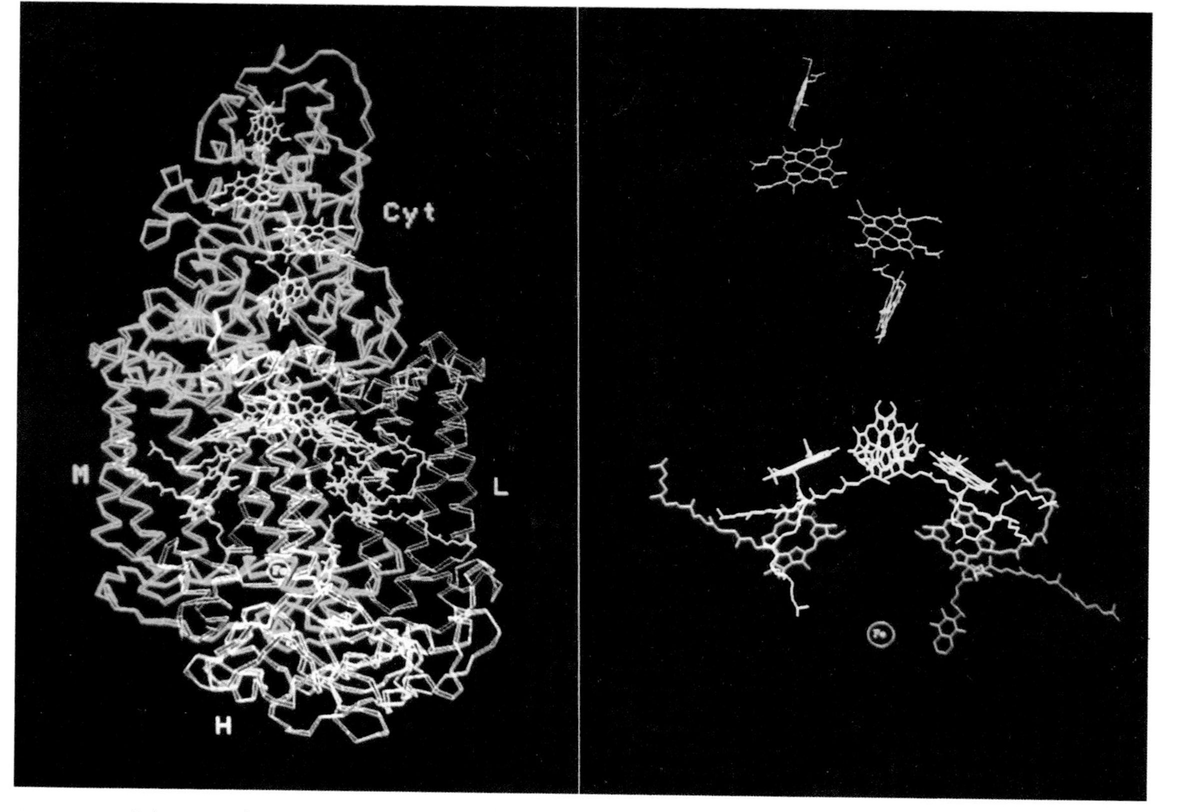

The structure of the reaction centre from *Rps viridis*. Left, the complete structure; right, the arrangement of the chromophores alone (orange = hemes; yellow = BChl b; blue = Bpheo b). See Chapter 4, Fig. 4.7B for details.

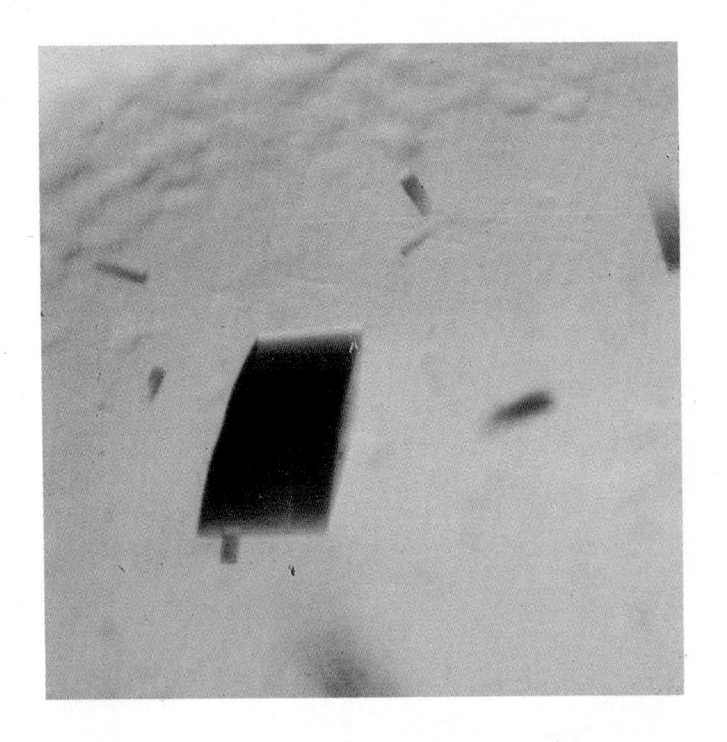

Crystals of reaction centres from *Rhodopseudomonas viridis*. See Chapter 4, Fig. 4.7A for details.

Plant Pigments

edited by

T. W. Goodwin
Emeritus Professor
Department of Biochemistry
University of Liverpool

1988

ACADEMIC PRESS
Harcourt Brace Jovanovich, Publishers
London San Diego New York Berkeley Boston
Sydney Tokyo Toronto

ACADEMIC PRESS LIMITED
24–28 Oval Road
London NW1 7DX

U.S. Edition published by
ACADEMIC PRESS INC.
San Diego, CA 92101

British Library Cataloguing in Publication Data

Goodwin, T.W. (Trevor Walworth)
Plant pigments.
I. Plants. Pigments. Chemical composition
I. Title
581.19′218

ISBN 0–12–289847–8

Typeset by Photo-Graphics
and printed in Great Britain by T.J. Press (Padstow) Ltd.,
Padstow, Cornwall

Contributors

George Britton *Department of Biochemistry, University of Liverpool, P.O. Box 147, Liverpool L69 3BX, U.K.*

Richard Cogdell *Department of Botany, University of Glasgow, Glasgow G12 8QQ, Scotland.*

T. W. Goodwin *9 Woodlands Close, Parkgate, S. Wirral, Cheshire L64 4RU, U.K.*

Jeffrey B. Harborne *Plant Science Laboratories, University of Reading, Reading, Berkshire RG6 2AH, U.K.*

W. Rau *Botanisches Institut, Universität München, Menzingerstr. 67, 8000 München 19, F.R.G.*

W. Rüdiger *Botanisches Institut, Universität München, Menzingerstr. 67, 8000 München 19, F.R.G.*

S. Schoch *Botanisches Institut, Universität München, Menzingerstr. 67, 8000 München 19, F.R.G.*

Harry Smith *Department of Botany, University of Leicester, University Road, Leicester LE1 7RH, U.K.*

Garry Whitelam *Department of Botany, University of Leicester, University Road, Leicester LE1 7RU, U.K.*

Preface

The first edition of *The Chemistry and Biochemistry of Plant Pigments* appeared in 1965. It stemmed from a Colloquium on this subject organized by the Biochemical Society and held in August 1962 in the Biochemistry Department of the University College of Wales, Aberystwyth. This meeting clearly revealed the demand for reviews of the latest developments by internationally accepted experts, to be collected together in one volume. The resulting book consisted of 19 chapters and 578 pages, and was sufficiently successful to warrant a second edition. This appeared in 1975 and consisted of two volumes comprising 23 chapters and 1243 pages; the second volume was devoted entirely to analytical methods. Ten years later the possibility of producing a third edition arose; undoubtedly this was needed but the upsurge in the rate of research activity, which was much greater than between 1965 and 1975, would have resulted in a publication of some 3500–4000 printed pages. This was clearly not commercially feasible and, after further thought, was judged not necessary scientifically because most fundamentals were already authoritatively enshrined in the second edition.

After much discussion it was eventually decided that an entirely new book was called for and here it is. It is selective, dealing only with a group of topics which the editor decided were the current 'growing points' of the subject. He took many soundings before coming to his decisions, which however are his own responsibility. The remit to authors was to produce critical rather than comprehensive reviews, which would have sufficient introductory material to make them self-contained. The authors have, in my view, achieved this aim admirably and I hope this volume will take its place with the previous two in the series as an effective critical, up-to-date contribution to the chemistry and biochemistry of plant pigments.

T. W. Goodwin

Contents

1
Chlorophylls

W. RÜDIGER and S. SCHOCH

Botanisches Institut Universität München Menzingerstr. 67, 8000 München 19, F.R.G.

1 INTRODUCTION

Chlorophylls belong to the great number of metallo-tetrapyrroles which have been called "pigments of life' (Battersby, 1985) because of their central importance in living systems. Two aspects can be considered when describing the vitally important function of these compounds:

(i) Being pigments these compounds absorb visible light. Harvesting sunlight and transforming its energy into biochemical energy is one

PLANT PIGMENTS
ISBN 0-12-289847-8

of the essentials for life on earth. Photosynthesis in its present form cannot be imagined without chlorophylls. One distinguishes antenna pigments which only harvest light energy and funnel this energy to the reaction centres. The reaction centre pigments transform this energy into biochemical energy, i.e. they are the starting points of the electron transport chains.

(ii) Complexation of the central metal ion by tetrapyrroles can modify its electronic and redox properties. By this means, a great variety of metal-containing enzymes and cofactors arise which are related via common biosynthetic precursors (scheme 1). It should be noted that the list of metals which are complexed by tetrapyrroles is by no means complete. However, no specific biological function is known for other metallo-tetrapyrroles, such as vanadium porphyrins found in various mineral oils or copper uroporphyrins found in feathers of the turaco bird. Of the many metal tetrapyrrole complexes which can be formed by chemical synthesis (Smith, 1975a) the zinc complexes should particularly be mentioned because they are very easily formed. Zinc-containing derivatives of chlorophylls can therefore occasionally be found as artifacts; traces of zinc ions can be derived from solvents, from extracted plant material or even from glass vessels.

It is interesting to note that the ability to produce chlorophyll seems to be restricted to photosynthetic organisms—bacteria and plants. Green "protective" colours of animals which apparently try to mimic the colour of chlorophyll-containing plants are due either to physical colouring (in amphibia and reptiles) or to other tetrapyrroles in combination with

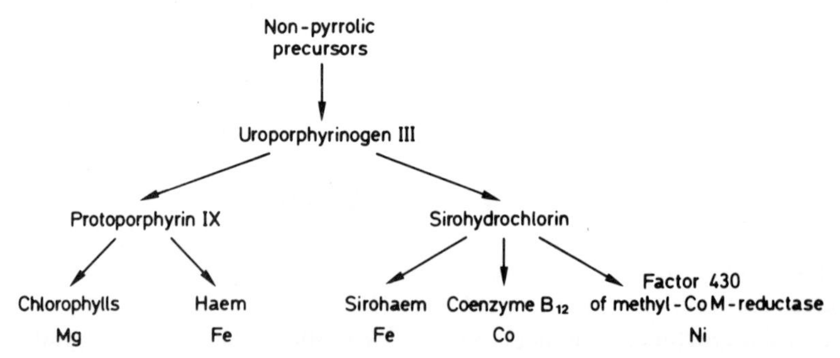

Scheme 1. Biosynthetic relationship between several metal-containing tetrapyrroles (after Thauer, 1985). Some important functions involve: chlorophylls in photosynthesis, haem in oxygen transport, in peroxidase and catalase reactions and in electron transport by cytochromes, sirohaem in sulphite and nitrite reduction, coenzyme B_{12} in carbon metabolism, factor 430 in the energy metabolism of methanogenic bacteria.

carotenoids, e.g. various bile pigments in insects (Rüdiger, 1970) or the sirohydrochlorin-related bonellin in *Bonellia viridis*.

In the following sections, chlorophylls will be described mainly from the viewpoint of their biosynthesis. Emphasis is given to chlorophylls of higher plants and algae. Since the first biosynthetic steps are identical or at least closely related to those of bacteriochlorophylls (i.e. chlorophylls of photosynthetic bacteria), data from these organisms will be included whenever applicable.

This article covers the literature until Spring 1987. Since a complete description or even a complete list of references dealing with chlorophylls is far beyond the scope of this chapter, many details had to be omitted. We tried, however, to select references from various laboratories so that laboratories working in the chlorophyll field are mentioned at least once in one or other connection. For more citations of older work, the reader is referred to the reviews of Bogorad (1976), Holden (1976), Jackson (1976), Schneider (1980), Castelfranco and Beale (1981, 1983), Porra and Meisch (1984) and Leeper (1985 a,b).

2 DISTRIBUTION AND CHEMICAL STRUCTURES

The number of known chlorophyll structures has steadily increased during the last few years. According to Scheer (1984), only three chlorophylls were known in 1960; this number increased to 20 in 1970 and to more than 100 in 1980, including all known structural variations. Some of these variations will be discussed later in connection with the corresponding biosynthetic pathway. Chlorophyll structures can be classified into three groups: porphyrins, dihydroporphyrins (= chlorins) and tetrahydroporphyrins (= bacteriochlorins). The basic structures are shown in scheme 2A. Porphyrinogens are colourless compounds, porphyrin precursors which are readily autoxidized to porphyrins. The IUPAC numbering system for porphyrins which will be used in this article is also indicated in scheme 2A.

The structures of chlorophylls a and b, the typical chlorophylls of higher plants, are presented in scheme 2B. Chlorophyll b is 7-desmethyl-7-formylchlorophyll a. According to Bonnett (1973), exchange of one substituent by another one in the same position can be indicated by the new substituent in square brackets. Following this proposal, chlorophyll b is [7-formyl]chlorophyll a. Analytical and synthetic work related to elucidation of these structures has been reviewed several times (Smith, 1975a; Jackson, 1976; Scheer, 1978) and will therefore not be treated here. Another chlorophyll of organisms carrying out oxygenic photosynthesis has

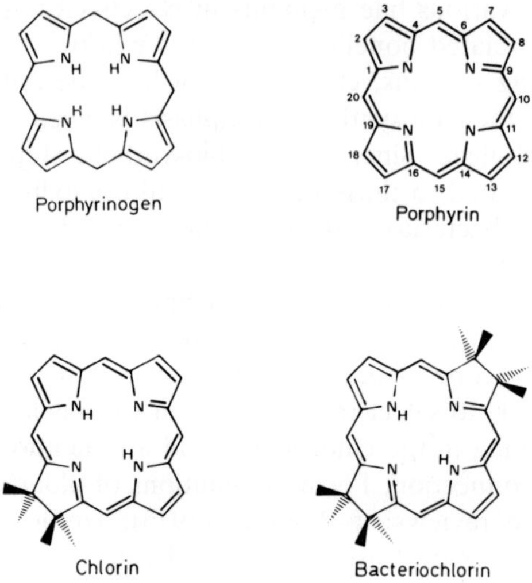

Porphyrinogen

Porphyrin

A

Chlorin

Bacteriochlorin

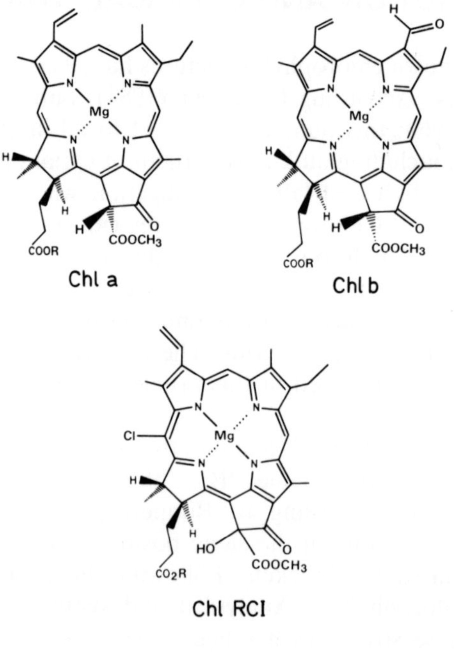

Chl a

Chl b

B

Chl RCI

Scheme 2

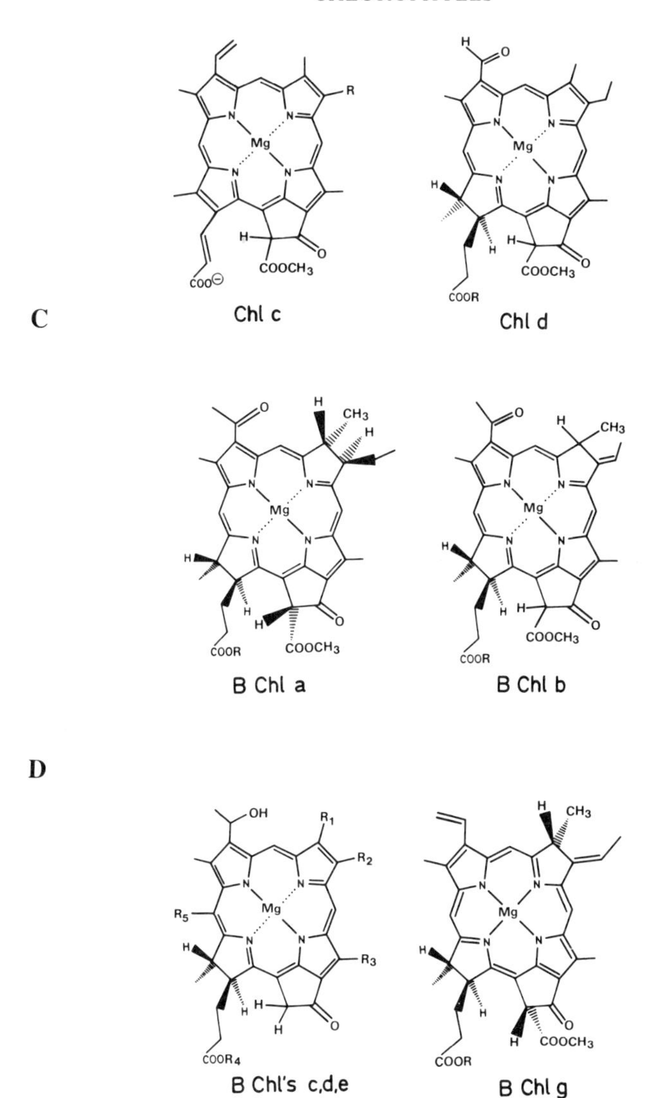

Scheme 2. Chemical structures of chlorophylls and related compounds. **A** Basic structures of macrocycles. **B** Chlorophylls of higher plants. Chl a, Chl b, Chl RCI = chlorophylls a, b, and RCI, respectively. **C** Chlorophylls of algae. R = ethyl: chlorophyll c_1 ("monovinyl compound"), R = vinyl: chlorophyll c_2 ("divinyl compound"). **D** Chlorophylls of bacteria. B Chl = bacteriochlorophyll; R = long-chain alcohol, generally phytol or geranylgeraniol; R_4 in B Chl c generally farnesol; for heterogeneity of alcohols see text; R_1 = methyl in B Chl c and d or formyl in B Chl e; R_2 = ethyl, n-propyl or isobutyl; R_3 = methyl or ethyl, R_5 = methyl in B Chl c; hydrogen in B Chl d and e. **B**–**D**:R = phytyl, if not otherwise stated.

only recently been detected; since it occurs in a 1:1 molar ratio with reaction centre I (= P_{700}) it has been called chlorophyll RCI (Dörnemann and Senger, 1981). Its structure has been established (Dörnemann and Senger, 1986; Scheer *et al.*, 1986) as 20-chloro-13^2-hydroxychlorophyll a (see scheme 2B). Both substituents, the 13^2-hydroxy group (Pennington *et al.*, 1967) and the 20-chloro group (Hynninen and Lötjönen, 1981) can easily be introduced artificially. The hypothesis that chlorophyll RCI might be an artifact has been tested and rejected, (Dörnemann and Senger, 1986; Scheer *et al.*, 1986) although it has not yet been ruled out. It is interesting to note that chlorophyll a′, a compound which can also artificially be produced from chlorophyll a, has also been described as a natural constituent of the reaction centre I (Watanabe *et al.*, 1985). Chlorophyll a′ is believed to be the 13^2-*R* epimer of chlorophyll a which has the 13^2-S configuration (Scheer, 1978).

Structures of typical algal chlorophylls are given in scheme 2C. Chlorophyll c belongs to the porphyrin group but chlorophyll d to the chlorin group of structures. Chlorophyll c is the only chlorophyll (besides biosynthetic chlorophyll precursors) which occurs as a free carboxylic acid rather than as an ester. Structural work on these pigments has been reviewed by Jackson (1976).

Structures of bacterial chlorophylls are shown in scheme 2D. The bacteriochlorophylls c and d (formerly called *Chlorobium* chlorophylls 660 and 650, respectively) and bacteriochlorophyll e belong to the chlorin type. Whereas bacteriochlorophyll a belongs to the bacteriochlorin type, bacteriochlorophylls b and g, having an exocyclic double bond at ring B, can be classified as intermediate between the chlorin and bacteriochlorin types. Work on the elucidation of the tetrapyrrole structure of bacteriochlorophylls a, c and d has been reviewed by Jackson (1976). The structures given in his review for bacteriochlorophylls c and d have to be corrected; the 13^2-carboxymethyl group must be replaced by hydrogen. Further work on structures of bacteriochlorophylls c and d has been published (Brockmann, 1976; Smith *et al.*, 1980, 1985). Structural elucidation of bacteriochlorophylls e (Risch *et al.*, 1978; Smith and Simpson, 1986) and g (Brockmann and Lipinski, 1983; Michalski *et al.*, 1987) has been described more recently.

The nature of the esterifying alcohol of bacteriochlorophylls deserves special attention. Bacteriochlorophylls a and b of a wide variety of photosynthetic bacteria contain phytol and geranylgeraniol ($\Delta^{2,6,10,14}$-phytatetraenol) (Künzler and Pfennig, 1973; Gloe and Pfennig, 1974) but not farnesol, as was erroneously reported (Brockmann and Knobloch, 1972). Another hitherto unknown alcohol, $\Delta^{2,10}$-phytadienol was found in bacteriochlorophyll b of *Ectothiorhodospira halochloris* (Steiner *et al.*, 1981). The main alcohol of bacteriochlorophyll c is farnesol; five minor compounds were identified in bacteriochlorophylls c from *Chlorobium limicola* as phytol, $\Delta^{2,6,10,14}$-phytatetraenol, $\Delta^{2,6}$-phytadienol, cis-9-hexadecan-1-ol and 4-undecyl-2-furanmethanol (Caple *et al.*, 1978).

The distribution of chlorophylls among photosynthetic organisms is shown in Table 1.1. Chlorophyll a is found in all oxygenic photosynthetic organisms. Since (a small amount of) chlorophyll RCI is also found in all these organisms, as far as they were analysed, it has been suggested that these chlorophylls both occur in the reaction centres. The bulk of chlorophyll a is however found in the antenna (or light-harvesting apparatus) of the respective organisms.

Bacteriochlorophyll a seems to be as widely distributed in organisms carrying out anoxygenic photosynthesis as is chlorophyll a in organisms carrying out oxygenic photosynthesis. However, bacteriochlorophyll a can be substituted entirely by bacteriochlorophyll b in several bacteria. Recent interest has been focused upon bacteriochlorophyll b-containing reaction centres of *Rhodopseudomonas viridis*, which were the first reaction

Table 1.1 Distribution of chlorophylls among photosynthetic organisms (according to Jackson, 1976; Jones, 1978, completed)

Organism	Chlorophyll					
	a	b	c	d	RCI	
Higher plants, ferns, mosses	+	+	−	−	+	
Algae						
Chlorophyta	+	+	−	−	+	
Chrysophyta	+	−	+	−	?	
Euglenophyta	+	+	−	−	?	
Pyrrophyta	+	−	+	−	?	
Phaeophyta	+	−	+	−	?	
Rhodophyta	+	−	−	+	?	
Cyanophyta (cyanobacteria)	+	−	−	−	+	
Prochlorophyta	+	+	−	−	+	
	Bacteriochlorophyll					
	a	b	c	d	e	g
Rhodospirillaceae	+ or	+				
Chromatiaceae	+ or	+				
Chlorobiaceae	(+)[1]		+[2] or	+[3] or	+[4]	
Heliobacterium chlorum	?					+

[1]Minor amounts of bacteriochlorophyll a are found in most *Chlorobium* spp.
[2]Bacteriochlorophyll c has been found to be a main component in most *Chlorobium* spp.
[3]Bacteriochlorophyll d has been found e.g. in *Chlorobium vibrioforme*.
[4]Bacteriochlorophyll e has been found e.g. in *Chlorobium phaeobacterioides* and *Chl. phaeovibriodes*.

Table 1.2 Spectral data of chlorophylls and phaeophytins determined in diethylether at room temperature. The structures of the compounds are to be found in scheme 2. The blue maximum is the Soret band, the red maximum has been described throughout as Q_x band. ϵ = molar extinction coefficient [cm^{-1} M^{-1}]

Compound	Absorption		Ref.	Emission	Ref.
	blue maximum in nm ($\epsilon \times 10^{-3}$)	red maximum in nm ($\epsilon \times 10^{-3}$)		red max. in nm	
Protochlide a	432 (200.)	624 (24.3)	1[6]	630	9
Protophaeide a	417 (151.9)	565 (14.9)	1	642	9
Chl a	430 (117.7)	662 (90.1)	1	665	2
Phae a	408 (115.0)	667 (55.5)	1	673	2
Chl b	455 (158.6)	644 (56.2)	1	649	2
Phae b	434 (191.2)	655 (37.3)	1	657	2
Chl RCI[1]	433 (100%)[5]	672 (71%	13	675	13
Phae RCI[1]	415 (100%)[5]	678 (52%)	13	682	13
Chl c	447 (138.5)	628 (13.4)	1	635	2
Phae c	421 (133.7)	590 (8.5)	1	650	2
Chl c[1]	442 (70.7)	628 (9.6)	3	631	12
Chl c[1]	446 (212.3)	629 (23.9)	4	632	12
Chl c[1]	445 (195.8)	630 (22.7)	4	632	12
Chl d	447 (87.4)	688 (98.3)	1	693	2
Phae d	421 (85.9)	692 (72.4)	1	701	2
B Chl a	359 (73.3)	773 (91.0)	2	782	10
B Phae a	358 (113.6)	749 (67.5)	2	760	2
B Chl b[1]	368 (94%)[5]	794 (100%)	5		
B Phae b	398 (237%)[5]	776 (100%)	6		
B Chl c[2]	429 (112)	660 (73)	7	667	11
B Phae c[2]	406 (86.2)	663 (46)	7		
B Chl d[3]	423 (117)	651 (88.3)	7		
B Phae d[3]	403 (84.7)	658 (44.1)	7		
B Chl e[1,4]	458 (100%)[5]	647 (34%)	8		
B Phae e[1,4]	439 (100%)[5]	654 (24%)	8		

[1]Measured in acetone.
[2]Mixture of several homologues; ϵ calculated with the substituents $R^1 = R^3 = R^5 = CH_3$; $R^2 = C_2H_5$; R^4 = farnesyl.
[3]Mixture of several homologues; ϵ calculated with the substituents $R^1 = R^3 = CH_3$; $R^2 = C_2H_5$; R^4 = farnesyl; R^5 = H.
[4]Mean of B Chl e fractions isolated from six different species; the peak varies by ± 2 nm; the relative intensities up to 50%.
[5]Arbitrary units.
[6]The molar extinction coefficients were calculated from the specific extinctions coefficients in reference 1.

centres to be crystallized and studied by high-resolution X-ray analysis (Deisenhofer *et al.*, 1985; Michel and Deisenhofer, 1987). This bacterial reaction centre contains, besides other components, four molecules of bacteriochlorophyll (a or b) and two molecules of bacteriophaeophytins (a or b). These compounds might differ in their esterifying alcohols, e.g. bacteriochlorophyll a_{GG} and bacteriophaeophytin a_P in the reaction centre of *Rhodospirillum rubrum* (Walter *et al.*, 1979). Bacteriochlorophylls c, d and e are mixtures of homologues (see scheme 2D). They are antenna pigments in the Chlorobiaceae.

3 ANALYTICAL METHODS

Isolation of chlorophylls requires at first detachment from the chlorophyll-binding proteins. This is achieved by polar solvents: 80% aqueous acetone for higher plants or methanol for bacteria and algae (Holden, 1976; Svec, 1978; Scheer, 1987). These solvents also penetrate cell membranes so that mechanical disruption of cells need not necessarily be complete. Sometimes other polar solvents have been used, e.g. dimethylformamide (Moran and Porath, 1980; Inskeep and Bloom, 1985; Bergweiler and Lütz, 1986) or dimethylsulphoxide (Hiscox and Israelstam, 1979). For quantitative determination of the single chlorophylls in such extracts by spectrophotometry, formulae have been published (Holden, 1976). For quantities > 0.1 mg of chlorophyll, the best purification step is precipitation from the acetone or methanol solution with dioxane (Iriyama and Shiraki, 1979). Further purification is achieved by column chromatography on powdered sugar (reviewed by Svec, 1978; see also Hynninen, 1979) on Sepharose CL-6B (Iriyama *et al.*, 1981; Omata and Murata, 1983), or DEAE-Sepharose (Omata and Murata, 1983; Araki *et al.*, 1984). Isocratic

References to Table 1.2

1. French, C.S. (1960). In: *Handbuch der Pflanzenphysiologie* (Ed. W. Ruhland) Bd. 5/1.
2. Smith, J. H. C. and Benitez, A. (1955). In: *Moderne Methoden der Pflanzenanalyse* (Eds. K. Paech, M. V. Tracey) Band IV, p. 142, Springer-Verlag, Berlin.
3. Jeffrey, S. W. (1969). *Biochim. Biophys. Acta* **177**, 456.
4. Jeffrey, S. W. (1972). *Biochim. Biophys. Acta* **279**, 15.
5. Baumgarten, D. (with Sauer, K.) (1970) *MS Thesis*, University of California, Berkeley.
6. Steiner, R. (1981). *Zulassungsarbeit*, Universität München.
7. Holt, A. S. (1965). In: *Chemistry and Biochemistry of Plants* (Ed. T. W. Goodwin) p. 3, Academic Press, New York.
8. Gloe, A. (1977). *Dissertation*. Universität Göttingen.
9. Frey, M. A., Alberte, R. S. and Schiff, J. A. (1979). *Biol. Bull.* **157**, 368.
10. Connolly, J. S., Smauel, E. B. and Janzen, F. (1982). *Photochem. Photobiol.* **36**, 565.
11. Goedheer, J. C. (1966). In: *The Chlorophylls* (Eds. L. P. Vernon, G. R. Seely) p. 147. Academic Press, New York.
12. Wilhelm, C. (1987). *Botanic Acta*, in press.
13. Dörnemann, D. and Senger, H. (1982). *Photochem. Photobiol.* **35**, 821.

high-performance liquid chromatography (HPLC) on silica gel has also been described (Watanabe *et al.*, 1984).

The isolated pigments can be characterized by their absorption or fluorescence spectra (Tables 1.2 and 1.3). For additional characterization,

Table 1.3 Fluorescence spectral data of chlorophylls ("monovinyl" compounds) and [8-vinyl]chlorophylls ("divinyl" compounds) in diethylether at 77K, according to Rebeiz *et al.* (1983). The λ_{max} values are expressed in nm.

	Monovinyl		Divinyl	
	Excitation	Emission	Excitation	Emission
Protochlide a	437	625	443	625
Protochl a	436	624	442	624
Chlide a	449	675	459	675
Chl a	448	673	458	673

Table 1.4 Chromatographic systems employed for the separation of chlorophylls and related compounds by thin-layer chromatography

Adsorbent	Compounds to be separated	Ref.
Cellulose	Chl a,b, Phae a,b, Pheide a,b,	1
Sucrose	Chl a,a′,b,b′, 13^2-OH Chl a, a′	2
Silica gel	Protochlide, Chlide a,b	3
Silica gel	Phaeide a,b, Pyrophaeide a, 13^2-OH Phaeide a	4
Cellulose + triglyceride	Chl a,a′,b,b′, Phae a, BChl a,a′	5
Reverse-phase silica gel (HPTLC-RP_8)	Phaeide a, Pyrophaeide a	6
Reverse-phase silica-gel (HPTLC-RP_8)	Chl a,b, Phae a,b, 13^2-OH Chl a,b	7

References

1. Sievers, G. and Hynninen, P. H. (1977). *J. Chromatogr.* **134**, 359.
2. Sahlberg, J. and Hynninen, P. H. (1984). *J. Chromatogr.* **291**, 331.
3. Duggan, J. X. and Rebeiz, C. A. (1982). *Biochim. Biophys. Acta* **714**, 248.
4. Endo, E., Hosoya, H., Koyama, T. and Ichioka, M. (1982). *Agric. Biol. Chem.* **46**, 2183.
5. Scholz, B., Willascheck, K. D., Müller, H. and Ballschmiter, K. (1981). *J. Chromatogr.* **208**, 156.
6. Takeda, Y., Saito, Y. and Uchiyama, M. (1983). *J. Chromatogr.* **280**, 188.
7. Schoch, S., Rüdiger, W., Lüthy, B. and Matile, P. (1984). *J. Plant Physiol.* **115**, 85.

Table 1.5 HPLC of chlorophylls and related compounds on reversed-phase columns (C_8 or C_{18})

Compounds	Retention time (min)				
	System 1	System 2	System 3	System 4	System[f] 5
Chlide b	3.8				
Chl c		6.4	1.5	2.6	
Chlide a	6.6	5.6	2.2	2.3	
Phaeide b	7.6				
Phaeide a	8.7	7.9	3.0	7.7	
Chl b	14.0	11.5	5.6	16.0	21.5
Chl a	16.0	12.7	7.1	17.1	23.0
Phae b	18	14.8	10.8	18.3	28
Phae a	22	17.3	14.7	19.3	32

[f]Excellent system to separate carotinoides and chlorophylls.

System 1: Column C_{18}; gradient (concave) from 75:25 methanol-water (A) to ethyl acetate (B). Final conditions 50% A and 50% B. Schwartz, S. J., Woo, S. L. and von Elbe, J. H. (1981). *J. Agric. Food Chem.* **29**, 533.

System 2: Column C_{18}; linear gradient from 10:10:80 ion-pairing reagent (1.5 g tetrabutyl-ammonium acetate + 7.7 g ammonium acetate made up to 100 ml water)–water-methanol (A) to 20:80 acetone–methanol (B). Final conditions 100% B. Mantoura, R. F. C. and Llewellyn, C. A. (1983). *Anal. Chim. Acta* **151**, 297.

System 3: Column C_8; two steps solvent programme, first 90:10 methanol–water (until phaeide a is eluted) second 98:2 methanol–water.
Falkowski, P. G. and Sucher, J. (1981). *J. Chromatogr.* **213**, 349.

System 4: Column C_{18}; linear gradient from 90:10 acetonitrile–water to 100% ethylacetate. Wright, S. W. and Shearer, J. D. (1984). *J. Chromatogr.* **294**, 281.

System 5: Column C_{18}; linear gradient from 25%[A]:75%[B] to 100%[B] A = water; B = 1:3 methanol–acetonitrile.
Braumann, T. and Grimme, L. H. (1981). *Biochim. Biophys. Acta* **637**, 8.

the original chlorophylls can be transformed into the respective phaeophytins with mineral acids (see Svec, 1978 and earlier papers cited therein), e.g. by shaking the ether solution with 5–10% aqueous HC1 until the green chlorophyll colour changes to the greyish phaeophytin colour. The spectral data of the phaeophytins are also found in Table 1.2. Characterization and determination of chlorophylls and [8-vinyl]chlorophylls can be achieved by determination of the fluorescence excitation spectra at low temperature (Table 1.3). The fluorescence emission is identical for mono- and divinyl compounds (Table 1.3).

Further characterization of the pigments and checking for purity can be performed by thin-layer chromatography (Table 1.4) or HPLC (Table 1.5). Many systems have been described to deal with manifold separation problems. The pigments are detected either by their absorption (colour) or their fluorescence. The tables contain only a few examples covering a

Table 1.6 HPLC of chlorophylls esterified with various alcohols. Retention times are given in minutes

	Proto Chl a System 1	Chl a System 1	Phae a System 2	Phae a System 3	B Chl a System 1	Chl a System 4[1]
GG[2]	12.1	8.0	10.2	4.1	4.4	27.3
DHGG[2]	14.1	9.3	11.6	8.1	4.9	32.6
THGG[2]	16.6	11.0	13.8	9.4	5.6	40.9
Phytol	19.7	13.0	16.0	10.6	6.5	49.2

[1]Excellent system for separating non-esterified chlorophylls.
[2]GG = geranylgeraniol, $\Delta^{2,6,10,14}$-phytatetraenol; DHGG = dihydrogeranylgeraniol, $\Delta^{2,10,14}$-phytatrienol; THGG = tetrahydrogeranylgeraniol, $\Delta^{2,14}$-phytadienol.

System 1: Column C_{18} separation with methanol at 40°C.
Shioi, Y., Fukae, R. and Sasa, T. (1983). *Biochim. Biophys. Acta* **722**, 72.
System 2: Column C_8; separation with a mixture of methanol–water (95:5) at room temperature.
Schoch, S. (1978). *Z. Naturforsch.* **33c**, 712.
System 3: Column C_{18}; separation with a mixture of 90:10 methanol–acetone at room temperature.
Schoch, S., Lempert, U., Wieschhoff, H. and Scheer, H. (1978). *J. Chromatogr.* **157**, 357.
System 4: Column C_{18}; separation with a mixture of 95:5 methanol–water containing 13 mM acetic acid (pH 4.2) at 40°C.
Shioi, Y., Doi, M. and Sasa, T. (1984). *J. Chromatogr.* **298**, 141.

large range of diverse problems and systems. Special separation problems may require modified systems. Hanamoto and Castelfranco (1983) were able to separate mono- and divinyl species of protochlorophyllide and chlorophyllide by reversed-phase HPLC using an ion-pairing reagent similar to system 2 in Table 1.5. In Table 1.6, systems are listed for separation of pigments esterified with various alcohols.

Mass spectrometry and nuclear magnetic resonance (NMR) spectroscopy have been used throughout for elucidation of chlorophyll structures. Application of these methods to chlorophylls and other porphyrins have been reviewed by Smith (1975b) and Scheer and Katz (1975).

4 BIOSYNTHESIS OF CHLOROPHYLLS

A Formation of 5-aminolaevulinic acid

The first specific precursor for formation of chlorophylls and other tetrapyrroles is 5-aminolaevulinic acid (ALA) which can be formed in two ways: the Shemin pathway and the C_5-pathway (scheme 3). Formation from glycine and succinyl coenzyme A (Shemin pathway) has been known

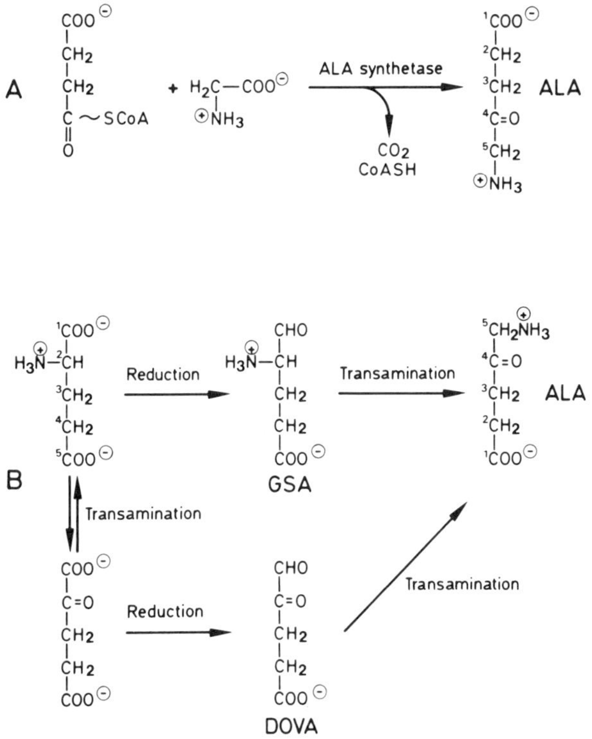

Scheme 3. Formation by ALA on two pathways. **A** Shemin pathway; **B** C_5 pathway including alternative routes.

for a long time in animals (for biosynthesis of haem) and in some photosynthetic bacteria, e.g. *Rhodopseudomonas sphaeroides* (for biosynthesis of bacteriochlorophyll). Formation of bacteriochlorophyll as in *R. sphaeroides* via this pathway has recently been confirmed by ^{13}C NMR studies (Oh-hama *et al.*, 1985). The C_5-pathway for ALA formation was first described by Beale and Castelfranco (1974) and Beale *et al.* (1975). It was found that several C_5 compounds (glutamate, glutamine, ketoglutarate) were more effectively incorporated into ALA than 2-^{14}C glycine in higher plants. Subsequently, several reports on *in vitro* ALA formation from α-ketoglutarate or glutamate appeared, summarized by Castelfranco and Beale (1981, 1983). The most convincing evidence for operation of the C_5 pathway in chlorophyll biosynthesis of higher plants and green algae comes from ^{13}C NMR studies (Oh-hama *et al.*, 1982: Porra *et al.*, 1983). These experiments provided conclusive evidence that the entire carbon skeleton of the C_5 precursors is incorporated into the chlorophyll molecule. Such ^{13}C NMR studies with [2-^{13}C]glycine and [1-^{13}C]glutamate demonstrated that bacteriochlorophylls a and c, respectively, are—in contrast to the situation in *R. sphaeroides*—formed via the C_5 pathway in the photosynthetic bacterium *Chromatium vinosum* (Oh-hama *et al.*, 1986a) and in the green sulphur bacteria *Prosthecochloris aestuarii* (Oh-hama *et al.*, 1986b) and *Chlorobium vibriforme* (Smith and Huster, 1987). Thus photosynthetic bacteria can contain either the C_5 or the Shemin pathway. Oh-hama *et al.* (1987a) suggested that the C_5 pathway is an ancestral form of ALA formation and that the Shemin pathway has been evolved together with enzymes of the tricarboxylic acid cycle, especially 2-oxoglutarate dehydrogenase.

Early reports on the implication of the Shemin pathway for chlorophyll biosynthesis in *higher plants* have been summarized by Bogorad (1976). One of the essential arguments for operation of this pathway is the specific incorporation of [2-^{14}C]glycine into chlorophylls and other porphyrins. It has been questioned whether this occurs at all in higher plants (Porra, 1986). Beale and Castlefranco (1974) had found that [1-^{14}C]glycine is incorporated as effectively as [2-^{14}C]glycine into chlorophyll of greening bean or barley leaves. Such an unspecific incorporation has to be considered in all experiments of this type. Meller and Gassmann (1982) studied the incorporation of various ^{14}C-labelled precursors into ALA in etiolated leaves of barley in continuous darkness and during greening in the light. Mainly from the differences of incorporation of [1-^{14}C]glutamate into ALA in the dark and in the light, the authors concluded that ALA formation occurred preferably by the Shemin pathway in the dark (presumably in mitochonodria) and by the C_5 pathway in the light (presumably in plastids). The results with other precursors (e.g. [5-^{14}C]glutamate, [2-^{14}C]glycine), however, do not provide convincing evidence for this view. Earlier reports on the operation of both the Shemin and C_5 pathway in higher plants (Klein and Porra, 1982) were later withdrawn to the advantage of the C_5 pathway (Porra, 1986). Recent incorporation experiments with [1-^{14}C]glutamate and [2-^{14}C]glycine point to the C_5 pathway in etiolated maize (Weinstein *et al.*, 1987a).

The situation is apparently not uniform in the field of *algae*. The C_5 pathway seems to be the only pathway in Cyanobacteria (Kipe-Nolt and Stevens, 1980) and in the eukaryotic alga *Cyanidium caldarium* (Troxler and Offner, 1979; Weinstein *et al.*, 1987a) whereas both pathways may operate in *Scenedesmus obliquus* (Klein and Senger, 1978a,b) and *Euglena gracilis* (Weinstein *et al.*, 1987a). In the latter case, the authors discuss a compartmentation of the pathways: the Shemin pathway in the mitochondria (for haem a) and the C_5 pathway in the chloroplast (for chlorophyll a). Whether the C_5 pathway in eukaryotes is restricted to plastids is not yet entirely clear. In one paper the direct incorporation of [1-^{14}C]glutamate into haem in duck blood was reported (Franck *et al.*, 1984).

Two possible routes have been and still are considered for the C_5 pathway, namely via glutamate-1-semialdehyde (GSA) or via 4,5-

dioxovalerate (DOVA) (scheme 3). The last step is a transamination in both routes—intramolecular with GSA and intermolecular with DOVA. The enzymic reaction of DOVA transaminase has been investigated in algae (Gassman *et al.*, 1968a; Klein and Senger, 1978b; Salvador, 1978) and higher plants (Hayashi and Moguchi, 1983). Foley and Beale (1982) concluded from comparison of K_M values that the true substrate for the enzyme from *Euglena gracilis* should be glyoxylate rather than DOVA. However, the K_M value for DOVA is smaller than that for glyoxylate in the enzyme from *Chlorella fusca* (Meisch *et al.*, 1983) and from *Scenedesmus obliquus* (Kah *et al.*, 1986). Contrary to the situation with *Euglena* extracts, DOVA transaminase and glyoxylate transaminase activities could be separated by chromatography in the latter case. A connection of DOVA with ALA biosynthesis has also been postulated because of parallel light regulation of DOVA transaminase and of ALA formation (Hayashi and Moguchi, 1983) and because of DOVA accumulation after inhibition of transamination with amino-oxyacetate (Kah *et al.*, 1986). It should be recalled, however, that DOVA transaminase activity has also been described in *Rhodopseudomonas sphaeroides* (Neuberger and Turner, 1963) which forms bacteriochlorophyll not via the C_5 pathway but via the Shemin pathway.

The proposal that GSA is an intermediate came from experiments with cell-free systems of higher plants (Gough and Kannangara, 1977; Kannangara and Gough, 1977). The enzyme activity of GSA amino-transferase was subsequently described (Kannangara and Gough, 1978); the enzyme is apparently inhibited by gabaculine (Kannangara and Schouboe, 1985). Whereas the diethylacetal of GSA was well characterized (Houen *et al.*, 1983), doubts concerning the existence of free GSA have been proposed (Meisch and Maus, 1983; Kah and Dörnemann, 1987). The synthetic compound was characterized only with attached protective groups but it polymerized immediately after cleavage of these groups. Stabilization of GSA might occur by masking the free aldehyde group either by binding to the enzyme (Meisch and Maus, 1983; see also Scheme 6), by hydration (G. Kannangare, personal communication) or by ring closure with the carboxy group (P. Jordan, personal communication).

A new aspect arose when it was detected that a low molecular weight RNA is involved in ALA formation from glutamate. This is true for barley (Kannangara *et al*, 1984), *Chlamydomonas* (Huang *et al.*, 1984), *Chlorella* (Weinstein and Beale, 1985), *Euglena* and *Cyanidium* (Weinstein *et al.*, 1987b), and also for *Methanobacterium* (Friedmann and Thauer, 1986). The RNA turned out to be a glutamate-specific tRNA, the sequence of which was determined for barley (Schön *et al.*, 1986; see scheme 4). This tRNA equals $tRNA^{Glu}$ from several sources but has the modified base 5-methylaminomethyl-2-thiouridine (mam^5s^2U) in the anticodon (Schön *et al.*, 1986). The nucleotide mam^5s^2U has so far been

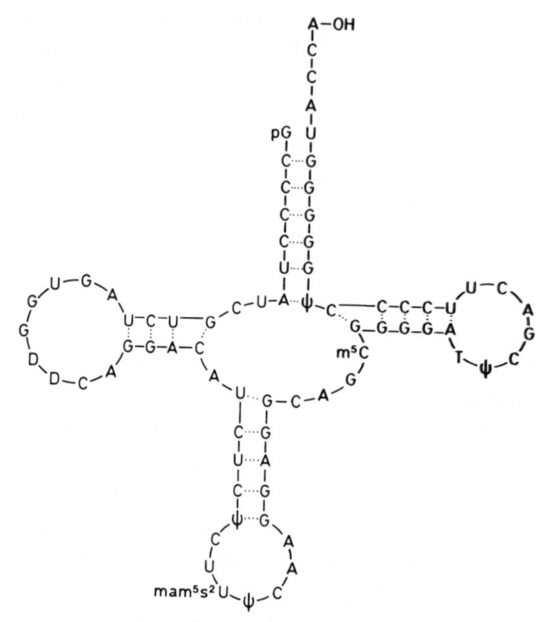

Scheme 4. Structure of glutamate-binding tRNA from barley which is specific for ALA formation, named $tRNA^{DALA}$ (after Schön *et al.*, 1986).

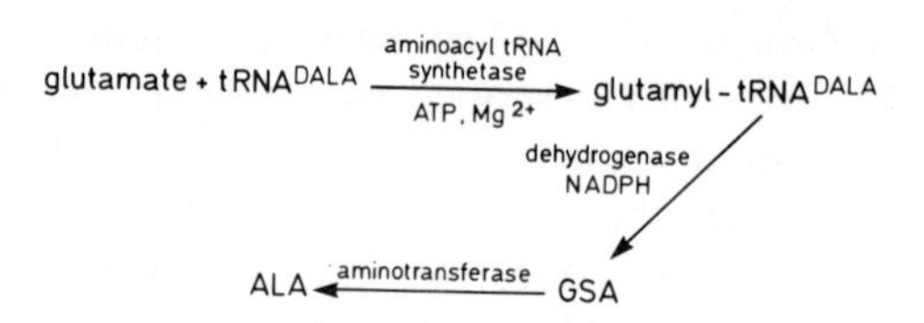

Scheme 5. Formation of ALA from glutamate, according to Kannangara *et al.* (1984), Schön *et al.* (1986) and Wang *et al.* (1987).

described only for tRNAs of procaryotes. Although it has been reported that the $tRNA^{Glu}$ species from *Escherichia coli* can substitute the tRNA species in *Chlamydomonas* extracts (Huang and Wang, 1986) no such substitution of the endogenous $tRNA^{DALA}$ was possible in barley (Kannangara *et al.*, 1984) or *Chlorella* (Weinstein and Beale, 1985).

This is one of the very rare examples of involvement of an aminoacyl-tRNA in a metabolic pathway other than protein biosynthesis, namely the reduction of tRNA-bound glutamate to the tRNA-bound semialdehyde (scheme 5). According to the present view (Schön *et al.*, 1986, Wang *et al.*, 1987), the first step specific for ALA formation is ligation of glutamate to the described $tRNA^{DALA}$. The next step is dehydrogenation of the aminoacyl-tRNA with NADPH (Wang *et al.*, 1984).

Schön *et al* (1986) assume—without presenting direct experimental evidence—that a hydrolase cleaves GSA from RNA[DALA] before the aminotransferase can form ALA. Considering the instability of free GSA (see above), one can alternatively assume that the aminotransferase reacts directly with the tRNA-bound GSA and that cleavage from the tRNA occurs concomitantly with ALA formation. If the aminotransferase contains a cofactor like pyridoxalphosphate, the first transamination step would lead to DOVA which is still bound to the tRNA (see scheme 6). The second step of the transamination would then be ALA formation concomitantly with the cleavage from the tRNA. This hypothetical mechanism, which includes bound GSA and bound DOVA within one sequence, would end the controversy over GSA or DOVA as alternative routes for ALA formation.

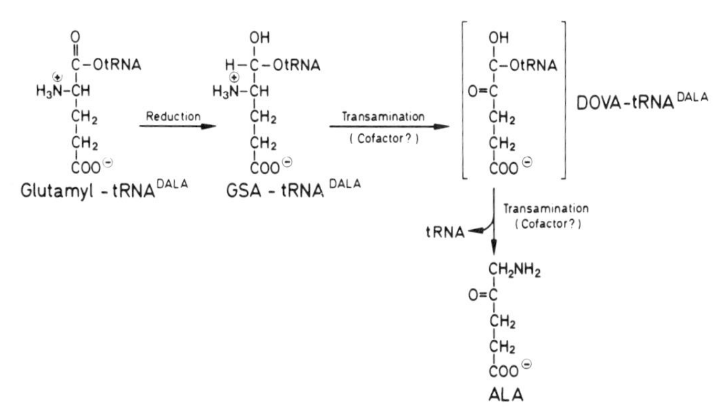

Scheme 6. Hypothetical pathway for ALA formation from glutamate which includes tRNA-bound glutamate semialdehyde and dioxovalerate in one sequence.

B 5-Aminolaevulinic acid dehydratase (porphobilinogen synthase)

Condensation of two molecules of ALA with (formal) elimination of two molecules of water leads to the first pyrrole compound in the pathway, prophobilinogen (PBG). The enzyme catalysing this reaction was first obtained from animal systems and from *Rhodospeudomonas sphaeroides* (summarized by Bogorad, 1976) and later also from higher plants (summarized by Schneider, 1980). An efficient purification method using monoclonal antibodies for the enzyme from spinach (Liedgens *et al.*, 1980) enabled these authors to determine the molecular properties of this enzyme in detail (Liedgens *et al.*, 1983). Whereas the enzyme from animal sources consists of eight subunits with the molecular weight of about 35 kDa each, the enzymes from photosynthetic organisms are hexameric with subunits of 40 kDa (*R. spheroides*) or about 50 kDa (spinach). The animal enzyme is a zinc protein containing probably four active centres per octamer (Jaffe and Hanes, 1986) and four zinc ions per octamer which, according to Hasnain *et al.* (1985), do not bind or activate the substrate directly. Whereas this enzyme is inhibited by metal chelators such as ethylenediamine tetraacetate or 8-quinolinol, no such inactivation is found for the enzyme from *R. sphaeroides*. The latter enzyme requires K ions for activation whereas the plant enzyme needs Mg ions instead. It is assumed that the respective metal ions serve as stabilizators of the enzyme rather than as participants in the enzymic reaction. The plant enzyme seems to be localized in plastids and eventually in the cytoplasm (not in mitochondria), even in non-green tissue like etiolated pea leaves and spadices of Arum (Smith, 1987).

A common feature of the enzymes from animals, photosynthetic bacteria and plants is the formation of a Schiff base between the carbonyl group of the substrate and an amino group of the enzyme. Although it had first been assumed that only the ALA molecule leading to formation of the acetic acid side-chain of PBG forms a Schiff base (Nandi and Shemin, 1968; Barnard *et al.*, 1977), Jordan and Seehra (1980) and Jordan and Gibbs (1985) were able to demonstrate that the ALA molecule leading to the propionic acid side-chain of PBG binds first to the enzyme (scheme 7, mechanism 2). For the mechanism, see also Jaffe and Markham (1987).

The free carboxylic acid group is probably necessary for binding to the enzyme since ALA methyl ester is a very poor substrate (Shemin, 1970). The carboxylate group of ALA is probably fixed by a cationic residue of the protein. Enzyme inactivation with the arginine-specific reagent butanedione and protection against this inactivation by the substrate, ALA, demonstrated involvement of one arginine residue per bound ALA (Liedgens *et al.*, 1983).

The ALA dehydratases from *R. sphaeroides* and from animals contain cysteines which are essential for the enzyme activity. The activity of these enzymes is inhibited by SH-reagents, such as iodoacetamide. It is supposed that these cysteines, which are different from the zinc-binding cysteines (Hasnain *et al.*, 1985), take part in the reaction as proton donors and acceptors, as well as a histidine residue. The spinach enzyme is not inhibited by iodoacetamide (Liedgens *et al.*, 1983; proton transfer is eventually catalysed by acidic amino acid residues which are abundant in the plant enzyme.

A mechanism has been proposed (Nandi and Shemin, 1968; Barnard *et al.*, 1977) in which a similarity with the aldolase or transaldolase mechanism is considered (scheme 8). While these authors assume that X in scheme 8 is oxygen, Jordan and Gibbs (1985) pointed out that X is enzyme-NH^+ but Y could be either oxygen or enzyme-NH^+. The last

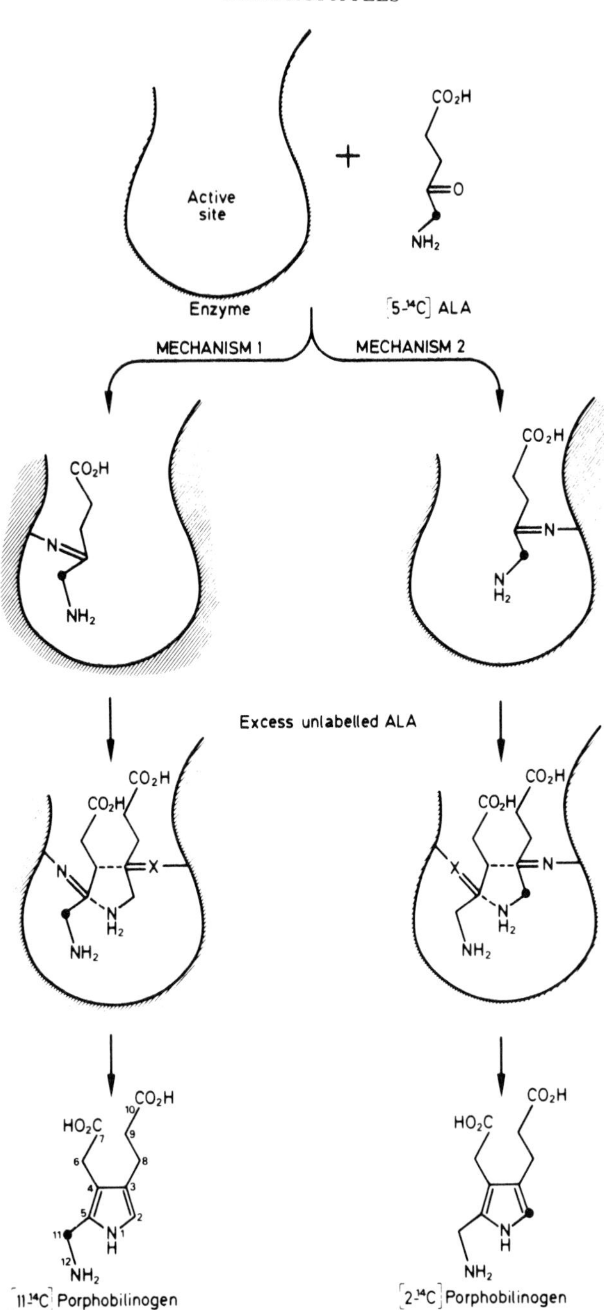

Scheme 7. Formation of PBG from ALA (according to Jordan and Gibbs, 1985).

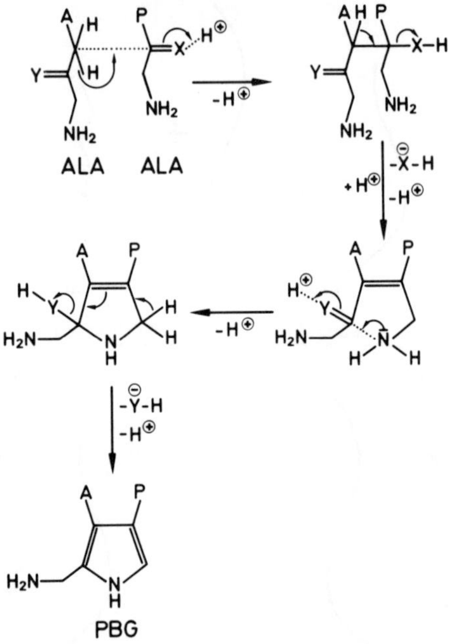

Scheme 8. Mechanism of 5-aminolaevulinate dehydratase, adopted from Nandi and Shemin (1968) and Barnard *et al.* (1977). A = acetic acid residue; P = propionic acid residue. X and Y are either enzyme-NH^+ (i.e. Schiff base of substrate) or oxygen (i.e. free substrate).

reaction step which forms the pyrrole has been discussed as a mechanism of freeing the product from the active site of the enzyme (Shemin, 1970). Several competitive inhibitors of ALA dehydratase have been described which contain the structural feature of the succinyl residue of ALA (scheme 9). They are also believed to form Schiff bases with the enzyme. The best known compound is laevulinic acid which has often been used for biosynthetic studies because inhibition of ALA dehydratase leads to accumulation of excess ALA. The succinyl residue seems to be essential (interaction of the carboxylate with the cationic group of the enzyme and formation of a Schiff base between carbonyl group and amino group of the enzyme) whereas the rest of the inhibitor molecule can vary (see scheme 9). It is interesting that DOVA also belongs to these inhibitors (Shioi *et al.*, 1985) because it has been considered as a possible biosynthetic precursor of ALA (see above). Since 2-ketoglutarate neither inhibits nor forms a Schiff base with the enzyme (Shemin, 1970), the variable residue is not allowed to carry a negative charge.

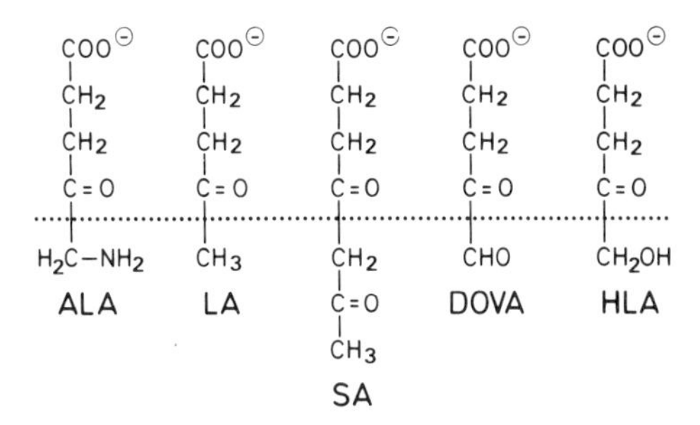

Scheme 9. Chemical structures of ALA and competitive inhibitors (after Meller and Gassman, 1981, completed). LA = laevulinic acid; SA = succinylacetone; DOVA = 4,5-dioxovalerate (see Shioi *et al.*, 1985); HLA = 5-hydroxylaevulinic acid (Yamasaki and Moriyama, 1970).

C Tetrapyrrole formation

The condensation of four molecules of PBG leads to the first detectable tetrapyrrole. The *chemical* (i.e. acid-catalysed) reaction yields a mixture of four uroporphyrinogens (urogens I to IV) which can also isomerize in hot acid after ring closure. It is interesting that only one of the four possible isomers predominates in equilibrium, namely urogen III (Mauzerall, 1960) which has been explained theoretically (von Dobeneck *et al.*, 1972). This may have relevance for prebiotic formation of tetrapyrroles since all natural tetrapyrroles with a known function are derived from urogen III. Rare cases in which urogen I or derivatives thereof are found in nature (e.g. porphyrin excretion in porphyria, porphyrin deposition in shells of marine organisms) may be malfunctions or deviations from the normal biosynthesis.

The *biochemical* reaction consists of two enzymic steps in which a product of isomer type I is formed at first, which is rearranged to a compound of type III only in the second step (scheme 10). The enzymes

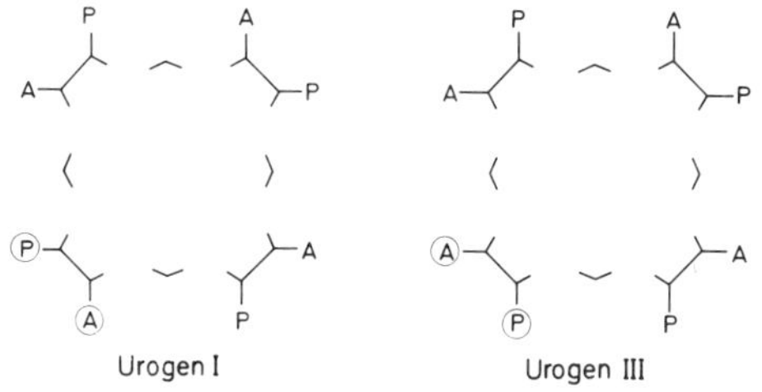

Urogen I

Urogen III

Scheme 10. Position of β-substituents in uroporphyrinogen isomers I and III. A = acetic acid residue; P = propionic acid residue.

which catalyse the reaction of PBG to urogen III are now called hydroxymethylbilane synthase (formerly porphobilinogen deaminase or urogen I synthase) and uroporphyrinogen III synthase (formerly urogen III cosynthase). It is the achievement of A.R. Battersby and his co-workers who have elucidated the exact nature and the mechanism of these two enzymic steps within more than 10 years, mainly by a combination of enzymological studies with careful (and extremely difficult) synthetic organic–chemical work and NMR spectroscopy (summarized by Battersby, 1985,1986 and Leeper, 1985a,b).

It has long been known that urogen I can be obtained by incubation of PBG with the first enzyme (then called urogen I synthase). However, Battersby and his colleagues demonstrated that the true product of the enzymic reaction is an open chain tetrapyrrole, namely a hydroxymethylbilane (scheme 11) and that this compound readily undergoes non-enzymic ring closure to yield urogen I. The enzyme has been studied from different sources—human erythrocytes (Anderson and Desnick, 1980), spinach (Higuchi and Bogorad, 1975), *Rhodopseudomonas sphaeroides* (Davies and Neuberger, 1973; Jordan and Shemin, 1973), *Chlorella regularis* (Shioi *et al.*, 1980) rat spleen (Williams, 1984), *Euglena gracilis* (Williams *et al.*, 1981) and *Escherichia coli* (Hart *et al.*, 1986). The first step of the enzyme reaction is the covalent binding of one PBG to the enzyme with elimination of NH_3. As binding residue, lysine (X = NH; Hart *et al.*, 1984), cysteine (X = S; Evans *et al.*, 1984, 1986) or an enzyme linked dipyrrole (P. Jordan, personal communication) have been discussed. Subsequently more PBGs are condensed (Anderson and Desnick, 1980; Berry *et al.*, 1981) with elimination of NH_3 and the final product released—eventually with the intermediary formation of an azafulvene which should still be enzyme-bound (Jones *et al.*, 1984; Neidhart *et al.*, 1985)—as hydroxymethylbilane. It should be pointed out that the PBG which at first binds to the enzyme is the one leading to ring A of the final porphyrin and furthermore, that the hydroxymethylbilane contains its side chains still in the order of head-to-tail condensation of the PBGs, i.e. without any rearrangements. Hence its chemical ring closure reaction yields exclusively urogen I.

Formation of urogen III is catalysed by the second enzyme, namely uroporphyrinogen III synthase. Purification and properties of the enzyme from *Euglena* have been described by Hart and Battersby (1985). The reaction involves ring closure and rearrangement of the tetrapyrrole. The enzyme mechanism is shown in scheme 12. The true substrate of this enzyme is the hydroxymethylbilane produced by the first enzyme. Early experiments with either PBG or synthetic di-, tri- or tetrapyrroles containing an α-aminomethyl group (Battersby *et al.*, 1982; Sburlati *et al.*, 1983) revealed that urogen III synthase does not accept these

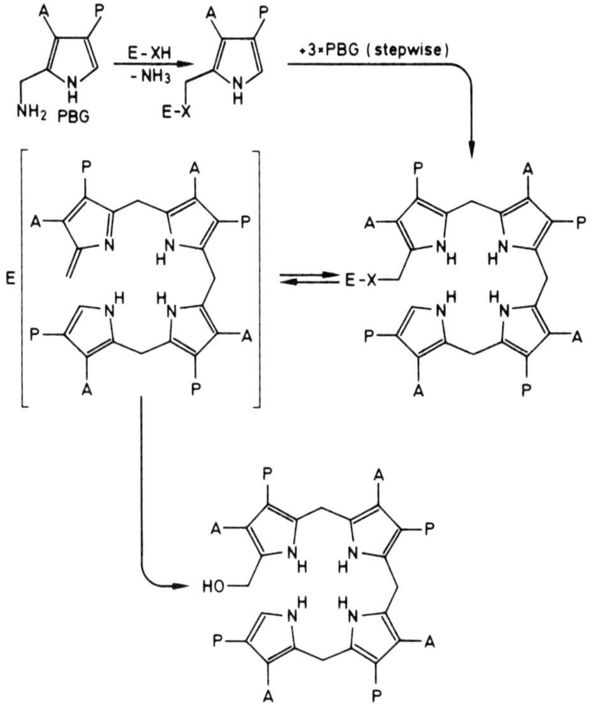

Scheme 11. Formation of hydroxymethylbilane from PBG. E-XH = hydroxymethylbilane synthase; X = NH, S or dipyrrole.

substrates as such but produces urogen III only if the first enzyme, hydroxymethylbilane synthase, is also present. The second enzyme was therefore called "urogen III cosynthase" because it was believed to act only in the presence of the first enzyme. It should be recalled here that the only known product of reaction of PBG with the first enzyme at that time was urogen I which is also not accepted by the second enzyme. It became clear only after elucidation of the hydroxymethylbilane structure that this compound is accepted by urogen III synthase without the presence of any other enzyme.

One of the key results of ^{13}C NMR spectroscopy with labelled precursors was the finding that all carbon atoms of the hydroxymethylbilane are retained in the product urogen III and that ring D is turned around but the methine carbons retain their position. Two possible mechanisms are discussed for this rearrangement reaction: one (labelled A in scheme 12) in which ring D is cleaved off the tetrapyrrole, transferred to the enzyme protein and condensed to the tripyrrole moiety in the opposite direction,

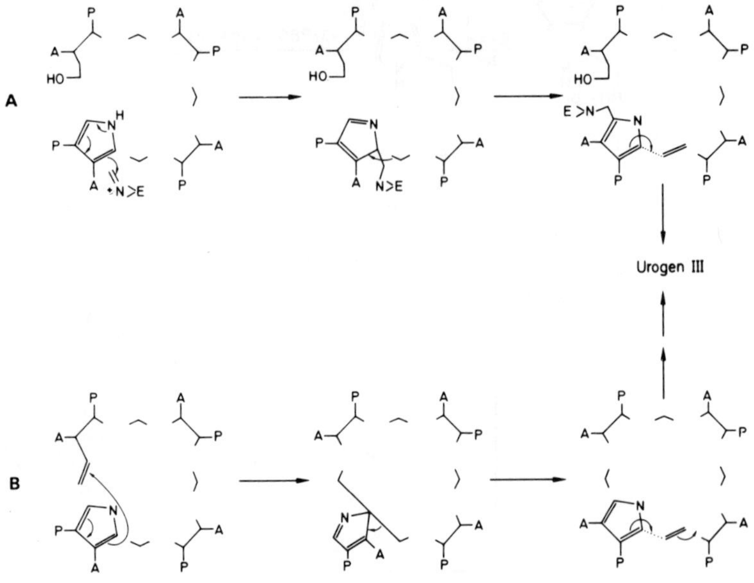

Scheme 12. Two proposed mechanisms for uroporphyrinogen III synthase. **A** ring D is released from and re-attached to the rest of the tetrapyrrole; **B** ring D remains as spiro compound throughout attached to the rest of the tetrapyrrole.

and the other (labelled B in scheme 12) in which ring D remains connected with the tripyrrole moiety and forms a spiro compound as enzyme-bound intermediate. The latter possibility is presently favoured (Battersby, 1985,1986).

D Formation of protoporphyrin

Urogen III is the last common intermediate for the biosynthesis of all tetrapyrroles (see scheme 1). One branch leading away from chlorophylls, namely to vitamin B_{12}, sirohaem, and factor F-430, is not considered here; reviews on this branch are available (Battersby, 1986; Leeper, 1985a,b; Thauer, 1985). In the chlorophyll and protohaem branch, the side chains of urogen III are at first modified to those of protogen. This compound is then dehydrogenated to protoporphyrin. Hence protoporphyrin is the first true pigment in this branch.

The side chains of urogen III are successively modified by two enzymes, namely urogen III decarboxylase and coprogen oxidase. The former enzyme transforms the four acetic acid side chains into methyl groups yielding coprogen III, a reaction which can also be achieved by treating urogen with hot acid. A reaction mechanism has been proposed (scheme 13) which is supported by incorporation experiments with stereospecifically labelled precursors (Barnard and Akhtar, 1979; Battersby *et al.*, 1981). Urogen III decarboxylase has best been isolated and investigated from human and animal sources (Mauzerall and Granick, 1958; de Verneuil *et al.*, 1983; Elder *et al.*, 1983; Kawanishi *et al.*, 1983; Straka and Kushner, 1983; Romeo *et al.*, 1986). The monomer size of the enzyme from several sources was found to be between 40 and 56 kDa. In any case, the first

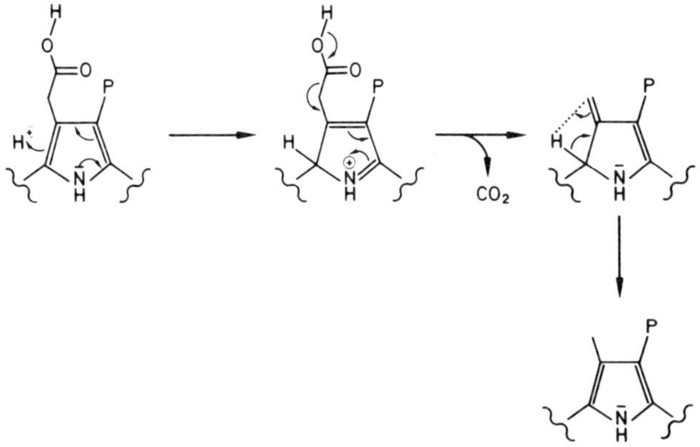

Scheme 13. Mechanism of uroporphyrinogen III decarboxylase (according to Leeper, 1985a, modified). P = propionic acid residue. The decarboxylation reaction is only presented for one ring but should occur consecutively at all four rings in the order D, A, B, C.

decarboxylation step occurs at the acetyl group on ring D. There is some medical interest in this enzyme because its deficiency leads to a certain type of porphyria (porphyria cutanea tarda) in which urogen will be produced in large quantities, partly excreted in the urine and autoxidized to uroporphyrin. Since the enzyme which contains essential thiol groups is inhibited by a number of compounds including thiol reagents, several metal ions (e.g. Cu^{2+}, Zn^{2+}, Hg^{2+}) and halogenated aromatics (e.g. PCBs), such a porphyria can also be produced artificially in mammals (Sinclair *et al.*, 1986 and papers cited therein). A corresponding reaction of plants or photosynthetic bacteria has not yet been described.

The next step is the successive oxidative decarboxylation of the propionic acid side chains of rings A and B (in this order, Elder *et al.*, 1978) to produce protogen IX from coprogen III. The enzyme coprogen

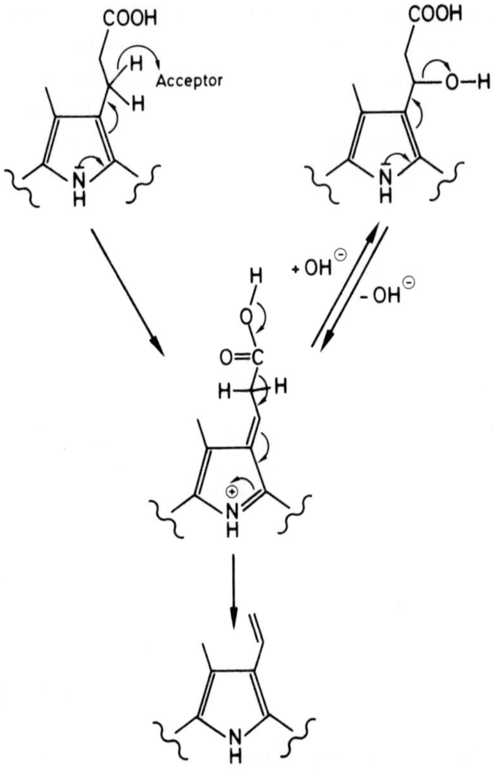

Scheme 14. Mechanism of coproporphyrinogen oxidase (according to Leeper, 1985a). The reaction occurs first at ring A and then at ring B.

oxidase has been isolated from bovine liver (Yoshinaga and Sano, 1980a) and yeast (Camadro *et al.*, 1986). A reaction mechanism has been proposed (Leeper, 1985a; scheme 14) which considers the facts that the hydroxylated species can be accepted as a substrate (Yoshinaga and Sano, 1980b) but that the enzyme does not show typical properties of hydroxylating enzymes (Seehra *et al.*, 1983). It has been demonstrated earlier by deuterium labelling experiments (Zaman *et al.*, 1972; Battersby, 1978) that decarboxylation cannot proceed via a β-keto or β-imino structure because only one deuterium atom (not two) is lost in the side chain in the course of decarboxylation. The mechanism involves abstraction of a hydride ion. Whereas the electron acceptor in aerobic organisms is molecular oxygen, anaerobes use a different electron acceptor, presumably NADP. The stereochemistry of the oxidative decarboxylation has been investigated (Zaman and Akhtar, 1976; Battersby, 1978; Yoshinaga and Sano, 1980b; Seehra *et al.*, 1983); the reaction involves trans-elimination of hydrogen and CO_2.

Investigation of the next enzyme, protogen oxidase, is hampered by the fact that all porphyrinogens are easily autoxidized to the corresponding porphyrin in the presence of air. Since molecular oxygen is the natural electron acceptor for the protogen oxidase reaction in aerobic organisms, special precautions have to be taken for the enzyme test (Camadro *et al.*, 1982; Jacobs and Jacobs, 1984). Whereas the enzymes from plants and mammals have been solubilized with detergents (Jacobs and Jacobs, 1984, 1987) this was not possible with the bacterial enzyme. In anaerobically grown microorganisms, oxidation is directly linked with the electron transport chain of membranes (Jacobs and Jacobs, 1981).

Dehydrogenation of porphyrinogens to porphyrins requires removal of six electrons (and six protons). Jones *et al.* (1984) concluded from incorporation of stereospecifically labelled PBG, namely $[2,11,S\text{-}^{3}H_2; 2,11\text{-}^{14}C_2]$ PBG into haem in a broken cell system from chicken blood, that the meso hydrogens of protogen are abstracted with high stereoselectivity in accordance with an enzymic reaction. Three of the hydrogens, one each at C-5, C-15 and C-20 respectively, are removed from one face of the macrocycle whereas the hydrogen at C-10 is removed from the other face. The authors suggested that the first mentioned hydrogens are abstracted as "hydride" together with protons from the pyrrole nitrogens and the last mentioned hydrogen as a proton via tautomerization (scheme 15). In the overall balance, "hydride" abstraction means removal of two electrons and two protons for each of the three steps, whereas tautomerization proceeds without overall gain or loss of protons.

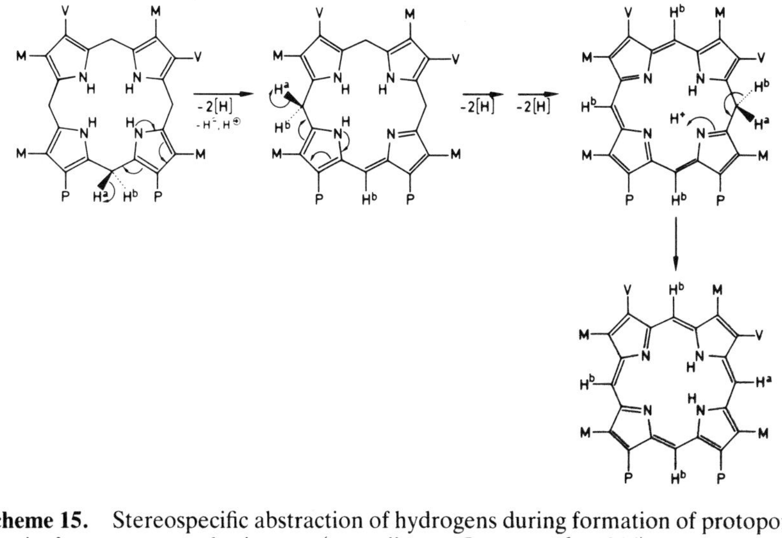

Scheme 15. Stereospecific abstraction of hydrogens during formation of protoporphyrin from protoporphyrinogen (according to Jones *et al.*, 1984).

E Pathway to protochlorophyllide

Protoporphyrin is the last common intermediate in two branches of tetrapyrrole biosynthesis, namely the iron branch leading to haem and hence to cytochromes, and the magnesium branch leading to the chlorophylls. The first few steps in the magnesium branch are depicted in scheme 16. They inlcude insertion of magnesium, formation of the monomethylester, formation of the isocyclic ring and eventually hydrogenation of one vinyl group. The resulting product, protochlorophyllide, has been obtained by incubation of either crude homogenates of etiolated seedlings or intact isolated chloroplasts with suitable precursors, such as glutamate, ALA or protoporphyrin *in vitro* (Ellsworth and Hervish, 1975; Daniell and Rebeiz, 1984; Fuesler *et al*., 1984a). It is the last true porphyrin on this pathway; the subsequent step, hydrogenation to the chlorin macrocycle, will be discussed in a separate section.

Whereas the first specific enzyme for the iron branch, ferrochelatase, has extensively been investigated (for reviews see Bogorad, 1976; Leeper, 1985a,b), few investigations are known on magnesium chelatase. Gorchein

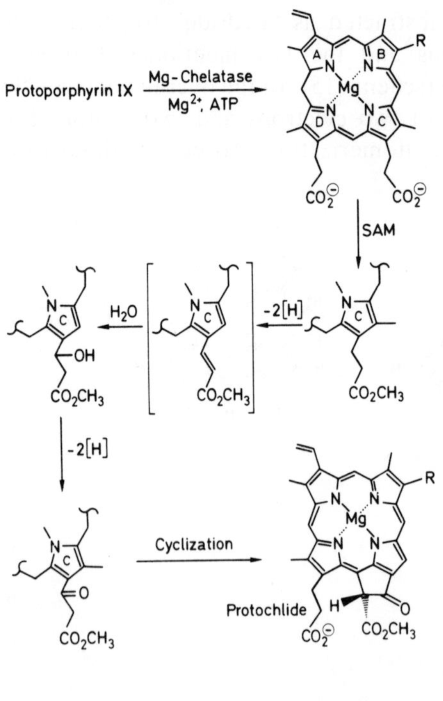

Scheme 16. Pathway of the Mg branch from protoporphyrin to protochlorophyllide. R = vinyl or ethyl, see p. 29.

(1972,1973) was able to demonstrate insertion of Mg into protoporphyrin which was administered to whole cells of *Rhodopseudomonas sphaeroides*, but not if the cells were broken up. Because the resulting product was Mg protoporphyrin monomethyl ester, Gorchein concluded that two enzymic steps (Mg chelation and methylation) are closely and obligatorily connected. However, the sequence of reactions, first Mg chelation and then methylation, was determined for developing cucumber etiochloroplasts by isolation of the respective products after incubation with the necessary cofactors (Fuesler *et al.*, 1982). After some preliminary observations on Mg protoporphyrin formation *in vitro* (Smith and Rebeiz, 1977; Castelfranco *et al.*, 1979), Pardo *et al.* (1980) were able to characterize Mg chelatase activity in isolated, developing chloroplasts from cucumber. The plastids had to be intact. The enzyme required ATP as well as the substrates, Mg^{2+} and protoporphyrin. It has been speculated that the role of ATP is to "squeeze out the water of hydration surrounding the Mg^{2+} ion" (Castelfranco and Beale, 1981). AMP proved to be a strong inhibitor, whereas ADP was only slightly inhibitory. Richter and Riemits (1982) found some Mg chelation of exogenous protoporphyrin in the presence of ATP also with lysed plastids from cucumber.

The next step, methylation of Mg protoporphyrin, is catalysed by a methyltransferase which uses S-adenosyl-L-methionine as methyl donor. This enzyme has been described by Ebbon and Tait (1969) for *Euglena gracilis* and by Ellsworth and Dullaghan (1972) for wheat seedlings. The enzyme S-adenosylmethionine:Mg-protoporphyrin O-methyltransferase has been purified by affinity chromatography starting with extracts from *Rhodopseudomonas sphaeroides* (Hinchigeri *et al.*, 1984), *E. gracilis* (Hinchigeri and Richards, 1982) and etiolated wheat (Hinchigeri *et al.*, 1981; Kwan *et al.*, 1986). The enzyme mechanism (scheme 17) seems to

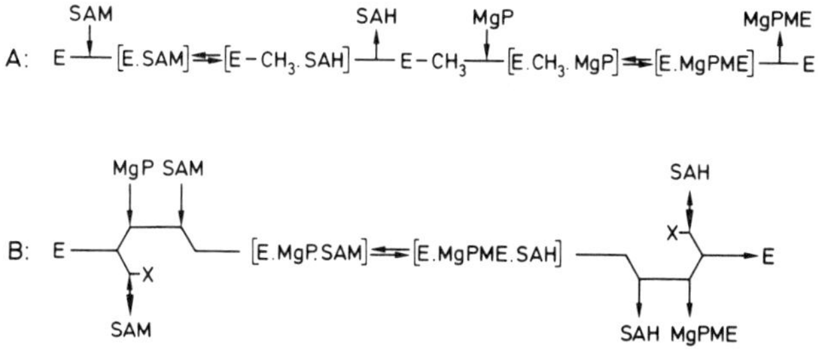

A: E —(SAM)— [E.SAM] ⇌ [E–CH_3.SAH] —(SAH)— E–CH_3 —(MgP)— [E.CH_3.MgP] ⇌ [E.MgPME] —(MgPME)— E

B: E —(MgP, SAM; SAM –X)— [E.MgP.SAM] ⇌ [E.MgPME.SAH] —(SAH, MgPME; X– SAH)— E

Scheme 17. Kinetic reaction mechanisms of S-adenosylmethionine:Mg-protoporphyrin O-methyltransferases. **A** from etiolated wheat ("ping-pong mechanism"); **B** from *R. sphaeroides* ("ordered mechanism"). The "random mechanism" of *Euglena gracilis* is similar to B, but the order of substrate binding is random (after Richards *et al.*, 1987).

be different for the enzyme from each of the three sources: a "ping-pong" mechanism was proposed for the enzyme from etiolated wheat (Ellsworth *et al.*, 1974; Hinchigeri *et al.*, 1981), a random mechanism for the enzyme from *Euglena* (Hinchigeri and Richards, 1982), and an ordered mechanism with Mg protoporphyrin as the first substrate for the enzyme from *R. sphaeroides* (Hinchigeri *et al.*, 1984). The magnesium complex of protoporphyrin is a much better substrate than protoporphyrin itself. This is in accordance with the proposed order of reactions (see scheme 16). Free protoporphyrin monomethylester might be derived by secondary removal of magnesium if the further metabolism of the Mg complex is blocked (Crawford and Wang, 1983). Interestingly, the enzyme is inhibited by chlorophyllide a, protochlorophyllide, Mg protoporphyrin monomethylester and the respective metal-free compounds (Ellsworth and St. Pierre, 1976). However, these authors concluded that the methyltransferase is not a regulatory enzyme of the chlorophyll biosynthetic pathway because neither allosteric regulation nor light induction was found.

Early proposals on formation of the isocyclic ring were derived from studies of *Chlorella* mutants by Ellsworth and Aronoff (1969) (summarized by Bogorad, 1976; Castelfranco and Beale, 1981; Leeper 1985a). These mutants excreted a number of compounds which were supposed to be chlorophyll precursors. According to the chemical structures, the compounds were believed to be derived from Mg protoporphyrin monomethylester by transformation of the propionic acid methyl ester side chain into an acrylic acid, β-hydroxypropionic acid, and β-ketopropionic acid methyl ester side chain. This was in accordance with the suggestion of a typical β-oxidation of the esterified side chain, although direct proof was still lacking. The idea that the β-keto ester side chain could easily form the isocyclic ring was supported by the finding that the synthetic β-keto compound underwent cyclization by chemical dehydrogenation (Cox *et al.*, 1969). Whereas the chemical cyclization is expected to produce a racemate at the new asymmetric centre at C-13^2, natural chlorophylls have the 13^2R configuration (Wolf *et al.*, 1967). It can be concluded that the cyclization is—as are all other steps—an enzymic reaction. A plausible mechanism for cyclization is given by Leeper (1985a).

A cell-free system for cyclization from etiolated cucumber seedlings has been described by Castelfranco and co-workers. Although intact developing plastids had to be used at first (Chereskin *et al.*, 1982; Fuesler *et al.*, 1984a), the reaction could later also be performed with sonicated plastids if both supernatant and 150 000 **g** pellet were combined (Wong and Castelfranco, 1984). The active factors of both fractions were heat-labile. This was clear evidence that at least two different enzymes were involved. The enzyme, which was called Mg protoporphyrin IX monomethylester (oxidative) cyclase, needs NADH or NADPH and molecular oxygen as cofactors. In an *in vitro* system from wheat seedlings, Ellsworth and Murphy (1979) had described biosynthesis of protochlorophyllide from Mg protoporphyrin monomethylester via two intermediates. Different cofactors were used in this experiment, namely NAD, ATP and coenzyme A. The compounds were only characterized by their fluorescence spectra, however. In several higher plants, iron deficiency led to lack of the cyclization reaction (Spiller *et al.*, 1982). However, the reaction was not inhibited by iron chelators *in vitro* (Chereskin *et al.*, 1982). Inhibition was found with methyl viologen and other inhibitors of the electron transport chain. Starting with Mg protoporphyrin monomethylester, the compound with the β-hydroxypropionate group was identified as an intermediate (Wong *et al.*, 1985). This compound and the corresponding β-ketopropionate derivate have been synthesized (Smith and Goff, 1986); both are converted to protochlorophyllide by the cell-free system.

Protoporphyrin is a 3,8-divinyl compound and the bulk of final chlorophylls has the 3-vinyl-8-ethyl structure, with the exception of a maize mutant which produces considerable amounts of [8-vinyl]chlorophylls a and b (see Bazzaz and Brereton, 1982; Wu and Rebeiz, 1985). Hydrogenation of the 8-vinyl group must therefore occur at some stage of the biosynthetic pathway. As shown in scheme 16, all intermediates between Mg protoporphyrin and protochlorophyllide have been found in *Chlorella* mutants as divinyl and as monovinyl, monoethyl derivatives (Ellsworth and Aronoff, 1968, 1969). These authors concluded that the enzymes of chlorophyll biosynthesis are not very specific with regard to the presence of vinyl or ethyl side chains at C-8. On the other hand, Rebeiz and co-workers (summarized by Rebeiz *et al.*, 1983) concluded from their results that well separated mono- and divinyl pathways should exist (Tripathy and Rebeiz, 1986). These authors were able to demonstrate, in a great number of papers, that chlorophylls and their precursors are heterogeneous with regard to their fluorescence properties. The compounds with longer wavelength fluorescence excitation maxima are the divinyl, while those with shorter wavelength fluorescence are the monovinyl–monethyl compounds (Wu and Rebeiz, 1984; see also Table 1.3). Although fluorescence excitation is ideal for detection of even minimal amounts of byproducts, it is not very useful for quantitation of these products. Nevertheless it is clear that a considerable part of protochlorophyllide in cucumber etioplasts is the divinyl compound*

* This compound has previously been called Mg-divinylphaeoporphyrin a_5. We use the name [8-vinyl]protochlorophyllide (see Bonnett, 1973).

(Hanamoto and Castelfranco, 1983; Wu and Rebeiz, 1984) whereas the bulk of protochlorophyllide in etiolated barley and maize is the monovinyl–monethyl compound (Belanger *et al.*, 1982; Bombart *et al.*, 1984). The relative proportions of divinyl and monovinyl compounds seem to vary not only with the plant species but also with the physiological state of one and the same species: whereas etiolated seedlings of both monocots and dicots contain protochlorophyllide (monovinyl pigment), reaccumulation in the dark after repeated light flashes yields nearly exclusively [8-vinyl]protochlorophyllide in dicots (Belanger *et al.*, 1982).

The enzyme "vinyl reductase' has not yet been investigated intensively. Ellsworth and Hsing (1973, 1974) described enzyme activity in extracts from etiolated wheat seedlings in which the 8-vinyl group of Mg protoporphyrin methyl ester was reduced with NADH as cosubstrate; [8-vinyl]protochlorophyllide was not a substrate. Richards *et al.* (1987) described two fractions, a soluble and a pelletable vinyl reductase in lysed wheat etioplasts. The soluble enzyme was enriched 70-fold by affinity chromatography. This enzyme accepted Mg protoporphyrin methylester as well as [8-vinyl]protochlorophyllide; the best incorporation of radioactive label was obtained with $(4R^{-3}H)$NADPH whereas $(4S^{-3}H)$NADPH and NADH were less effective. No reduction of the vinyl group was found in the experiments of Castelfranco and co-workers (e.g. Wong *et al.*, 1985; Huang and Castelfranco, 1986) in which formation of the isocyclic ring was investigated in intact or broken plastids from cucumber (see above). This is remarkable because the incubation mixture also contained NADPH or NADH in these cases. Duggan and Rebeiz (1982a) have accumulated evidence that reduction of the 8-vinyl group occurs only at the stage of chlorophyllide or even esterified chlorophyll in cucumber. Differences between different plant species, especially monocots and dicots, may be reflected here. Emery and Akhtar (1985) presented evidence that the reduction of the 8-vinyl group occurs via the formal trans-addition of hydrogen in the course of bacteriochlorophyll a biosynthesis.

F Formation of the chlorin chromophore: reduction of ring D

Angiosperm seedlings do not become green if germinated in total darkness. These pale yellow, etiolated seedlings accumulate chlorophyll precursors to the extent of $<1\%$ of the amount of chlorophylls in light-grown controls. These precursors consist mainly of protochlorophyllide, esterified protochlorophyll (5-15% of the precursor pool; Schoch *et al.*,

1977; Shioi and Sasa, 1983a) and traces of other precursors like Mg protoporphyrin ester (Rebeiz *et al.*, 1983). Since most of the accumulated protochlorophyllide is immediately transformed into chlorophyllide a by a short light pulse, a light-dependent enzyme, protochlorophyllide reductase, must be present in etiolated seedlings. This enzyme will be discussed below. Light regulation of chlorophyll biosynthesis is mainly achieved by this enzyme.

A second, light-independent pathway of chlorophyll biosynthesis exists, of which the significance and distribution may sometimes be underestimated. It has been known for a long time that algae, mosses, some ferns and gymnosperm seedlings can form chlorophyll in the dark (Michel-Wolwertz, 1977 and papers cited therein), but it became clear only recently that light-independent chlorophyll formation also occurs in preilluminated angiosperms (Adamson *et al.*, 1980, 1985a,b; Maitra and Mukherji, 1983) and dark-grown tobacco cell cultures (Ikegami *et al.*, 1984). Because of its limited capacity and differences in the incorporation of precursors into chlorophyll in the light and in the dark (Packer and Adamson, 1986), it has been suggested that this pathway may either allow a further (but limited) thylakoid development to continue at night (Adamson *et al.*, 1985b; Packer and Adamson, 1986) or play a chlorophyll repair role at night in angiosperm (Tripathy and Rebeiz, 1987).

It now appears that the light-dependent and light-independent pathways of chlorophyll biosynthesis are not mutually exclusive but generally occur in one and the same organism. This is true not only for angiosperms, but also for gymnosperms which form distinctly more chlorophyll in the light than in the dark (Kasemir and Mohr, 1981). In accordance with this view, light-dependent protochlorophyllide reductase has also been found in gymnosperms (Griffiths and Mapleston, 1978). Several mutants of algae have been described which form chlorophyll only in the light, whereas the parent wild type was able to synthesize chlorophyll in the dark (reviewed by Castelfranco and Beale, 1981). This phenomenon can easily be understood if one assumes that both light-dependent and light-independent pathways are present in the wild type and that only the light-independent pathway is lost in the mutation.

The two enzymic pathways have their counterparts in chemistry: hydrogenation of ring D can be achieved either by chemical or by photochemical reduction. In a review of these reactions, Scheer (1978) stated that "chlorins derived from highly symmetric model porphyrins can be prepared in high yields by chemical reduction, but less symmetric chlorins like the ones related to chlorophylls are not accessible in this way". Asymmetric chlorins can be prepared from porphyrins, however, which possess considerable steric hindrance between neighbouring coplanar groups. Woodward (1960) used this principle of the "overcrowded periphery" for his total synthesis of chlorophyll. The final reaction product has the correct stereochemistry at ring D, namely the 17,18-trans configuration. The reaction is not a direct reduction of ring D, however, but proceeds via several intermediates. The photochemical reduction, also known as the "Krasnowskii reaction", seems to lead directly to a chlorin (besides other hydrogenation products of the porphyrin) but the product has the 17,18-cis configuration in this case (Wolf and Scheer, 1973; see also Scheer, 1978).

Whereas only preliminary investigations on the light-independent enzymic reduction of ring D are known (Adamson *et al.*, 1987) the light-dependent protochlorophyllide reductase has been investigated for a long time (reviewed by Bogorad, 1976; Schneider, 1980; Castelfranco and Beale, 1981,1983; Virgin, 1981). It catalyses the stereoselective reduction of ring D which is a key reaction of chlorophyll biosynthesis (scheme 18). The enzyme specificity was investigated by Griffiths and co-workers (see

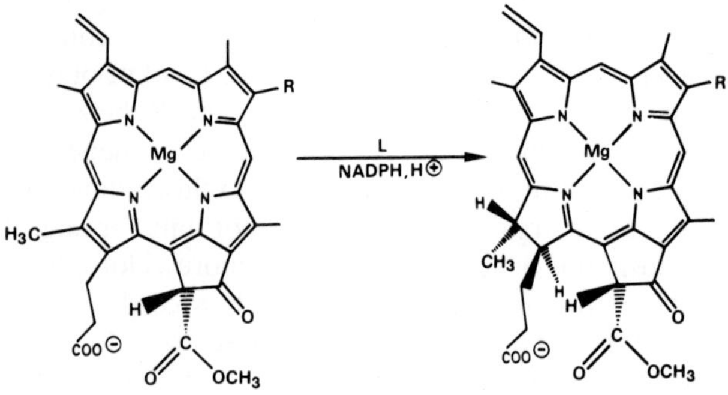

Scheme 18. Reaction of light-dependent NADPH: protochlorophyllide oxidoreductase.

Griffiths, 1978 and earlier papers cited therein). It needs NADPH; NADH and other reductants are ineffectual. The enzyme accepts exogenous protochlorophyllide and [8-vinyl]protochlorophyllide but not the esterified species, protochlorophyll. Whereas the Mg-free protophaeophorbide is no substrate, its zinc complex is accepted by the enzyme (Griffiths, 1980). Most of the protochlorophyllide which accumulates in angiosperms in the dark (see above) is already bound to this enzyme. The complex, which has often been called "protochlorophyll(ide) holochrome", can be considered as a preformed enzyme–substrate complex; immediate formation of chlorophyllide a is observed *in vitro* on irradiation of the isolated complex.

The enzyme protein has been identified, isolated from etiolated plants and investigated in several laboratories (e.g. Apel *et al.*, 1980; Oliver and Griffiths, 1981; Ikeuchi and Murakami, 1982). It is the main protein of prolamellar bodies (Ryberg and Sundquist, 1982; Dehesh and Ryberg, 1985). It consists of only one polypeptide chain; the molecular weight is between 36 and 38 kDa depending on the plant species. Earlier reports on the existence of several enzymically active polypeptides are due to partial proteolysis either *in vitro* during the extraction (Röper *et al.*, 1983; Röper, 1985) or *in vivo* during senescence of the etiolated plants (Bergweiler *et al.*, 1983). After irradiation, i.e. after photoconversion of protochlorophyllide *in vivo*, most of the enzyme disappears by proteolysis (Kay and Griffiths, 1983; Griffiths and Walker, 1987).

In 1957 Shibata first described the spectral changes occurring just after protochlorophyll(ide) phototransformation in etiolated leaves at room temperature. The long wavelength chlorophyll(ide) formed within 30 s, absorbing around 684 nm, and changed gradually to chlorophyll(ide) species, absorbing first around 672 nm (observed within 10–20 min) and then around 678 nm (observed after several hours). The shift from 684 to 672 nm has conventionally been named the Shibata shift. A great number of investigations on such spectral shifts followed (reviewed by Bogorad, 1976; Kasemir, 1983 a,b). The present view for *in vivo* phototransformation (scheme 19A) includes a chlorophyll(ide) species absorbing at 678 nm, observed after less than 1 s at room temperature (Gassman *et al.*, 1968b) and a non-fluorescent intermediate absorbing at around 690 nm. This intermediate, X_{690}, is photoreversible (Litvin *et al.*, 1981). It is stable at liquid nitrogen temperature (Dujardin and Sironval, 1977; Dujardin and Correia, 1979) and decays at room termperature in less than 8 ms (Franck *et al.*, 1980). If only a minor part (5–10%) of the protochlorophyllide present in etiolated leaves was phototransformed by

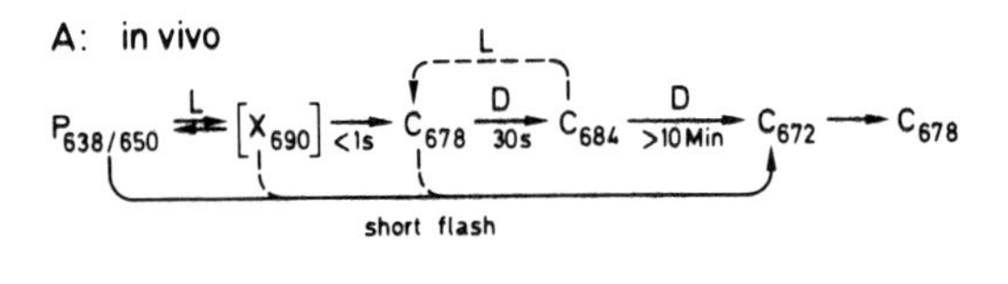

Scheme 19. Spectral shifts observed after the first phototransformation of protochlorophyllide to chlorophyllide a. **A** *in vivo*; **B** *in vitro*. Times are indicated for the shifts at room temperature.

a 15-ms light pulse (Klockare and Virgin, 1983) or by low-intensity light (Ogawa and Konishi, 1977), the Shibata shift (684 to 672 nm) was not observed; the newly formed chlorophyll(ide) already had its absorption maximum at 672 nm within 10 s after the flash (Klockare and Virgin, 1983). Scheme 19A also shows the observation of Franck and Inoue (1984) that red light can shift C_{684} back to C_{678}. This seems to be related to the light-dependent changes of the NADPH:NADP ratio in plants (Mapleston and Griffiths, 1978) rather than to photochemical reactions of the pigment (Franck and Schmid, 1985).

The same spectral shifts have also been found *in vitro* with etioplast membrane preparations or purified "protochlorophyllide holochrome" (see scheme 19B). The short-lived intermediate X_{690} (Franck and Mathis, 1980; Inoue *et al.*, 1981) appeared instantly after the laser flash and decayed with a rate constant of 3.4×10^5 s^{-1}. In earlier work (cited in Bogorad, 1976), phototransformation stopped at the stage of C_{678}. Griffiths (1978) detected a further shift to 682 nm by addition of NADPH. The classic Shibata shift (682 to 672 nm) could be reproduced *in vitro* by addition of protochlorophyllide which apparently displaces chlorophyllide from the enzyme (Oliver and Griffiths, 1982).

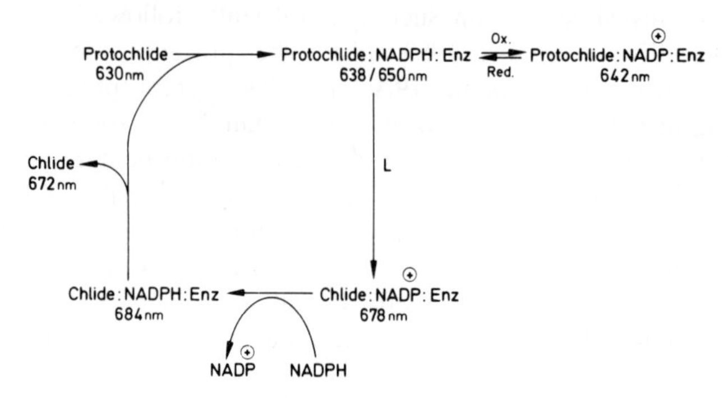

Scheme 20. Reactions observed *in vitro* which explain the spectral shifts indicated in scheme 19B (after Oliver and Griffiths, 1982). Enz = NADPH:protochlorophyllide oxidoreductase.

Oliver and Griffiths (1982) presented a scheme (scheme 20) which explains the spectral shifts observed *in vivo* and *in vitro* at the molecular level. An essential point of this scheme is that a ternary complex of protochlorophyllide, NADPH, and the enzyme has to be formed before the phototransformation can take place. Free protochlorophyllide has an absorption peak at above 630 nm. In the ternary complex with NADPH, the absorption peak is shifted to 638/650 nm. After oxidation of the enzyme or the presence of excess $NADP^+$, the absorption peak is shifted to about 642 nm but the complex is not phototransformable (El Hamouri and Sironval, 1980; El Hamouri *et al.*, 1983). This together with free protochlorophyllide may constitute the "non-phototransformable pool" of protochlorophyllide found in etiolated seedlings. The first stable photoproduct (absorbing at 678 nm) is chlorophyllide a bound to the oxidized (= $NADP^+$-containing) enzyme. This form is stable *in vitro*, especially in the presence of $NADP^+$. In the presence of excess NADPH, a form absorbing at 684 nm accumulates. The next step, a slow shift to 672 nm (Shibata shift) can be simulated *in vitro* under conditions (addition of thiol reagents (Oliver and Griffiths, 1982) or detergents, freeze-thawing) which release chlorophyllide from the enzyme. The form with an absorption peak at 672nm is therefore considered to be free chlorophyll(ide). It is assumed that chlorophyllide is released *in vivo* because it is replaced by protochlorophyllide (Oliver and Griffiths, 1982). The last shift from 672 to 678 nm observed *in vivo* (see scheme 19A) is not observed *in vitro*. Bogorad (1976) pointed out that esterification of chlorophyllide a takes place during the last shift, but it should be remembered that other events, e.g. rearrangement of the plastid structure and relocalization of protochlorophyllide reductase, also take place at the same time (Ryberg and Dehesh, 1986).

G Biosynthesis of phytol and phytylation of chlorophyllide

An essential part of the complete chlorophyll molecule is the phytol residue which, although it makes up about one-third of the molecular weight of chlorophyll, is often neglected in chlorophyll reviews. Phytol belongs to the class of isoprenoids. The specific C_5-precursor of the isoprenoid pathway, isopentenyl diphosphate (IPP), is incorporated in the cytoplasm of plants mainly into the C_{15}-compound farnesyl diphosphate, but in plastids mainly into the C_{20}-compound geranylgeranyl diphosphate (GGPP; see scheme 21). Prenyltransferase activity leading to this C_{20}-compound was demonstrated in plastids either indirectly (Benz and Rüdiger, 1981a; Kreuz and Kleinig, 1981) or directly (Block *et al.*, 1980). The disphosphate form is apparently a substrate for a hydrogenase of plastids which catalyses the formation of phytyl diphosphate; cofactor is NADPH (Soll and Schultz, 1981; Soll *et al.*, 1983). Whereas GGPP is preferentially in the supernatant fraction of plastid lysates or tissue extracts (Benz *et al.*, 1981, 1983), phytyl diphosphate is preferentially membrane-bound (Rüdiger and Benz, 1984).

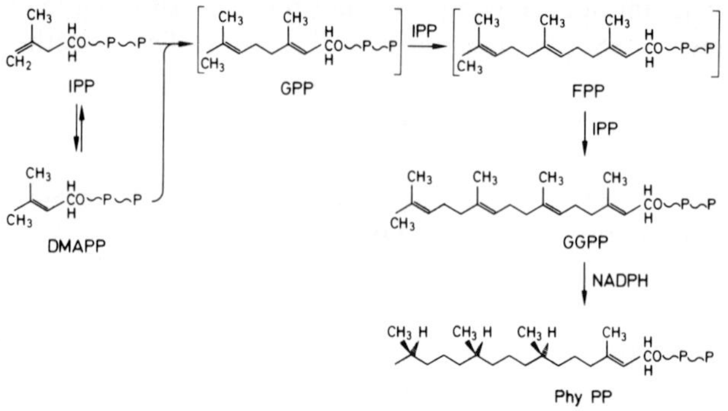

Scheme 21. Biosynthesis of phytyl diphosphate. IPP = isopentenyl diphosphate; DMAPP = dimethylallyl diphosphate; GPP = geranyl diphosphate; FPP = farnesyl diphosphate; GGPP = geranylgeranyl diphosphate.

Although esterification of chlorophyllide with free phytol by the enzyme chlorophyllase has formerly been discussed as a possible step in chlorophyll biosynthesis (reviewed by Bogorad, 1976), such a possibility is unlikely according to more recent experimental results. Chlorophyllase catalyses hydrolysis of chlorophylls into chlorophyllides and free alcohols under normal conditions; the reverse reaction requires a large excess of free alcohols. The required high concentrations of phytol could not be detected in greening oat, wheat and barley seedlings even under optimum conditions of chlorophyll biosynthesis (Steffens *et al.*, 1976). Akhtar and co-workers (1984) considered possible mechanisms for esterification of chlorophylls (scheme 22). These authors detected ^{18}O-label in the phytol or geranylgeraniol obtained by chemical hydrolysis of bacteriochlorophylls after incorporation of $[1\text{-}^{18}O_2]$-ALA (see also Emery and Akhtar, 1987). The label is expected in the carboxyl groups of the newly formed chlorophyllides. According to scheme 22, ^{18}O-label of the carboxy group of ring D is not expected in the hydroxy group of the alcohol after chlorophyllase reaction (mechanism 1) or after esterification via an activated carboxylic

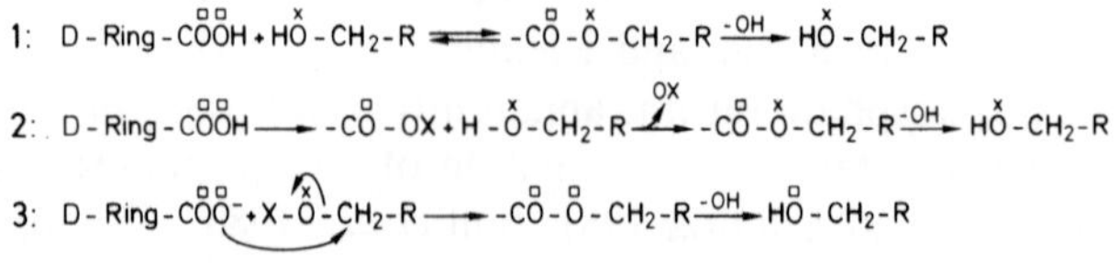

Scheme 22. Possible mechanisms for esterification in chlorophyll and bacteriochlorophyll biosynthesis (after Akhtar *et al.*, 1984). Only in mechanism 3 can the label from the carboxy oxygens be found in the alcohol after alkaline hydrolysis.

acid derivative (mechanism 2). The label is however expected in the alcoholic hydroxy group if the esterification occurs via prenylation of the free carboxy group, i.e. via an activated alcohol (mechanism 3).

An enzyme activity which meets these requirements was detected in 1980 and named chlorophyll synthetase (Rüdiger *et al.*, 1980). This enzyme accepts the diphosphates of phytol or geranylgeraniol; the monophosphates or the free alcohols react only in the presence of excess ATP which is able to phosphorylate these compounds in the enzyme preparations.

The enzyme preparation from etioplasts prefers GGPP (or geranylgeraniol + ATP) to phytyl diphosphate (or phytol + ATP; Rüdiger *et al.*, 1980) whereas the enzyme from green chloroplasts has a preference for phytyl diphosphate (Soll *et al.*, 1983). The C_{15} compound farnesol is a less suitable substrate whereas the C_{10} compound geraniol (see scheme 21) and non-iosoprenoid alcohols like n-pentadecanol are not substrates. Farnesyl diphosphate (or farnesol + ATP) leads to small amounts of farnesylchlorophyllide if washed plastid membranes are used as the enzyme source. In the presence of IPP, however, only geranylgeranylchlorophyllide is obtained (Benz and Rüdiger, 1981a). This is apparently due to prenyltransferase activity in the membrane fraction which may effectively compete with chlorophyll synthetase for farnesyl diphosphate. The specificity of chlorophyll synthetase for the tetrapyrrole unit was investigated by Benz and Rüdiger (1981b). Whereas modifications at rings A (to 3-acetylchlorophyllide) and C (to pyrochlorophyllide) are tolerated by the enzyme, the structure of rings B and D seem to be critical because neither bacteriochlorophyllide nor protochlorophyllide is a suitable substrate. The Mg-free phaeophorbide is also not accepted by chlorophyll synthetase. These properties are different from those of chlorophyllase which has no specificity with regard to the alcohol and also accepts bacteriochlorophyllide and phaeophorbide (in the synthetic reaction). The enzymic esterification can also be observed in etiolated leaves which had been infiltrated with exogenous chlorophyllide a (Vezitskii and Walter, 1981).

Esterification of newly formed chlorophyllide has also been studied *in vivo*. Protochlorophyllide which had accumulated in *etiolated* seedlings in the dark can be transformed into chlorophyllide a by a light pulse or light flash. Esterification of this newly formed chlorophyllide starts immediately after phototransformation and is completed after 1–2 h at room temperature. Analysis of newly formed chlorophyll in oat (Schoch *et al.*, 1977) and bean (Schoch, 1978) seedlings revealed the presence of several isoprenoid alcohols. It was concluded from kinetic analysis that geranylgeraniol was the first alcohol to be linked to the newly formed

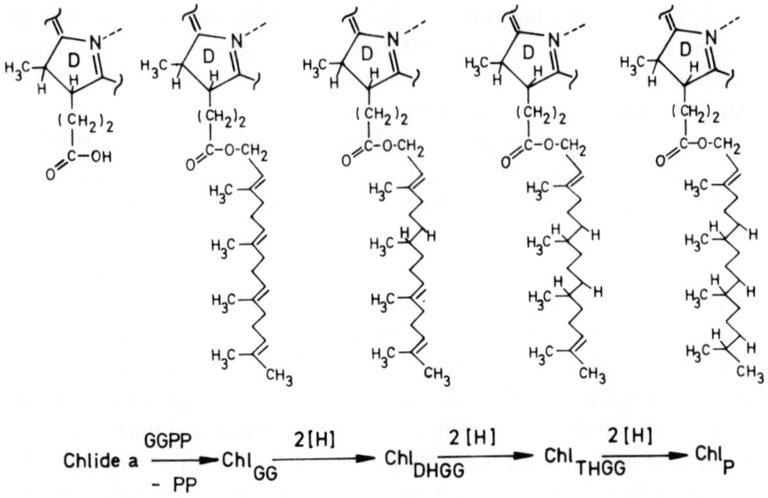

$$\text{Chlide a} \xrightarrow[-\text{PP}]{\text{GGPP}} \text{Chl}_{GG} \xrightarrow{2[\text{H}]} \text{Chl}_{DHGG} \xrightarrow{2[\text{H}]} \text{Chl}_{THGG} \xrightarrow{2[\text{H}]} \text{Chl}_{P}$$

Scheme 23. Chlorophyll precursors containing various alcohols detected in etiolated seedlings after short irradiation (after Rüdiger and Benz, 1984).

chlorophyllide and that hydrogenation occurs stepwise to the final product phytylchlorophyllide (scheme 23). The structures of the intermediates between geranylgeraniol and phytol were derived from mass spectroscopy (Rüdiger *et al.*, 1976; Schoch and Schäfer, 1978). It has been demonstrated that this hydrogenation activity can be lost *in vivo* without parallel loss of chlorophyll synthetase activity by treatment of the seedlings with aminotriazole (Rüdiger and Benz, 1979) or with S-ethyl-dipropylthiocarbamate (Wilkinson, 1985) or by anaerobic pretreatment of seedlings (Schoch *et al.*, 1980).

In order to label newly formed chlorophyll in *green* seedings, shoots of light-grown barley seedlings were incubated with ^{14}C-ALA in the dark for 2 h and then transferred into the light for 15 min. Since the ^{14}C-label was only found in phytylchlorophyllide and not in geranylgeranylchlorophyllide (Dehesh, Schoch, Lempert and Rüdiger, unpublished results), it was concluded that chlorophyllide of green seedlings is esterified directly with phytyl diphosphate and not with geranylgeranyl diphosphate, which would imply only subsequent hydrogenation to phytylchlorophyllide. Thus it seems that two pathways exist for the esterification. In etiolated seedlings, the pathway via Chl_{GG} predominates, whereas in green seedlings the pathway via phytyldiphosphate seems to be preferred.

H Parallel pathways of chlorophyll biosynthesis

The finding of two possible pathways for esterification of chlorophyllide a does not yet mean that such pathways must be operating well separated from each other within the plastids. One possible explanation could be the availability of substrates, e.g. if more GGPP were available in etioplasts and more phytyl diphosphate in chloroplasts, the results could be explained since chlorophyll synthetase takes both of these substrates. Nevertheless, there might be different chlorophyll synthetases in etioplasts and chloroplasts because of differences in the specifity for substrates (see above).

Esterification must also occur to a certain extent on the stage of protochlorophyllide. As mentioned above, some esterified protochlorophyll is always found besides protochlorophyllide in etiolated plants. The same four alcohols as in newly formed chlorophyll (see scheme 23) are found in protochlorophyll (Shioi and Sasa, 1983b). While the absolute amount of esterified protochlorophyll increases with the age of etiolated seedlings (Shioi and Sasa, 1983a), the percentage of the esterified pigment decreases because more (unesterified) protochlorophyllide is accumulated with time. In 2-day-old, dark grown bean seedlings in which the plastids are still more similar to proplastids than to well-developed etioplasts, and in greening *Euglena*, as much as 50% of total protochlorophyll(ide) is esterified (Lancer *et al.*, 1976; Kindman *et al.*, 1978). It is not yet known whether another enzyme is responsible for esterification in this case or whether a certain "leakiness" of chlorophyll synthetase (Benz and Rüdiger, 1981b) allows esterification of protochlorophyllide by this enzyme. According to Lancer *et al.* (1976), esterified protochlorophyll is directly phototransformed into chlorophyll upon irradiation in these young bean seedlings. This certainly constitutes a parallel pathway of chlorophyll biosynthesis. Rebeiz *et al.* (1983) also found small amounts of esterified species of earlier chlorophyll precursors (protoporphyrin, Mg-protoporphyrin). These authors found heterogeneity of the alcohol residues of these esterified precursors according to GC/MS analysis but the alcohols have not been identified so far. Nevertheless, the data allowed the conclusion that two different types of alcohols, isoprenoid and non-isoprenoid exist. Rebeiz *et al.* (1983) assume separate pathways for tetrapyrrole isoprenoid esters, non-isoprenoid esters and carboxylic acids. This division, combined with the possibilities of mono- and divinyl pathways, leads to a scheme with six parallel pathways. It should be recalled, however, that the carboxylic acid pathway makes up for more that 90% chlorophyll biosynthesis according to these authors. As mentioned above, the divinyl pathway seems to be preferred in dicots and the monovinyl pathway in monocots.

I Chlorophyll b and other chlorophylls

Chlorophyll b is a constituent of the light-harvesting system in higher plants, green algae, Euglenaceae and Prochlorophyta (see Burger-Wiersma *et al.*, 1986); its amount is usually one-third (or less) of that of chlorophyll a. Chlorophyll b contains a formyl group at ring B (see scheme 2) where chlorophyll a and its precursors contain a methyl substituent. It is obvious therefore to assume an oxygenation reaction at this methyl group; a hitherto unknown intermediate could be a compound with a hydroxymethyl substituent. At which stage of chlorophyll biosynthesis does such a reaction occur? Since an enzyme for this reaction has not yet been described, current ideas about chlorophyll b formation are mostly based on indirect evidence.

It has been mentioned above that etiolated plants form chlorophyllide a upon short irradiation, which is then esterified to chlorophyll a in the dark. Under these conditions, chlorophyll b is also formed in the dark at the cost of chlorophyll(ide) a. As pointed out by Klockare and Virgin (1984), chlorophyll b formation can still be observed in wheat seedlings under these conditions when all chlorophyllide a has already been esterified to chlorophyll a. Since no chlorophyll b formation occurs at 0°C under these conditions, the authors assume enzymic transformation of chlorophyll a into chlorophyll b. This assumption is in accordance with widely accepted conclusions from earlier feeding experiments (summarized by Shlyk, 1971; Bogorad, 1976) in which radioactively labelled chlorophyll a or precursors were transformed into chlorophyll b. However, Bogorad also mentioned the great difficulty of bringing chlorophylls to radiopurity. This implies non-chlorophyllous compounds in the chlorophyll fractions and can be extended to artifactual products formed from chlorophylls during the purifiication procedures. Such artifacts might be found in pigment fractions different from those of the parent pigments. Therefore, some of the early (eventually conflicting) data from tracer experiments must be interpreted with caution.

During studies on the light-dependency of chlorophyll b formation, Virgin (1977) came to the conclusion that chlorophyll b is formed from newly synthesized chlorophyll a molecules. Whereas Virgin believed that this newly synthesized chlorophyll a could be C_{672} (see scheme 19A), Rüdiger (1987) suggested that it could as well be geranylgeranylchlorophyllide a, which precedes phytylchlorophyllide a in greening leaves.

Oelze-Karow and Mohr (1978) observed that chlorophyll b formation in greening mustard seedlings is greatly enhanced by red light pretreatment acting through phytochrome. Although these authors could not define the immediate precursor of chlorophyll b they stated: "it is tempting to suggest chlorophyllide a as the common precursor for chlorophyll a and chlorophyll b". This would imply chlorophyllide b as the direct precursor of chlorophyll b. Trace amounts of chlorophyllide b have been found in greening pea (Aronoff, 1981) and cucumber seedlings (Duggan and Rebeiz, 1982b; Shioi and Sasa, 1983c). These laboratories came to the conclusion that not all of the detected chlorophyllide b could have arisen from chlorophyllase action on chlorophyll b during the extraction. The possibility that chlorophyllide b is a normal intermediate is supported by the findings that it is a good substrate for chlorophyll synthetase (Benz and Rüdiger, 1981b) and that the newly formed chlorophyll b contains the alcohols geranylgeraniol, dihydrogeranylgeraniol and tetrahydrogeranylgeraniol in addition to phytol (Shioi and Sasa, 1983c).

The direct formation of chlorophyllide b from protochlorophyllide (not via chlorophyllide a) is suggested for the mutant strain Y-1 of *Chlamydomonas reinhardtii* (Bednarik and Hoober, 1985a,b). This mutant, which needs light for normal chlorophyll biosynthesis, excretes several chlorophyllides into the medium upon transfer from 25°C (normal growth temperature) to 38°C. This is protochlorophyllide in the dark or chlorophyllide a in the light. Surprisingly, chlorophyllide b (but not chlorophyllide a) is excreted in the dark after treatment of the cells with heterocyclic bases like phenanthroline or phenanthridine. Chlorophyllide b is formed under these conditions at the expense of protochlorophyllide. Although these authors do not speculate about the details of this direct conversion of protochlorophyllide to chlorophyllide b, the results suggest oxygenation of the methyl group at the stage of protochlorophyllide. This would eventually lead to a "protochlorophyllide b". Early reports on the detection of such pigment were later withdrawn (discussed in Bogorad, 1976). It remains to be seen whether an unknown protochlorophyllide which is accumulated by a mutant strain C-2A' of *Scenedesmus obliquus* in the dark (Kotzabasis and Senger, 1986a) is "protochlorophyllide b"; chlorophyllide b is among the chlorophyllides which are promptly formed upon irradiation (Kotzabasis and Senger, 1986b).

In summary, the question of chlorophyll b biosynthesis is not yet settled. Oxygenation of the methyl group of ring B could occur either at the stage of chlorophyll a or chlorophyllide a or protochlorophyllide a. As discussed above, experimental results in favour of each of these possibilities have been obtained. Since the conflicting data are from various plants, ranging from higher plants to algae, it cannot be excluded

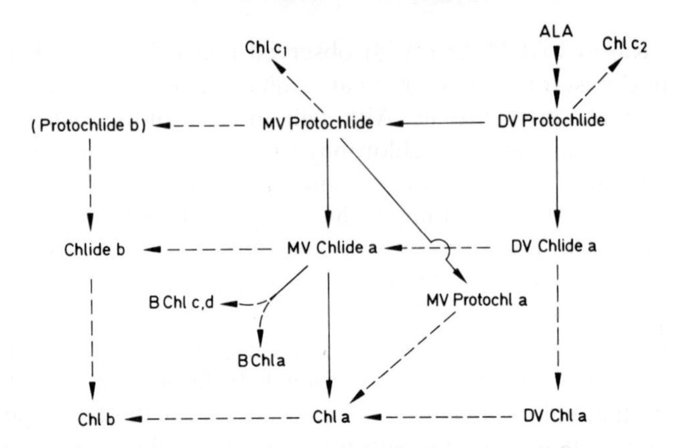

Scheme 24. Possible pathways of biosynthesis of chlorophylls as derived from their structural relationship.

that the reaction occurs at different biosynthetic stages in various plants.

Owing to the lack of direct experimental evidence, present ideas on biosynthetic routes to other chlorophylls are mostly based on the relationships of their chemical structures (see scheme 24).

Chlorophyll RCI should be derived from chlorophyll a by introduction of a hydroxy group at C-13^2 and by a chlorine substituent at C-20. Both reactions can easily be performed as chemical reactions (Hynninen and Lötjönen, 1981; Scheer *et al.*, 1986) but its biochemical counterparts are still unknown. The modifying reactions can also occur at the chlorophyllide stage (Kotzabasis and Senger, 1986b).

The structures of *chlorophylls* c_1 and c_2 are identical with those of protochlorophyllide and [8-vinyl]protochlorophyllide, respectively, except for the acrylate side chain at ring D. Since modification of the propionate side chain at ring C is believed to proceed via formation of an acrylate side chain (see scheme 16), such a transformation is suggested to occur in chlorophyll c-containing organisms which also have the propionate side chain at ring D.

Likewise, *chlorophyll d* should arise from the chlorophyll a structure by oxidation of the 3-vinyl group. but it is unknown at which stage of chlorophyll biosynthesis this occurs.

As mentioned in the introduction, early investigations on mutants of photosynthetic bacteria revealed that biosynthesis of *bacteriochlorophylls* follows the same general pathway as biosynthesis of chlorophyll a in higher plants; this was substantiated by isolation of several enzymes of this pathway from photosynthetic bacteria. The discussion of the steps up to chlorophyllide a (see above) therefore also contained results obtained from photosynthetic bacteria.

Bogorad (1976) and Jones (1978) have summarized the later steps between chlorophyllide a and bacteriochlorophylls (see scheme 24) as derived from such mutant studies. According to these studies, modification of the vinyl group at ring A to the acetyl group via a hydroxyethyl intermediate occurs before hydrogenation of ring B in the pathway of bacteriochlorophyll a biosynthesis. The structure of the hydroxyethyl compound has been elucidated by Houghton *et al.* (1983). Many of the intermediates have also been observed *in situ* by fluorescence and optically detected magnetic resonance studies (Beck and Drews, 1982; Beck *et al.*, 1983). These authors claim the occurrence of Mg-free derivatives (phaeophorbides) besides chlorophyll(ides). This conclusion has to be accepted with caution, however, because only protein-bound pigments were observed. In any case, protein-binding causes varying spectral shifts. Phaeophorbides and phaeophytins are usually believed to be demetallation artifacts which are formed during the extraction procedure. The finding of bacteriophaeophytin a as well as bacteriochlorophyll a in bacterial reaction centres which contain different esterifying alcohols (e.g. bacteriochlorophyll a_{GG} and bacteriophaeophytin a_P in *Rhodospirillum rubrum*; see Walter *et al.*, 1979) points to the possibility of separate pathways for the Mg-containing and Mg-free compounds.

Formation of the other bacteriochlorophylls is even less well known. From a chemical point of view, the ethylidene side chain at ring B of bacteriochlorophylls b and g could arise by 1,4 hydrogenation of a vinyl group at an unsaturated ring B or by elimination of a leaving group (e.g. hydroxy or sulphydryl) in the alpha position of the side chain attached to a saturated ring B.

5 COMPARTMENTATION AND REGULATION OF CHLOROPHYLL BIOSYNTHESIS

Chlorophylls of eukaryotes are localized within chloroplasts. The enzymes for ALA formation via the C_5 pathway (see section 4A) and the enzyme for GGPP formation (see section 4G) have also been found in chloroplasts. The discussion of compartmentation can therefore be restricted to the question of in which compartment within the chloroplast the single steps of chlorophyll biosynthesis occur.

It is usually considered that the soluble enzymes of this pathway occur in the stroma of plastids. This is certainly true for the enzymes of ALA and PBG formation which have been investigated in plants . The enzymes catalysing the subsequent steps to protoporphyrinogen have not yet been intensively investigated in higher plants. It cannot be excluded at present that these enzymes are localized in another compartment of the plant cell, e.g. the cytoplasm (see Smith, 1987). The enzyme protogen oxidase (see section 4D) has been found in chloroplast (and mitochondrial) membranes (Jacobs and Jacobs, 1987). The next enzyme, Mg chelatase, has been localized in developing chloroplasts (see section 4E). Since this enzyme was inhibited by p-chloromercuribenzene sulphonate, which does not penetrate into chloroplasts, Fuesler *et al.* (1984b) assume that Mg chelatase is localized in the chloroplast envelope. The same reagent did not inhibit activity of Mg-protoporphyrin monomethyl ester cyclase, although thiol reagents which can penetrate into plastids inhibit this activity. Since this reaction requires both a soluble and an insoluble enzyme (Wong and Castelfranco, 1984), these are assumed to be localized in the stroma and an inner plastid membrane. Similar conclusions were drawn by Nastulhag-Boyce *et al.* (1987). The above-mentioned localization of Mg chelatase contradicts to a certain extent the finding that Mg-protoporphyrin is preferentially found in the prothylakoid fraction of etioplasts after pretreatment of plants with 8-hydroxyquinoline and ALA (Ryberg and Sundquist, 1984). The possibility cannot be excluded however, that Mg-protoporphyrin is relocated in the prothylakoids after its formation in the envelope mebrane.

Interesting observations have been made with the next enzyme of the pathway, the light-dependent protochlorophyllide reductase (see section 4F). This enzyme is the main protein component of prolamellar bodies in etioplasts (Dehesh and Ryberg, 1985). It is also found (among many other proteins) to a certain extent in prothylakoids (Lütz *et al.*, 1981; Ryberg and Sundquist, 1984). After irradiation, it is relocalized in prothylakoids (Ryberg and Dehesh, 1986) and then proteolytically degraded. The remaining enzyme activity, however, is still sufficient to account for chlorophyll biosynthesis in the light. In mature chloroplasts, trace amounts of protochlorophyllide have been detected in the envelope (Pineau *et al.*, 1986a). Since the pigment is photoconverted to chlorophyllide a in this fraction (Pineau *et al.*, 1986b), protochlorophyllide reductase must be present in this membrane. The enzyme (or an immunologically cross-reacting protein) has also been detected outside the plastids (Dehesh *et al.*, 1987). The significance of these different types of enzyme localization, which seem to depend on plastid development, is not yet clear.

The last enzyme of chlorophyll synthesis, chlorophyll synthetase, has been found only in the thylakoid fraction of chloroplasts and the prothylakoid/prolamellar body fraction of etioplasts but not in the envelope (Blank-Huber, 1986). Since phytyl diphosphate, the only substrate for chlorophyll synthesis in mature chloroplasts (see section 4G) is formed exclusively in the envelope and not in either the stroma or the thylakoids (Soll *et al.*, 1983), some way of intermembrane transport must exist in the chloroplast, the details of which are still unknown.

The question of regulation of chlorophyll biosynthesis in plants and photosynthetic bacteria has been excellently reviewed and discussed in detail several times (Bogorad, 1976; Lascelles, 1978; Castelfranco and Beale, 1981; 1983; Kasemir, 1983a,b; Sironval and Brouers, 1984; Akoyunoglou and Senger, 1986).Therefore only some general aspects will be discussed hcre.

Two types of regulation of chlorophyll biosynthesis are known: (i) light regulation in higher plants and (ii) regulation by oxygen and light in photosynthetic bacteria. As discussed in section 4F, chlorophyll biosynthesis is mostly light-dependent but can also be light-independent in higher plants. Likewise, bacteriochlorophyll synthesis occurs in most photosynthetic bacteria only under anaerobic conditions (Lascelles, 1978; Oelze, 1981). Exceptions to this rule, however, are some bacteria which produce bacteriochlorophyll under aerobic conditions (Sato, 1978; Shiba *et al.*, 1979; Pierson *et al.*, 1985).

The mechanisms for these types of regulation are not yet known in detail. The best known light-regulated step is the reduction of protochlorophyllide to chlorophyllide, discussed in section 4F. Oh-hama *et al.* (1987b) report that this step can be inhibited by higher ("non-permissive") temperatures in a *Scenedesmus* mutant. It is generally assumed that protochlorophyllide reduction is light-regulated, not only in greening etiolated plants (in which the enzyme can easily be detected), but also in light-grown plants, although most of the enzyme and its mRNA are degraded in the light (see Dehesh *et al.*, 1987, and earlier papers cited therein). The light-dependent disappearance of the mRNA encoding this enzyme is a phytochrome effect (Batschauer and Apel, 1984).

Another phytochrome effect is the elimination of the "lag-phase" of chlorophyll accumulation which occurs at the beginning of greening of etiolated seedlings. The phytochrome effect might be due in part to prevention of chlorophyll photodecomposition (Oelze-Karow *et al.*, 1983) but certainly also to increased formation of chlorophyll synthesizing enzymes. Since prefeeding of etiolated tissue with ALA decreases or abolishes the lag-phase (discussed by Castelfranco and Beale, 1981), ALA formation is considered to be the main regulatory step under these conditions. It is not yet known exactly which is the key enzyme of this step but there are indications that the glutamyl-tRNA-reductase (see section 4A) may be the site of regulation (see below).

Incubation of plants with ALA in the dark leads to accumulation of large amounts of protochlorophyllide. It is generally assumed that the normally much smaller accumulation of protochlorophyllide is due to feedback inhibition of ALA formation by this pigment itself (see Stobart and Ameen-Bukhari, 1986). However, glutamyl-tRNA-reductase is inhibited by haem but not by protochlorophyllide (Wang *et al.*, 1987). As pointed out in several reviews (see above), increased accumulation of protochlorophyllide after incubation of plants with iron chelators (phenanthrolin, 2,2′-dipyridyl) has also been interpreted as due to decreased haem formation and hence lack of feedback control by haem. Haem also inhibits bacteriochlorophyll and ALA formation in photosynthetic bacteria (Lascelles, 1978). Control of bacteriochlorophyll synthesis by oxygen is mediated via ALA synthesis, whereas light may control the pathway in a different way (Oelze and Arnheim, 1983). Haem and oxygen do not necessarily have to act on the level of enzyme activity because it has been shown that coproporphyrinogen oxidase formation is inhibited in yeast by these factors on the pretranslational level (Zagorec and Labbe-Bois, 1986).

Castelfranco and Beale (1983) discussed other possible regulatory steps of chlorophyll synthesis, including Mg chelation, Mg protoporphyrin methylester esterase activity, isocyclic ring formation, turnover of chlorophyll precursors and protochlorophyllide reductase. These authors state that their conclusions are still largely hypothetical.

Many investigators, when studying "regulation of chlorophyll biosynthesis", in reality determine chlorophyll accumulation rates rather than synthesis rates. Stable accumulation of chlorophylls requires concomitant production of several components, including chlorophyll-binding proteins. A complex pattern of the overall regulation of accumulation may arise if these components interact with each other. Such interaction has been described, e.g. as stabilization by association of chlorophyll with the light-harvesting chlorophyll a/b protein (Apel and Kloppstech, 1980) or as inhibition of mRNA formation of this protein by chlorophyll precursors (Johanningmeier and Howell, 1984). Post-translational control of formation of the P700 chlorophyll a protein (Kreuz *et al.*, 1986) is eventually also due to chlorophyll precursors.

6 DEGRADATION OF CHLOROPHYLLS

Chlorophyll is degraded to a large extent each autumn. Surprisingly little is known about the fate of the pigment under the conditions of senescence. Chlorophyll degradation has been reviewed e.g. by Simpson *et al.* (1976)

Kufner (1980) and Hendry *et al.* (1987). Chlorophyll breakdown has been induced in intact plant tissues by prolonged dark periods, e.g. in *Euglena* (Schoch *et al.*, 1981), pea and barley (Maunders *et al.*, 1983), tobacco cell cultures (Schoch and Vielwerth, 1983) and parsley (Amir-Shapira *et al.*, 1987). Products of chlorophyll catabolism were also studied in senescing marine phytoplankton (Hallegraeff and Jeffrey, 1985) and *Rhodopseudomonas sphaeroides* (Haidl *et al.*, 1985). Apparently, the same reactions must have occurred in all systems: finding of chlorophyllides points to cleavage of the phytol residue; phaeophytins indicate release of Mg; 13^2-hydroxychlorophylls must have been formed by hydroxylation. Further catabolism seems to occur in some microalgae. Pyrophaeophytin has been isolated from *Euglena* (Schoch *et al.*, 1981) and open-chain tetrapyrroles have been detected in the dinoflagellate *Pyrocystis lunula* (Dunlap *et al.*, 1981). Photo-oxidation of porphyrins which carry alkyl substituents at a methine bridge (e.g. at C-20) has been shown to produce open-chain tetrapyrroles (Troxler *et al.*, 1980; Risch *et al.*, 1984).

Chlorophyllase, the enzyme which splits chlorophyll *in vitro* into chlorophyllide and phytol, has been known since the classic work of Willstätter. The enzyme, which is localized in chloroplast membranes, has been solubilized and purified from several sources. The enzyme from the diatom *Phaeodactylum tricornutum* is a glycoprotein consisting of 30 kDa subunits (Terpstra, 1981; Lambers *et al.*, 1985). The enzyme from rye seedlings has a molecular size of 39 kDa (Tanaka *et al.*, 1982). Besides chlorophylls a and b and the corresponding phaeophytins, bacteriochlorophylls (Tanaka *et al.*, 1983) and even protein-bound chlorophylls (Schoch and Brown, 1986) are accepted as substrate by the chlorophyllase. The enzyme activity is enhanced by membrane lipids and Mg ions (Terpstra and Lambers, 1983).

As mentioned in section 3, phaeophytins are readily found as artifacts of chlorophyll isolation. There are, however, some observations on the existence of an Mg-releasing enzyme (Owens and Falkowski, 1982; Ziegler *et al.*, 1982a,b).

Another possible artifact during pigment isolation is the hydroxylation of chlorophyll at C-13^2 (see section 2). This reaction can be achieved by addition of linolenic acid to a preparation of plastid membranes within a few minutes (Schoch *et al.*, 1984). The enzyme involved in this reaction has been called chlorophyll oxidase (Lüthy *et al.*, 1984; 1986).

Enzymes for further steps of catabolism have not yet been described. In summary, neither accumulation of large amounts of chlorophyll degradation products nor enzymes leading to such products have been described so far. Degradation products of chlorophyll can be photosensitizing pigments (Endo *et al.*, 1982). It may be essential for plant cells to eliminate such products rapidly because otherwise cells become damaged under strong light. A similar situation has been discussed by Castelfranco and Beale (1983) for chlorophyll precursors.

REFERENCES

Adamson, H., Hiller, R.G. and Vesk, M. (1980). *Planta* **150**, 269–74.
Adamson, H., Griffiths, T., Packer, N. and Sutherland, M. (1985a). *Physiol. Plant* **64**, 345–52.
Adamson, H., Packer, N. and Gregory J. (1985b). *Planta* **165**, 469–76.
Adamson, H., Walker, C., Bees, A. and Griffiths, T. (1987). In: *Progress in Photosynthesis Research* (Ed. J. Biggins) vol. 4, pp. 483–6. Martinus Nijhoff, Dordrecht.
Akhtar, M., Ajaz, A.A. and Corina, D.L. (1984). *Biochem. J.* **224**, 187–94.
Akoyunoglou, G. and Senger, H. (Eds.) (1986). *Regulation of Chloroplast Differentiation*. Alan R. Liss, New York.
Amir-Shapira, D., Goldschmidt, E.E. and Altman, A. (1987). *Proc. Natl. Acad. Sci.* **84**, 1901–5.
Anderson, P.M. and Desnick, R.J. (1980). *J. Biol. Chem.* **225**, 1993–9.
Apel, K. and Kloppstech, K. (1980). *Planta* **150**, 426–30.
Apel, K., Santel, H., Redlinger, T.E. and Falk, H. (1980). *Eur. J. Biochem.* **111**, 251–8.
Araki, S., Oohusa, T., Omata, T. and Murata, N. (1984). *Plant Cell Physiol.* **25**, 841–3.
Aronoff, S. (1981). *Biochem. Biophys. Res. Commun.* **102**, 108–12.
Barnard, G.F. and Akhtar, M. (1979). *J. Chem. Soc., Perkin Trans.* **I**, 2354–60.
Barnard, G.F., Itoh, R., Hohberger, L.H. and Shemin, D. (1977). *J. Biol. Chem.* **252**, 8964–74.
Batschauer, A. and Apel, K. (1984). *Europ. J. Biochem* **143**, 593–7.
Battersby, A.R. (1978). *Experientia* **34**, 1–13.
Battersby, A.R. (1985). *Proc. R. Soc. Lond. B* **225**, 1–26.
Battersby, A.R. (1986). *Acc. Chem. Res.* **19**, 147–52.
Battersby, A.R., Gutman, A.L., Fookes, C.J.R., Günther, H. and Simon, H. (1981). *J. Chem. Soc. Chem. Commun.* **1981**, 645–7.
Battersby, A.R., Fookes, C.J.R., Gustafson-Potter, K.E., McDonald, E. and Matcham, G.W.J. (1982). *J. Chem. Soc., Perkin Trans.* **I**, 2413–26.
Bazzaz, M.B. and Brereton, R.G. (1982). *FEBS Lett.* **138**, 104–8.
Beale, S.I. and Castelfranco, P.A. (1974). *Plant Physiol.* **53**, 297–303.
Beale, S.I., Gough, S.P. and Granick, S. (1975). *Proc. Natl. Acad. Sci. USA* **72**, 2719–2723.
Beck, J. and Drews, G. (1982). *Z. Naturforsch.* **37c**, 199–204.
Beck, J., Schütz v., J.U. and Wolf, H.C. (1983). *Z. Naturforsch.* **38c**, 220–9.
Bednarik, D.P. and Hoober, K.J. (1985a). *Arch. Biochem. Biophys.* **240**, 369–79.
Bednarik, D.P. and Hoober, K.J. (1985b). *Science* **230**, 450–3.
Belanger, F.C., Duggan, J.X. and Rebeiz, C.A. (1982). *J. Biol. Chem.* **257**, 4849–58.
Benz, J. and Rüdiger, W. (1981a). *Z. Pflanzenphysiol.* **102**, 95–100.

Benz, J. and Rüdiger, W. (1981b). *Z. Naturforsch.* **36c**, 51–7.
Benz, J., Hampp, R. and Rüdiger, W. (1981). *Planta* **152**, 54–8.
Benz, J., Haser, A. and Rüdiger, W. (1983). *Z. Pflanzenphysiol.* **11**, 349–56.
Bergweiler, P. and Lütz, C. (1986). *Environmental and Experim. Botany,* **26**, 207–10.
Bergweiler, P., Röper, U. and Lütz, C. (1983). *Physiol. Plant* **60**, 395–400.
Berry, A., Jordan, P.M. and Seehra, J.S. (1981). *FEBS Lett.* **129**, 220–4.
Blank-Huber, M. (1986). *Dissertation.* Univ. München.
Block, M.A., Joyard, J. and Douce, R. (1980). *Biochim. Biophys. Acta* **631**, 210–19.
Bogorad, L. (1976). In: *Chemistry and Biochemistry of Plant Pigments* (Ed. T.W. Goodwin). vol. 1, pp. 64–148. Academic Press, London.
Bombart, P., Dujardin, E., Grandjean, J., Laszlo, P. and Sironval, C. (1984). In: *Advances in Photosynthesis Research* (Ed. C. Sybesma) vol. IV. 6, pp. 753–6. Martinus Nijhoff/Dr. W. Junk, The Hague.
Bonnett, R. (1973). *Ann. N.Y. Acad. Sci.* **206**, 745.
Brockmann, Jr. H. (1976). *Phil. Trans. R. Soc. London B* **273**, 277–85.
Brockmann, Jr. H. and Knobloch, G. (1972). *Arch. Mikrobiol.* **85**, 123–6.
Brockmann, Jr. H. and Lipinski, H. (1983). *Arch. Microbiol.* **136**, 17–19.
Burger-Wiersma, T., Veenhuis, M., Korthals, H.J., Van de Wiel, C.C.M. and Mur, L.R. (1986). *Nature* **320**, 262–4.
Camadro, J.-M., Urban-Grimal, D. and Labbe, P. (1982). *Biochem. Biophys. Res. Commun.* **106**, 724–30.
Camadro, J.M., Chambon, H., Jolles, J. and Labbe, P. (1986). *Eur. J. Biochem.* **156**, 579–87.
Caple, M.B., Chow, H. and Strouse, C.E. (1978). *J. Biol. Chem.* **253**, 6730–6.
Castelfranco, P.A. and Beale, S.I. (1981). In: *The Biochemistry of Plants.* (Eds. P.K. Stumpf and E.E. Conn) vol. 8, pp. 375–421. Academic Press, New York.
Castelfranco, P.A. and Beale, S.I. (1983). *Ann. Rev. Plant Physiol.* **34**, 241–78.
Castelfranco, P.A., Weinstein, J.D., Schwarcz, S., Pardo, A.D. and Wezelman, B.E. (1979). *Arch. Biochem. Biophys.* **192**, 592–8.
Chereskin, B.M., Wong, Y.-S. and Castelfranco, P.A. (1982). *Plant Physiol.* **70**, 987–93.
Cox, M.T., Howarth, T.T., Jackson, A.K. and Kenner, G.W. (1969). *J. Amer. Chem. Soc.* **91**, 1232–3.
Crawford, M.S. and Wang, W.-Y. (1983). *Plant Physiol.* **71**, 303–6.
Daniell, H. and Rebeiz, C.A. (1984). *Biotechnol. Bioeng.* **26**, 481–7.
Davies, R.C. and Neuberger, A. (1973). *Biochem. J.* **133**, 471–92.
Dehesh, K. and Ryberg, M. (1985). *Planta* **164**, 396–9.
Dehesh, K. Kreuz, K. and Apel, K. (1987). *Physiol. Plant.* **69**, 173–81.
Deisenhofer, J., Epp, O., Miki, K., Huber, R. and Michel, H. (1985). *Nature* **318**, 618–24.
Dobeneck v., H., Hansen, B. and Vollmann, E. (1972). *Z.Naturforsch.* **27b**, 922–4.
Dörnemann, D. and Senger, H. (1981). *FEBS Lett.* **126**, 323–7.
Dörnemann, D. and Senger, H. (1986). *Photochem. Photobiol.* **43**, 573–81.
Duggan, J.X. and Rebeiz, C.A. (1982a). *Plant Sci. Lett.* **27**, 137–45.
Duggan, J.X. and Rebeiz, C.A. (1982b). *Biochim. Biophys. Acta* **714**, 248–60.
Dujardin, E. and Correia, M. (1979). *Photobiochem. Photobiophys.* **1**, 25–32.
Dujardin, E. and Sironval, C. (1977). *Plant Sci. Lett.* **10**, 347–55.
Dunlap, J.C., Hastings, J.W. and Shimomura, O. (1981). *FEBS Lett.* **135**. 273–6.

Ebbon, J.G. and Tait, G.H. (1969). *Biochem. J.* **111**, 473–82.
Elder, G.H., Evans, J.O., Jackson, J.R. and Jackson, A.H. (1978). *Biochem. J.* **169**, 215–23.
Elder, G.H., Tovey, J.A. and Sheppard, D.M. (1983). *Biochem. J.* **215**, 45–55.
El Hamouri, B. and Sironval, C. (1980). *Photobiochem. Photobiophys.* **1**, 219–23.
El Hamouri, B., Oliver, R.P. and Griffiths, W.T. (1983). *Photobiochem. Photobiophys.* **6**, 305–9.
Ellsworth, R.K. and Aronoff, S. (1968). *Arch. Biochem. Biophys.* **125**, 269–77.
Ellsworth, R.K. and Aronoff, S. (1969). *Arch. Biochem. Biophys.* **130**, 374–83.
Ellsworth, R.K. and Dullaghan, J.P. (1972). *Biochim. Biophys. Acta* **268**, 327–33.
Ellsworth, R.K. and Hervish, P.V. (1975). *Photosynthetica* **9**, 125–39.
Ellsworth, R.K. and Hsing, A.S. (1973). *Biochim. Biophys. Acta* **313**, 119–29.
Ellsworth, R.K. and Hsing, A.S. (1974). *Photosynthetica* **8**, 228–34.
Ellsworth, R.K. and Murphy, S.J. (1979). *Photosynthetica* **13**, 392–400.
Ellsworth, R.K. and St. Pierre, M.E. (1976). *Photosynthetica* **10**, 291–301.
Ellsworth, R.K., Dullaghan, J.P. and St. Pierre, M.E. (1974). *Photosynthetica* **8**, 375–83.
Emery, V.C. and Akhtar, M. (1987). *Biochemistry* **26**, 1200–8.
H., Hosoya, H., Koyama, T. and Ichioka, M. (1982). *Agric. Biol. Chem.* **46**, 2183–93.
Evans, J.N.S., Fagerness, P.E., Mackenzie, N.E. and Scott, A.I. (1984). *J. Am. Chem. Soc.* **106**, 5738–40.
Evans, J.N.S., Burton, G., Fagerness, P.E., Mackenzie, N.E. and Scott A.I. (1986). *Biochemistry* **25**, 905–12.
Foley, T. and Beale, S.L. (1982). *Plant Physiol.* **70**, 1495–502.
Franck, F. and Inoue, Y. (1984). *Photobiochem. Photobiophys.* **8**, 85–96.
Franck, F. and Mathis, P. (1980). *Photochem. Photobiol.* **32**, 799–803.
Franck, F. and Schmid, G.H. (1985). *Z. Naturforsch.* **40c**, 832–8.
Franck, F., Dujardin, E. and Sironval, C. (1980). *Plant Sci. Lett.* **18**, 375–80.
Franck, B., Bruse, M. and Dahmer, J. (1984). *Angew. Chemie* **96**, 1000–1.
Friedmann, H.C. and Thauer, R.K. (1986). *FEBS Lett.* **207**, 84–8.
Fuesler, T.P., Hanamoto, C.M. and Castelfranco, P.A. (1982). *Plant Physiol.* **69**, 421–3.
Fuesler, T.P., Castelfranco, P.A. and Wong, Y.-S. (1984a). *Plant Physiol.* **74**, 928–33.
Fuesler, T.P., Wong, Y.-S. and Castelfranco, P. A. (1984b). *Plant Physiol.* **75**, 662–4.
Gassman, M., Pluscec, J. and Bogorad, L. (1968a). *Plant Physiol.* **43**, 1411–14.
Gassman, M., Granick, S. and Mauzerall, D. (1968b). *Biochem. Biophys. Res. Commun.* **32**, 295–300.
Gloe, A. and Pfennig, N. (1974). *Arch. Microbiol.* **96**, 93–101.
Gorchein, A. (1972). *Biochem. J.* **127**, 97–106.
Gorchein, A. (1973). *Biochem. J.* **134**, 833–45.
Gough, S.P. and Kannangara, C.G. (1977). *Carlsberg Res. Commun.* **42**, 459–64.
Griffiths, W.T. (1978). *Biochem. J.* **174**, 681–92.
Griffiths, W.T. (1980). *Biochem. J.* **186**, 267–78.
Griffiths, W.T. and Mapleston, R.E. (1978). In: *Chloroplast Development* (Eds. G. Akoyunoglou and J.H. Argyroudi-Akoyunoglou) pp. 99–104. Elsevier-North Holland, Amsterdam.
Griffiths, W.T. and Walker, C.J. (1987). In: *Progress in Photosynthesis Research* (Ed. J. Biggins) vol. 4, pp. 469–74, Martinus Nijhoff, Dordrecht.

Haidl, H., Knödlmayr, K., Rüdiger, W., Scheer, H., Schoch, S. and Ulrich, J. (1985). *Z. Naturforsch.* **40c**, 685–92.
Hallegraeff, G.M. and Jeffrey, S.W. (1985). *Deep Sea Res.* **32**, 697–705.
Hanamoto, C.M. and Castelfranco, P.A. (1983). *Plant Physiol.* **73**, 79–81.
Hart, G.J. and Battersby, A.R. (1985). *Biochem. J.* **232**, 151–160.
Hart, G.J., Leeper, F.J. and Battersby, A.R. (1984). *Biochem. J.* **222**, 93–102.
Hart, G.J., Abell, C. and Battersby, A.R. (1986). *Biochem. J.* **240**, 273–6.
Hasnain, S.S., Wardell, E.M., Garner, C.D., Schlösser, M. and Beyersmann, D. (1985). *Biochem. J.* **230**, 625–33.
Hayashi, S. and Moguchi, T. (1983). *J. Biol. Chem.* **258**, 13693–6.
Hendry, G.A.F., Houghton, J.D. and Brown, S.B. (1987) *New Phytol.* **107**, 255–302.
Higuchi, N. and Bogorad, L. (1975). *Ann. N.Y. Acad. Sci.* **224**, 401–18.
Hinchigeri, S.B. and Richards, W.R. (1982). *Photosynthetica* **16**, 554–60.
Hinchigeri, S.B., Chan, J.C.-S. and Richards, W.R. (1981). *Photosynthetica* **15**, 351–9.
Hinchigeri, S.B., Nelson, D.W. and Richards, W.R. (1984). *Photosynthetica* **18**, 168–78.
Hiscox, J.D. and Israelstam, G.F. (1979). *Can. J. Bot.* **57**, 1332–4.
Holden, M. (1976). In: *Chemistry and Biochemistry of Plant Pigments* (Ed. T.W. Goodwin) vol. 2, pp. 1–37. Academic Press, London.
Houen, G., Gough, S.P. and Kannangara, C.G. (1983). *Carlsberg Res. Commun.* **48**, 567–72.
Houghton, J.D., Jones, O.T.G., Quirke, J.M.E., Murray, M. and Honeybourne, C.L. (1983). *Tetrahedron Lett.* **24**, 5703–6.
Huang, L. and Castelfranco, P.A. (1986). *Plant Physiol.* **82**, 285–8.
Huang, D.-D. and Wang, W.-Y. (1986). *J. Biol. Chem.* **261**, 13451–5.
Huang, D.-D., Wang, W.-Y., Gough, S.P. and Kannangara, C.G. (1984). *Science* **225**, 1482–4.
Hynninen, P.H. (1979). *J. Chromatogr.* **175**, 75–88.
Hynninen, P.H. and Lötjönen, S. (1981). *Tetr. Lett.* **22**, 1845–6.
Ikegami, J., Kamiya, A. and Hase, E. (1984). *Plant Cell Physiol.* **25**, 343–8.
Ikeuchi, M. and Murakami, S. (1982). *Plant Cell Physiol.* **23**, 1089–99.
Inoue, Y., Kobayashi, T., Ogawa, T. and Shibata, K. (1981). *Plant Cell Physiol.* **22**, 197–204.
Inskeep, W.P. and Bloom, P.R. (1985). *Plant Physiol.* **77**, 483–5.
Iriyama, K. and Shiraki, M. (1979). *J. Liq. Chromatogr.* **2**, 255–76.
Iriyama, K. Yosiwra, M., Ishii, T. and Shiraki, M. (1981). *J. Liq. Chromatogr.* **4**, 533–8.
Jackson, A.H. (1976). In: *Chemistry and Biochemistry of Plant Pigments* (Ed. T.W. Goodwin) Vol. I, pp. 1–63. Academic Press, London.
Jacobs, N.J. and Jacobs, J.M. (1981). *Arch. Biochem. Biophys.* **211**, 305–11.
Jacobs, J.M. and Jacobs, N.J. (1984). *Arch. Biochem. Biophys.* **229**, 312–19.
Jacobs, N.J. and Jacobs, J.M. (1987) *Biochemistry* **244**, 219–24.
Jaffe, E.K. and Hanes, D. (1986). *J. Biol. Chem.* **261**, 9348–53.
Jaffe, E.K. and Markham, G.D. (1987). *Biochemistry* **26**, 4258–64.
Johanningmeier, U. and Howell, S.H. (1984). *J. Biol. Chem.* **259**, 13541–9.
Jones, O.T.G. (1978). In: *The Photosynthetic Bacteria* (Eds. R.K. Clayton and W.R. Sistrom) pp. 751–7. Plenum Press, New York.
Jones, C., Jordan, P.M. and Akhtar, M. (1984). *J. Chem. Soc. Perkin Trans.* I **1984**, 2625–33.

Jordan, P.M. and Gibbs, P.N.B. (1985). *Biochem. J.* **227**, 1015–20.
Jordan, P.M. and Seehra, J.S. (1980). *FEBS Lett.* **114**, 283–6.
Jordan, P.M. and Shemin, D. (1973). *J. Biol. Chem.* **248**, 1019–24.
Kah, A. and Dörnemann, D. (1987). *Z. Naturforsch.* **42**, 209–14.
Kah, A., Dörnemann, D. and Senger, H. (1986). In: *Regulation of Chloroplast Differentiation* (Eds. Akoyunoglou, G., Senger, H.) pp. 35–42, Alan Liss, New York.
Kannangara, C.G. and Gough, S.P. (1977). *Carlsberg Res. Commun.* **42**, 441–57.
Kannangara, C.G. and Gough, S.P. (1978). *Carlsberg Res. Commun.* **43**, 185–94.
Kannangara, C.G. and Schouboe, A. (1985). *Carlsberg Res. Commun.* **50**, 179–91.
Kannangara, C.G., Gough, S.P., Oliver, R.P. and Rasmussen, S.K. (1984). *Carlsberg Res. Commun.* **49**, 417–37.
Kasemir, H. (1983a). *Photochem. Photobiol.* **37**, 701–8.
Kasemir, H. (1983b). In: *Enclyclopedia of Plant Physiology* (Eds. W. Shropshire jr. and H. Mohr) vol. 16, pp. 662–86. Springer Verlag, Berlin.
Kasemir, H. and Mohr, H. (1981). *Planta* **152**, 369–73.
Kawanishi, S., Seki, Y. and Sano, S. (1983). *J. Biol. Chem.* **258**, 4285–92.
Kay, S.A. and Griffiths, W.T. (1983). *Plant Physiol.* **72**, 229–36.
Kindman, L.A., Cohen, C.E., Zeldin, M.H., Ben-Shaul, Y. and Schiff, J.A. (1978). *Photochem. Photobiol.* **27**, 787–94.
Kipe-Nolt, J.A. and Stevens, S.E. jr. (1980). *Plant Physiol.* **65**, 126–8.
Klein, O. and Porra, R.J. (1982). *Hoppe-Seyler's Z. Physiol. Chem.* **363**, 551–62.
Klein, O. and Senger, H. (1978a). *Photochem. Photobiol.* **27**, 203–8.
Klein, O. and Senger, H. (1978b). *Plant Physiol.* **62**, 10–13.
Klockare, B. and Virgin, H.I. (1983). *Physiol. Plant* **57**, 28–34.
Klockare, B. and Virgin, H.I. (1984). *Israel J. of Bot.* **33**, 175–83.
Kotzabasis, K. and Senger, H. (1986a). *Z. Naturforsch.* **41c**, 1001–3.
Kotzabasis, K. and Senger, H. (1986b). *Naturwiss.* **73**, 681–2.
Kreuz, K. and Kleinig, H. (1981). *Plant Cell Rep.* **1**, 40–2.
Kreuz, K., Dehesh, K. and Apel, K. (1986). *Europ. J. Biochem.* **159**, 459–67.
Kufner, R.B. (1980). In: *Pigments in Plants* (Ed. F.C. Czygan) 2nd edn., pp. 308–13. Gustav Fischer, Stuttgart.
Künzler, A. and Pfennig, N. (1973). *Arch. Mikrobiol.* **91**, 83–6.
Kwan, L.Y.-M., Darling, D.L. and Richards, W.R. (1986). In: *Regulation of Chloroplast Differentiation* (Eds. G. Akoyunoglou and H. Senger) pp. 57–62. Alan R. Liss, New York.
Lambers, J.W.J., Velthuis, H.W. and Terpstra, W. (1985). *Biochim. Biophys. Acta* **831**, 213–24.
Lancer, H.A., Cohen, C.E. and Schiff, J.A. (1976). *Plant Physiol.* **57**, 369–74.
Lascelles, J. (1978). In: *The Phototrophic Bacteria* (Eds. R.K. Clayton and W.R. Sistrom) pp. 795–808, Plenum Press, London.
Leeper, F.J. (1985a). *Nat. Prod. Rep.* **2**, 19–47.
Leeper, F.J. (1985b). *Nat. Prod. Rep.* **2**, 561–80.
Liedgens, W., Grützmann, R. and Schneider, H.A.W. (1980). *Z. Naturforsch.* **35c**, 958–62.
Liedgens, W., Lütz, C. and Schneider, H.A.W. (1983). *Europ. J. Biochem.* **135**, 75–9.
Litvin, F.F., Ignatov, N.V. and Belyaeva, O.B. (1981). *Photobiochem. Photobiophys.* **2**, 233–7.
Lüthy, B., Martinoia, E., Matile, P. and Thomas, H. (1984). *Z. Pflanzenphysiol* **113**, 423–34.

Lüthy, B., Matile, P. and Thomas, H. (1986). *J. Plant Physiol.* **123**, 169–80.
Lütz, C., Röper, U., Beer, N.S. and Griffiths, W.T. (1981). *Europ. J. Biochem.* **118**, 347–53.
Maitra, P. and Mukherji, S. (1983). *Biochem. Physiol. Pflanzen* **178**, 207–11.
Mapleston, R.E. and Griffiths, W.T. (1978). *FEBS Lett.* **92**, 168–72.
Maunders, M.J., Brown, S.B. and Woolhouse, H.W. (1983). *Phytochem.* **22**, 2443–6.
Mauzerall, D. (1960). *J. Am. Chem. Soc.* **82**, 2605–9.
Mauzerall, D. and Granick, S. (1958). *J. Biol. Chem.* **232**, 1141–62.
Meisch, H.-U. and Maus, R. (1983). *Z. Naturforsch.* **38c**, 563–70.
Meisch, H.U., Hoffmann, H. and Reinle, E. (1983). *Biochim. Biophys. Acta* **743**, 281–9.
Meller, E. and Gassmann, M.L. (1981). *Plant Physiol.* **67**, 728–32.
Meller, E. and Gassman, M.L. (1982). *Plant Sci. Lett.* **26**, 23–9.
Michalski, T.J., Hunt, J.E., Bowman, M.K., Smith, U., Bardeen, K., Gest, H., Norris, J.R. and Katz, J.J. (1987). *Proc. Natl. Acad. Sci. USA,* **84**, 2570–4.
Michel, H. and Deisenhofer, J. (1987). In: *Progress in Photosynthesis Research* (Ed. J. Biggins) vol. 1, pp. 353–62. Martinus Nijhoff, Dordrecht.
Michel-Wolwertz, M.R. (1977). *Plant Sci. Lett.* **8**, 125–34.
Moran, R. and Porath, D. (1980). *Plant Physiol.* **65**, 478–9.
Nandi, D.L., and Shemin, D. (1968). *J. Biol. Chem.* **243**, 1236–42.
Nasrulhaq-Boyce, A., Griffiths, W.T. and Jones, O.T.G. (1987). *Biochem. J.* **243**, 23–9.
Neidhart, W., Anderson, P.C., Hart, G.J. and Battersby, A.R. (1985). *J. Chem. Soc. Chem. Commun.* **1985**, 924–9.
Neuberger, A. and Turner, J.M. (1963). *Biochim. Biophys. Acta* **67**, 342–5.
Oelze, J. (1981). In: *Subcellular Biochemistry* (Ed. D.B. Roodyn) vol. 8, pp. 1–73. Plenum Press, London.
Oelze, J. and Arnheim, K. (1983). *FEMS Microbiol. Lett.* **19**, 197–9.
Oelze-Karow, H. and Mohr, H. (1978). *Photochem. Photobiol.* **27**, 189–93.
Oelze-Karow, H., Rösch, H. and Mohr, H. (1983). *Photochem. Photobiol.* **37**, 565–9.
Ogawa, M. and Konishi, M. (1977). *Plant Cell Physiol.* **18**, 303–7.
Oh-hama, T., Seto, H., Otake, N. and Miyachi, S. (1982), *Biochem. Biophys. Res. Commun.* **105**, 647–52.
Oh-hama, T., Seto, H. and Miyachi, S. (1985). *Arch. Biochem. Biophys.* **237**, 72–9.
Oh-hama, T., Seto, H. and Miyachi, S. (1986a). *Arch. Biochem. Biophys.* **246**, 192–8.
Oh-hama, T., Seto, H. and Miyachi, S. (1986b) *Eur. J. Biochem.* **159**, 189–194.
Oh-hama, T., Seto, H. and Miyachi, S. (1987a). In: *Progress in Photosynthesis Research* (Ed. J. Biggins) vol. 4, pp. 445–8. Martinus Nijhoff, Dordrecht.
Oh-hama, T., Kotzabasis, K. and Senger, H. (1987b). *Physiol. Plant*, **69**, 29–34.
Oliver, R.T. and Griffiths, W.T. (1981). *Biochem. J.* **195**, 93–101.
Oliver, R.P. and Griffiths, W.T. (1982). *Plant Physiol.* **70**, 1019–25.
Omata, T. and Murata, N. (1983). *Plant Cell Physiol.* **24**, 1093–1100.
Owens, T.G. and Falkowski, P.G. (1982). *Phytochem.* **21**, 979–84.
Packer, N. and Adamson, H. (1986). *Physiol. Plant.* **68**, 222–30.
Pardo, A.D., Chereskin, B.M., Castelfranco, P.A., Franceschi, V.R. and Wezelman, B.E. (1980). *Plant Physiol.* **65**, 956–60.

Pennington, F.C., Strain, H.H., Svec, W.A. and Katz, J.J. (1967). *J.Am. Chem. Soc.* **89**, 3875–80.

Pierson, B.K., Giovannoni, S.J., Stahl, D.A. and Castenholz, R.W. (1985). *Arch. Microbiol.* **142**, 164–7.

Pineau, B., Dubertret, G., Joyard, J. and Douce, R. (1986a). *J. Biol. Chem.* **261**, 9210–5.

Pineau, B., Dubertret, G., Joyard, J. and Douce, R. (1986b). *Abstr. Eur. Congr. Photobiol.* Grenoble, p. 78.

Porra, R.J. (1986). *Eur. J. Biochem.* **156**, 111–21.

Porra, R.J. and Meisch, H.-U. (1984). *TIBS* **9**, 99–104.

Porra, R.J., Klein, O. and Wright, P.E. (1983). *Eur. J. Biochem.* **130**, 509–16.

Rebeiz, C.A., Wu, S.M., S.M. Kuhadja, M., Daniell, H. and Perkins, E.J. (1983). *Mol. Cell. Biochem.* **57**, 97–125.

Richards, W.R., Fung, M., Wessler, A.N. and Hinchingery, S.B. (1987). In: *Progress in Photosynthesis Research* (Ed. J. Biggins) vol. 4, pp. 475–82. Martinus Nijhoff, Dordrecht.

Richter, M.L. and Riemits, K.G. (1982). *Biochim. Biophys. Acta* **717**, 255–64.

Risch, N., Kemmer, T. and Brockmann, H. (1978). *Liebigs Ann. Chem.* **1978**, 585–94.

Risch, N., Schormann, A. and Brockmann, H. (1984). *Tetrahedr. Lett.* **25**, 5993–6.

Romeo, P.-H., Raich, N., Dubart, A. *et al.* (1986). *J. Biol. Chem.* **261**, 9825–31.

Röper, U. (1985). *Thesis.* University of Köln.

Röper, U., Bergweiler, P. and Lütz, C. (1983). *Z. Pflanzenphysiol.* **112**, 89–93.

Rüdiger, W., (1970). *Naturwiss.* **57**, 331–7.

Rüdiger, W. (1987). In: *Progress in Photosynthesis Research* (Ed. J. Biggins) vol. 4, pp. 461–7. Martinus Nijhoff, Dordrecht.

Rüdiger, W. and Benz, J. (1979). *Z. Naturforsch.* **34c**, 1055–7.

Rüdiger, W. and Benz, J. (1984). In: *Chloroplast Biogenesis* (Ed. R.J. Ellis) pp. 225–44. Cambridge University Press, Cambridge.

Rüdiger, W., Benz, J., Lempert, U., Schoch, S. and Steffens, D. (1976). *Z. Pflanzenphysiol.* **80**, 131–43.

Rüdiger, W., Benz, J. and Guthoff, C. (1980). *Eur. J. Biochem.* **109**, 193–200.

Ryberg, M. and Dehesh, K. (1986). *Physiol. Plant.* **66**, 616–24.

Ryberg, M. and Sundquist, C. (1982). *Physiol. Plant* **56**, 125–32.

Ryberg, M. and Sundquist, Ch. (1984). In: *Protochlorophyllide Reduction and Greening* (Eds. C. Sironval and M. Broners) pp. 69–84. Martinus Nijhoff/Dr. W. Junk, The Hague.

Salvador, G.F. (1978). *Plant Sci. Lett.* **13**(4), 351–5.

Sato, K. (1978). *FEBS Lett.* **85**, 207–10.

Sburlati, A., Frydman, R.B., Valasinas, A., Rose, S., Priestap, H.A. and Frydman, B. (1983). *Biochemistry* **22**, 4006–13.

Scheer, H. (1978). In: *The Porphyrins* (Ed. D. Dolphin) vol. II, Part B, pp. 1–43. Academic Press, New York.

Scheer, H. (1984). In: *Spectroscopy of Biological Molecules* (Eds. Sandorfy, C. and Theophanides, T.) pp. 409–45. D. Reidel Publishing Company.

Scheer, H. (1987). In: *Handbook of Chromatography, Plant Pigments* (Ed. H.P. Köst) vol. 1, in press; CRC Press, Boca Raton, Florida.

Scheer, H. and Katz, J.J. (1975). In: *Porphyrins and Metalloporphyrins* (Ed. K.M. Smith) pp. 399–524. Elsevier, Amsterdam.

Scheer, H., Gross, E., Nitsche, B., Cmiel, E., Schneider, S., Schäfer, W., Schiebel, H.M. and Schulten, H.R. (1986). *Photochem. Photobiol.* **43**, 559–71.
Schiff, J.A. (1978). In: *Chloroplast Development* (Eds. G. Akoyunoglou and J.H. Argyroudi-Akoyunoglou) pp. 747–67. Elsevier-North Holland, Amsterdam.
Schneider, H.A.W. (1980). In: *Pigments in Plants* (Ed. F.-C. Czygan) 2nd edn. pp. 237–307. Gustav Fischer, Stuttgart.
Schoch, S. (1978). *Z. Naturforsch.* **33c**, 712–14.
Schoch, S. and Brown, J. (1986). *J. Plant Physiol.* **126**, 483–94.
Schoch, S. and Schäfer, W. (1978). *Z. Naturforsch.* **33c**, 408–12.
Schoch, S. and Vielwerth, F.X. (1983). *Z. Pflanzenphysiol.* **110**, 309–17.
Schoch, S., Lempert, U. and Rüdiger, W. (1977). *Z. Pflanzenphysiol.* **83**, 427–36.
Schoch, S., Hehlein, C. and Rüdiger, W. (1980). *Plant Physiol.* **66**, 576–9.
Schoch, S., Scheer, H., Schiff, J.A., Rüdiger, W. and Siegelman, H.W. (1981). *Z. Naturforsch.* **36c**, 827–33.
Schoch, S., Rüdiger, W., Lüthy, B. and Matile, P. (1984). *J. Plant Physiol.* **115**, 85–9.
Schön, A., Krupp, G., Gough, S., Berry-Lowe, S., Kannangara, C.G. and Söll, D. (1986). *Nature* **322**, 281–4.
Seehra, J.S., Jordan, P.M. and Akhtar, M. (1983). *Biochem. J.* **209**, 709–18.
Shemin, D. (1970). *Naturwiss.* **57**, 185–90.
Shiba, T., Simidu, U. and Taga, N. (1979). *Appl. Environ. Microbiol.* **38**, 43–5.
Shibata, K. (1957). *J. Biochem.* (Tokyo) **44**, 147–73.
Shioi, Y. and Sasa, T. (1983a). *Arch. Biochem. Biophys.* **220**, 286–92.
Shioi, Y. and Sasa, T. (1983b). *Plant Cell Physiol.* **24**, 835–40.
Shioi, Y. and Sasa, T. (1983c). *Biochim. Biophys. Acta* **756**, 127–31.
Shioi, Y., Nagamine, M., Kuraki, M. and Sassa, T. (1980). *Biochim. Biophys. Acta* **616**, 303–9.
Shioi, Y., Doi, M. and Sasa, T. (1985). *Plant Cell Physiol.* **26**, 379–82.
Shlyk, A.A. (1971). *Ann. Rev. Plant Physiol.* **22**, 169–84.
Simpson, K.L., Lee, T.G., Rodriguez, D.B. and Chichester, C.D. (1976). In: *Chemistry and Biochemistry of Plant Pigments* (Ed. T.W. Goodwin) vol. 2, pp. 779–842. Academic Press, London.
Sinclair, P.R., Bement, W.J., Bonkovsky, H.L. *et al.* (1986). *Biochem. J.* **237**, 63–71.
Sironval, C. and Brouers, M. (eds.) (1984). *Protochlorophyllide Reduction and Greening*. Martinus Nijhoff/Dr. W. Junk Publ., The Hague.
Smith, A.G. (1987). In: *Progress in Photosynthesis Research* (Ed. J. Biggins) vol. 4, pp. 453–6. Martinus Nijhoff, Dordrecht.
Smith, B.B. and Rebeiz, C.A. (1977). *Arch. Biochem. Biophys.* **180**, 178–85.
Smith, K.M. (1975a). *Porphyrins and Metalloporphyrins*. Elsevier, Amsterdam.
Smith, K.M. (1975b). In: *Porphyrins and Metalloporphyrins* (Ed. K.M. Smith) pp. 381–97. Elsevier, Amsterdam.
Smith, K.M. and Goff, D.A. (1986). *J. Org. Chem.* **51**, 657–66.
Smith, K.M. and Huster, M.S. (1987). *J. Chem. Soc. Chem. Commun.* **1987**, 14–6.
Smith, K.M. and Simpson, D.J. (1986). *J. Chem. Soc. Chem. Commun.* **1986**, 1682–4.
Smith, K.M., Bushell, M.J., Rimmer, J. and Unsworth, J.F. (1980). *J. Amer. Chem. Soc.* **102**, 2437–48.
Smith, K.M., Goff, D.A. and Simpson, D.J. (1985). *J. Amer. Chem. Soc.* **107**, 4946–54.

Soll, J. and Schultz, G. (1981). *Biochem. Biophys. Res. Commun.* **99**, 907–12.
Soll, J., Schultz, G., Rüdiger, W. and Benz, J. (1983). *Plant Physiol.* **71**, 849–54.
Spiller, S.C., Castelfranco, A.M. and Castelfranco, P.A. (1982). *Plant Physiol.* **69**, 107–11.
Steffens, D., Blos, I., Schoch, S. and Rüdiger, W. (1976). *Planta* **130**, 151–8.
Steiner, R., Schäfer, W., Blos, I., Wieschhoff, H. and Scheer, H. (1981). *Z. Naturforsch.* **36c**, 417–20.
Stobart, A.K. and Ameen-Bukhari, I. (1986). *Biochem. J.* **236**, 741–8.
Straka, J.G. and Kushner, J.P. (1983). *Biochemistry* **22**, 4664–72.
Svec, W.A. (1978). In: *The Porphyrins* (Ed. D. Dolphin) vol. V, part C, pp. 341–99. Academic Press, London.
Tanaka, K., Kakumo, T., Yamashita, J. and Horio, T. (1982). *J. Biochem. (Tokyo)* **92**, 1763–73.
Tanaka, K., Kakumo, T., Yamashita, J. and Horio, T. (1983). *J. Biochem. (Tokyo)* **93**, 159–67.
Terpstra, W. (1981). *FEBS Lett.* **126**, 231–5.
Terpstra, W. and Lambers, J.W.J. (1983). *Biochim. Biophys. Acta* **746**, 23–31.
Thauer, R.K. (1985). *Biol. Chem. Hoppe-Seyler* **366**, 103–12.
Tripathy, B.C. and Rebeiz, C.A. (1986). *J. Biol. Chem.* **261**, 13556–64.
Tripathy, B.C. and Rebeiz, C.A. (1987). In: *Progress in Photosynthesis Research* (Ed. J. Biggins) vol. 4, pp. 439–43. Martinus Nijhoff, Dordrecht.
Troxler, R.F. and Offner, G.D. (1979). *Arch. Biochem. Biophys.* **195**, 53–65.
Troxler, R.F., Smith, K.M. and Brown, S.B. (1980). *Tetrahedr. Lett.* **21**, 491–4.
Verneuil de, H., Sassa, S. and Kappas, A. (1983). *J. Biol. Chem.* **258**, 2454–60.
Vezitskii, A.Y. and Walter, G. (1981). *Photosynthetica* **15**, 104–8.
Virgin, H. (1977). *Physiol. Plant.* **40**, 45–9.
Virgin, H.J. (1981). *Ann. Rev. Plant Physiol.* **32**, 451–63.
Walter, E., Schreiber, J., Zass, E. and Eschenmoser, A. (1979). *Helv. Chim. Acta* **62**, 899–920.
Wang, W.-Y., Huang, D.-D., Stachon, D., Gough, S.P. and Kannangara, C.G. (1984). *Plant Physiol.* **74**, 569–75.
Wang, W.-Y., Huang, D.-D., Chang, T.-E., Stachon, D. and Wegmann, B. (1987). In: *Progress in Photosynthesis Research* (Ed. J. Biggins) vol. 4, pp. 423–30. Martinus Nijhoff, Dordrecht.
Watanabe, T., Hongu, A., Honda, K., Nakazato, M., Konno, M. and Saitoh, S. (1984). *Anal. Chem.* **56**, 251–6.
Watanabe, T., Kobayashi, M., Hongu, A., Nakazato, M., Hiyama, T. and Murata, N. (1985). *FEBS Lett.* **191**, 252–6.
Weinstein, J.D. and Beale, S.I. (1985). *Arch. Biochem. Biophys.* **239**, 87–93.
Weinstein, J.D., Schneegart, M.A. and Beale, S.I. (1987a). In: *Progress in Photosynthesis Research* (Ed. J. Biggins) vol. 4, pp. 431–4. Martinus Nijhoff, Dordrecht.
Weinstein, J.D., Mayer, S.M. and Beale, S.I. (1987b) *ibid.* 435–8.
Wilkinson, R.E. (1985). *Pestic. Biochem. Physiol.* **23**, 289–93.
Williams, D.C. (1984). *Biochem. J.* **217**, 675–83.
Williams, D.C., Morgan, G.S., McDonald, E. and Battersby, A.R. (1981). *Biochem. J*, **193**, 301–10.
Wolf, H. and Scheer, H. (1973). *Liebigs Ann. Chem.* **1973**, 1741–9.
Wolf, H., Brockmann, H., Biere, H. and Inhoffen, K.H. (1967). *Justus Liebigs Ann. Chem.* **704**, 208–25.

Wong, Y.-S. and Castelfranco, P.A. (1984). *Plant Physiol.* **75**, 658–61.
Wong, Y.-S., Castelfranco, P.A., Goff, D.A. and Smith, K.M. (1985). *Plant Physiol.* **79**, 725–9.
Woodward, R.B. (1960). *Angew. Chem.* **72**, 651–62.
Wu, S.M. and Rebeiz, C.A. (1984). *Tetrahedron* **40**, 659–64.
Wu, S.-M. and Rebeiz, C.A. (1985). *J. Biol. Chem.* **260**, 3632–4.
Yamasaki, H. and Moriyama, T. (1970). *Biochem. Biophys. Res. Commun.* **38**, 638–43.
Yoshinaga, T. and Sano, S. (1980a). *J. Biol. Chem.* **255**, 4722–6.
Yoshinaga, T. and Sano, S. (1980b). *J. Biol. Chem.* **255**, 4727–31.
Zagorec, M. and Labbe-Bois, R. (1986). *J. Biol. Chem.* **261**, 2506–9.
Zaman, Z. and Akhtar, M. (1976). *Eur. J. Biochem.* **61**, 215–23.
Zaman, Z., Abboud, M.M. and Akhtar, M. (1972). *J. Chem. Soc. Chem. Commun.* **1972**, 1263–4.
Ziegler, R., Dresken, H., Herl, B., Menth, M., Schneider, H. and Zange, P. (1982a). *Deutsche Botanische Gesellschaft (Freiburg) Book of Abstracts.* 472.
Ziegler, R., Guha, N. and Schnell, B. (1982b). *Deutsche Botanische Gesellschaft (Freiburg) Book of Abstracts.* 473.

2
Distribution and Analysis of Carotenoids

TREVOR W. GOODWIN and GEORGE BRITTON

Department of Biochemistry, University of Liverpool, P.O. Box 147, Liverpool L69 3BX, U.K.

Part 1: Distribution

Part 2: Analysis

PLANT PIGMENTS
ISBN 0-12-289847-8

PART 1: DISTRIBUTION

1 INTRODUCTION

Full background information on the distribution of carotenoids in plants has been given by Goodwin (1976, 1980). In the present chapter later data will be assessed in relation to this previous information. The general structural features, nomenclature and numbering of the carotenoids will be assumed. The full semi-systematic name of pigments will be given the first time they are mentioned.

2 HIGHER PLANTS

A Photosynthetic tissues

1 NATURE

Carotenoids accumulate in the chloroplasts of all green leaves and it is striking that all the many plants examined contain the same major carotenoids, namely β-carotene (β,β-carotene, I), lutein [(3*R*,3′*R*,6′*R*)-β,ε-carotene-3,3′-diol, II], violaxanthin [(3*S*,5*R*,6*S*,3′*S*,5′*R*,6′*S*)-5,6,5′,6′-diepoxy-5,6,5′,6′-tetrahydro-β,β-carotene-3,3′-diol, III] and neoxanthin [(3*S*,5*R*,6*R*,3′*S*,5′*R*,6′*S*)-5′,6′-epoxy-6 ,7-didehydro-5 ,6,5′,6′-tetrahydro-β,β-carotene-3,5,3′-triol, IV]. α-Carotene [(6′*R*)-β,ε-carotene, V], β-cryptoxanthin [(3*R*)-β,β-caroten-3-ol, VI], zeaxanthin [(3*R*,3′*R*)-β,β-

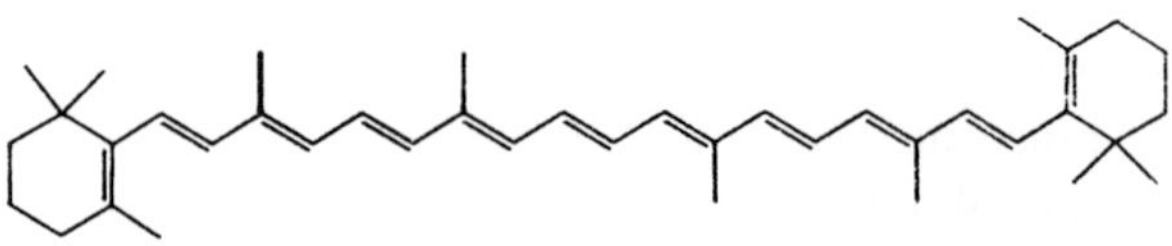

I β-Carotene

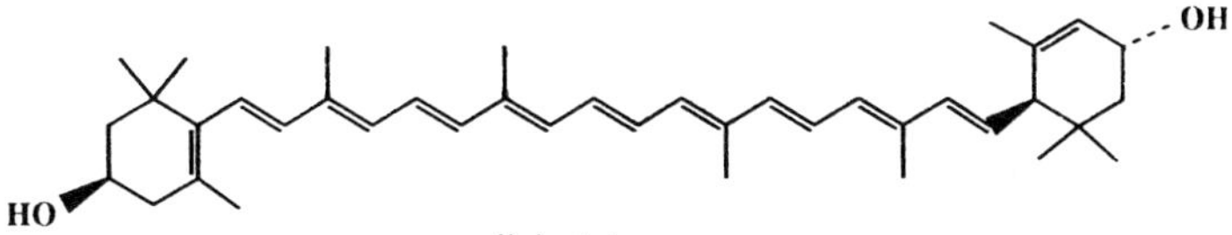

II Lutein

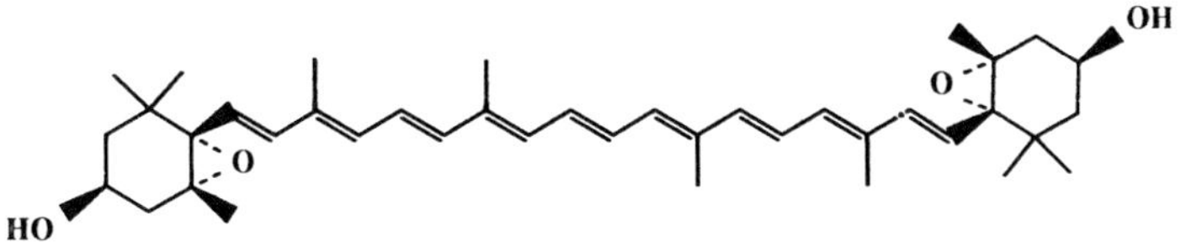

III Violaxanthin

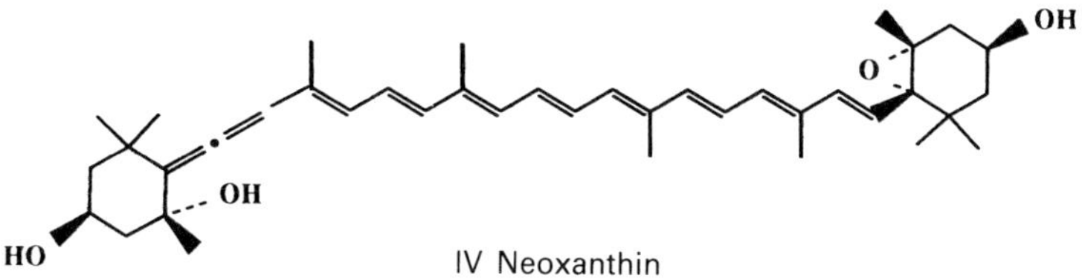

IV Neoxanthin

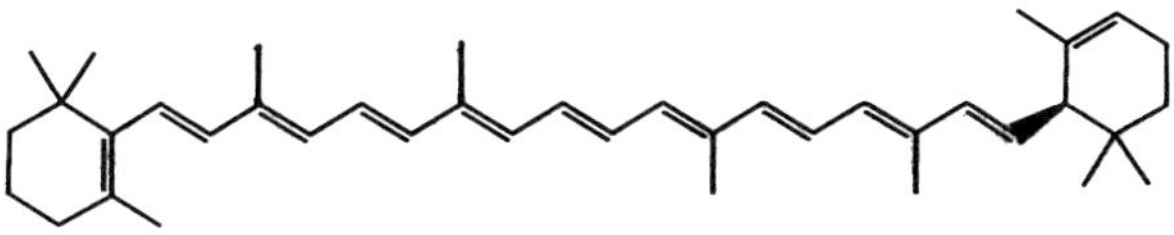

V α-Carotene

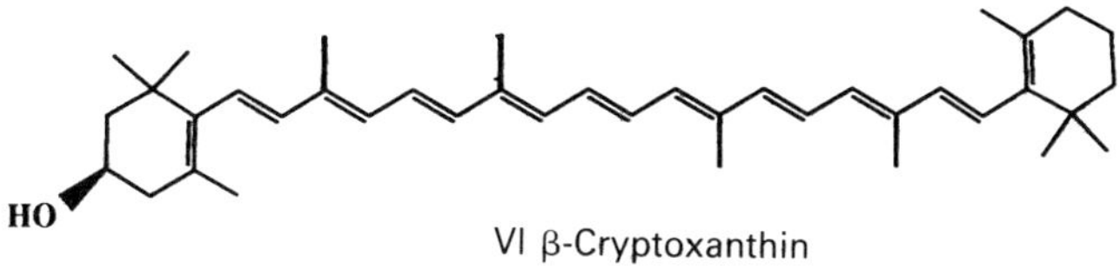

VI β-Cryptoxanthin

carotene-3,3′-diol, VII], antheraxanthin [(3*S*,5*R*,6*S*,3′*R*)-5,6-epoxy-5,6-dihydro-β,β-carotene-3,3′-diol, VIII], and lutein-5,6-epoxide [(3*S*, 5*R*, 6*S*, 3′*R*, 6′*R*)-5,6-epoxy-5,6-dihydro-β,ε-carotene-3,3′-diol, IX] are commonly also present as minor components. Except in the senescent autumn leaves of some deciduous trees, the xanthophylls are unesterified. Quantitative differences between species are encountered but, in general, carotenes

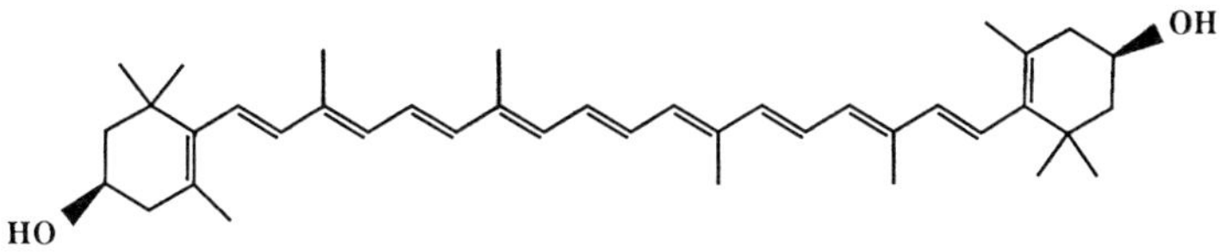

VII Zeaxanthin

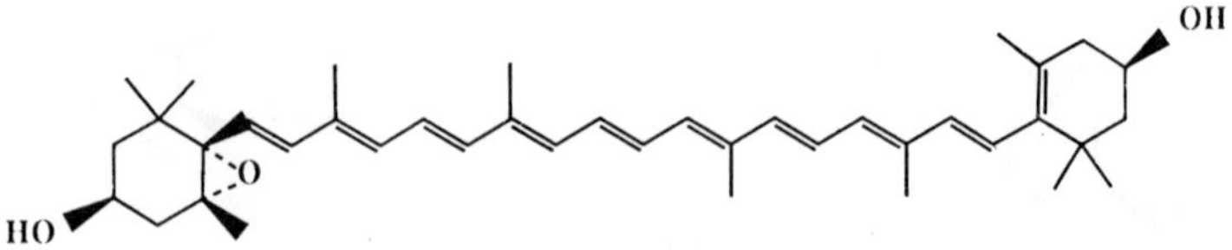

VIII Antheraxanthin

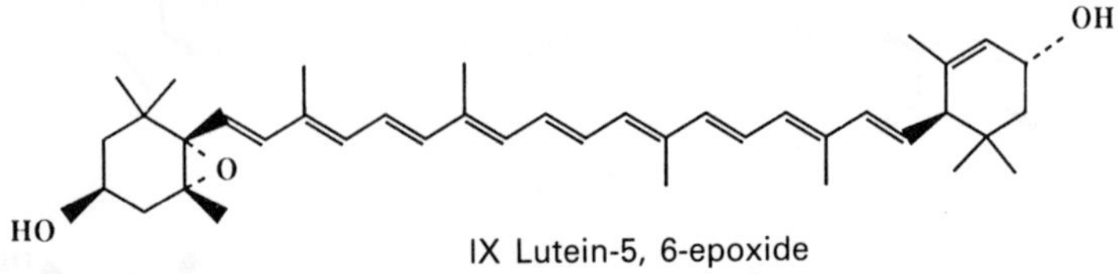

IX Lutein-5, 6-epoxide

(hydrocarbons) represent about 25% and lutein, the main xanthophyll, about 45% of the total carotenoid present. The absolute configuration of neoxanthin (Eugster, 1982) confirms its biosynthetic origin from the β,β-ring carotenoids, not from lutein.

Recently a new xanthophyll, lactucaxanthin [(3*R*, 6*R*, 3′*R*, 6′*R*)-ϵ,ϵ-carotene-3,3′-diol, X] has been isolated from certain composites, including lettuce (*Lactuca sativa*) (Siefermann-Harms *et al.*, 1981). This is the first report of an ϵ,ϵ-carotenoid in higher plants, although they are found in algae (see part 1, section 3) and fish (Goodwin, 1984).

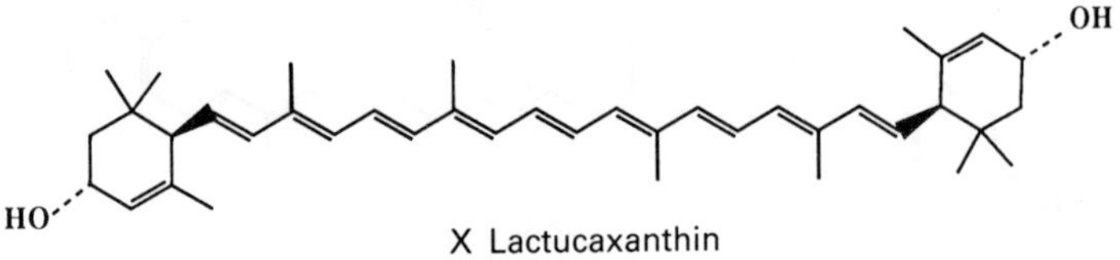

X Lactucaxanthin

The leaves or needles of many gymnosperms, especially in the autumn and winter, contain substantial amounts of ketocarotenoids. Czeczuga (1986) analysed the carotenoids of 26 species and reported the presence of adonixanthin (3,3′-dihydroxy-β,β-caroten-4-one, XI) in seven species from the Pinaceae and of rhodoxanthin (4′,5′-didehydro-4,5′-*retro*-β,β-carotene-3,3′-dione, XII) in all species of the Cupressaceae (four),

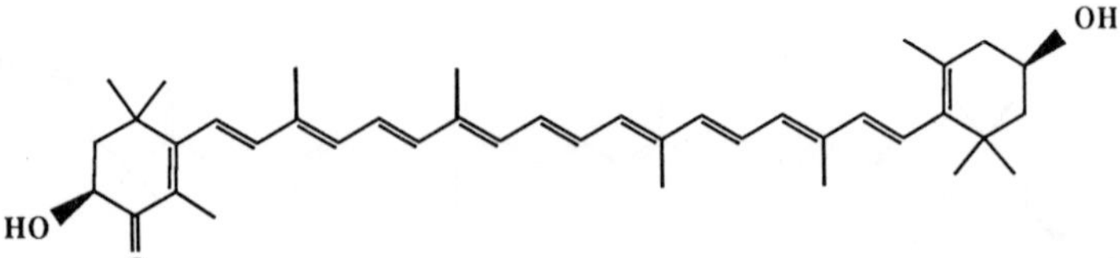

XI Adonixanthin

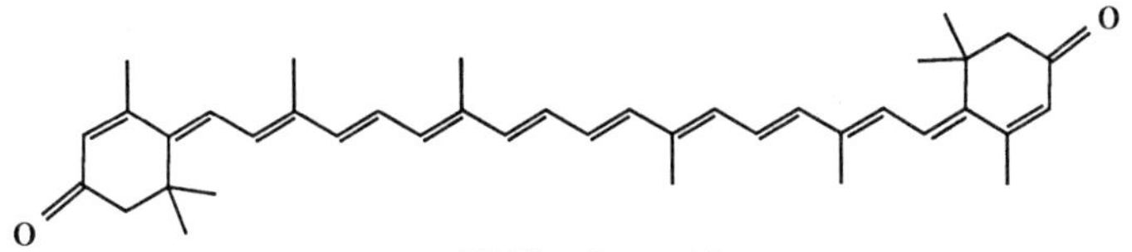

XII Rhodoxanthin

Taxaceae (one) and Taxodiaceae (one) examined, in agreement with an earlier report by Ida (1981) from a study of gymnosperms in Japan. Another unusual carotenoid, considered by Czeczuga to be lycoxanthin (ψ,ψ-caroten-16-ol, XIII) was present in seven of the species examined.

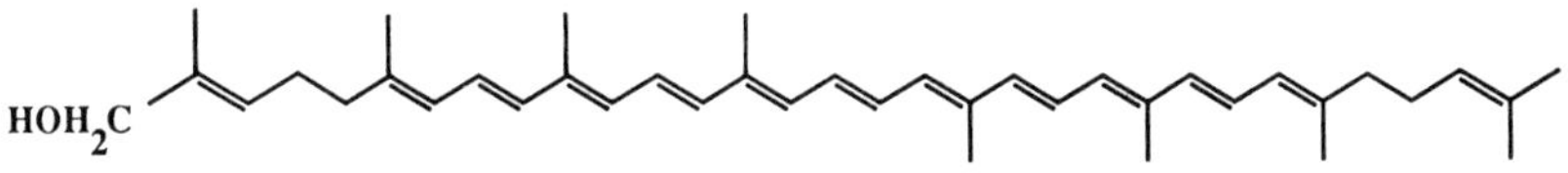

HOH_2C

XIII Lycoxanthin

Unfortunately, in this work the identifications were based only on chromatographic behaviour and absorption spectra, and re-examination by modern high-performance liquid chromatography (HPLC) and spectroscopic methods would be worthwhile. Such an examination has confirmed the identification of rhodoxanthin (25% of the total carotenoid) in bronze winter leaves of *Thuja occidentalis* var. Rheingold, largely as the 6,6-di-*trans* isomer, though some of the 6-*trans*-6′-*cis* and 6,6′-di-*cis* isomers were also present (A. J. Young and G. Britton, unpublished results).

The unusual secocarotenoid, semi-β-carotenone (5,6-seco-β,β-carotene-5,6-dione, XIV) has been found as the main pigment (70% of the total)

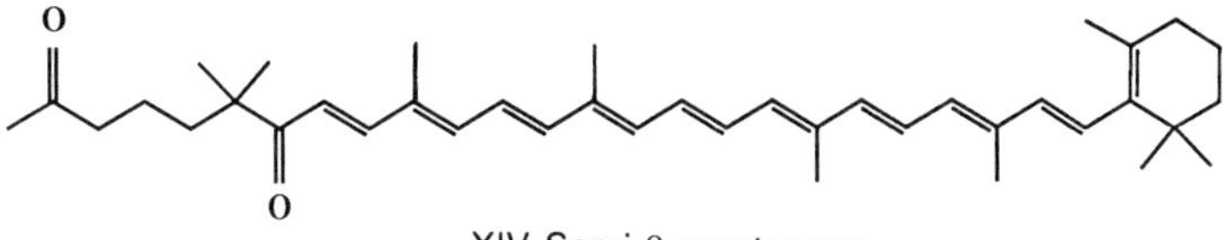

XIV Semi-β-carotenone

in young red leaflets of two cycads, *Ceratozamia fuscoviridis* and *C. kuesteriana* and identified by full spectroscopic characterization (Cardini *et al.*, 1987). Smaller amounts of several related hydroxylated secocarotenoids were also isolated, and one of these has been shown to have the novel 5-hydroxy-6-oxo-seco-β end-group (5-hydroxy-5,6-seco-β,β-caroten-6-one, XV) (F. Cardini and G. Britton, unpublished results).

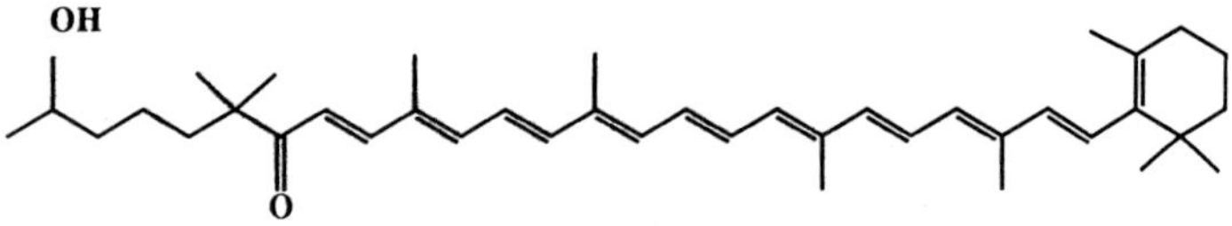

XV 5-Hydroxy-5,6-seco-β,β-caroten-6-one

2 LOCALIZATION

The carotenoids in chloroplasts exist in pigment–protein complexes but as they are not covalently bound to the acceptor proteins the detergents used in isolating the complexes frequently produce degraded complexes and release "free" carotenoids. The tortuous development of the methodology in this field has recently been thoroughly surveyed by Siefermann-Harms (1985). The development and use of non-denaturing solubilization methods have revealed the carotenoid disposition in the two photosystems involved in photosynthesis in higher plants and algae (oxygen-evolving photosynthesis). Photosystem I (PSI) consists of a core complex I (CCI) and a light-harvesting complex I (LHCI) (nomenclature of Thornber, 1986). CCI contains one β-carotene molecule per 40 chlorophyll a molecules whereas LHCI is a chlorophyll a/b-containing protein with which are associated the main leaf xanthophylls, lutein, violaxanthin and neoxanthin. Photosystem II consists of CCII and LHCIIα and LHCIIβ; the smallest of the five subunits of CCII contains about 50 chlorophyll and 7 β-carotene molecules (Tang and Satoh, 1985); in LHCIIβ one molecule of apoprotein contains 4 chlorophyll a molecules, 3 chlorophyll b molecules and 1–2 xanthophyll molecules (Thornber *et al.*, 1979; Lichtenthaler *et al.*, 1982). Detailed discussion of these and other complexes is provided by Siefermann-Harms (1985) and Cogdell (Chapter 4, this volume).

The ketocarotenoids referred to above, i.e. rhodoxanthin etc. in gymnosperms and semi-β-carotenone and related compounds in cycads, are not present in the chloroplasts but in oil droplets similar in form to plastoglobuli.

B Reproductive tissues

1 ANTHERS

The carotenoids in the anthers of lilies, first examined by Karrer in his classical chemical studies of carotenoids from 1930 to 1950 and from which antheraxanthin was first isolated, have been re-examined in detail by Eugster's group. A new xanthophyll lilixanthin [(3*S*, 4*S*, 3′*S*, 5′*R*)-3,4,3′-trihydroxy-β,κ-caroten-6′-one, XVI] was obtained from *Lilium tigrinum* cv. Red Night (Märki-Fischer and Eugster, 1985a), as was karpoxanthin [(3*S*, 5*R*, 6*R*, 3′*R*)-5,6-dihydro-β,β-carotene-3,5,6-3′-tetraol, XVII] first isolated in 1985 from rosehips (see below) (Märki-Fischer and Eugster, 1985b); epikarpoxanthin (XVIII), the 6-epimer of karpoxanthin, was also found in the anthers. Capsanthin [(3*R*, 3′*S*, 5′*R*)-3,3′-dihydroxy-β,κ-caroten-6′-one, XIX], first reported in anthers of lilies by Karrer in the 1950s has now been shown to be a mixture of four components, the (9*Z*)-, (9′*Z*), (13*Z*) and (13′*Z*) geometrical isomers [*Z* and *E* correspond to *cis* and *trans* in the old nomenclature]; (Baranyai *et al.*, 1981; Märki-Fischer and Eugster, 1985a).

The absolute configuration of antheraxanthin (VIII) is now established (Märki-Fischer *et al.*, 1982). The "*cis*-antheraxanthin' from *Lilium* spp. was assigned the 9*Z* configuration by the original workers (Toth *et al.*, 1978); this has been confirmed by nuclear magnetic resonance (NMR) studies (G. Englert, quoted by Eugster, 1982).

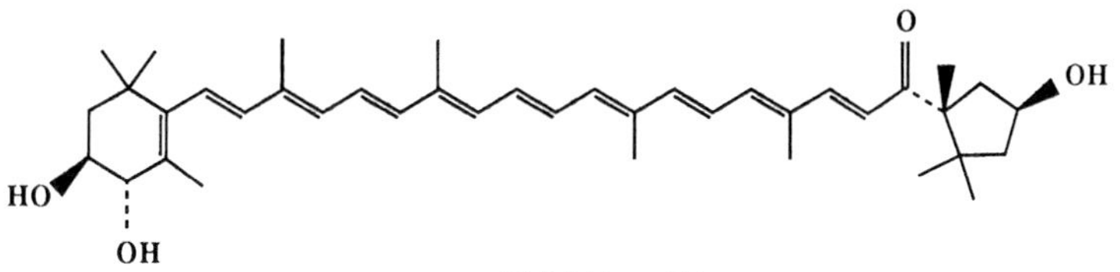

XVI Lilixanthin

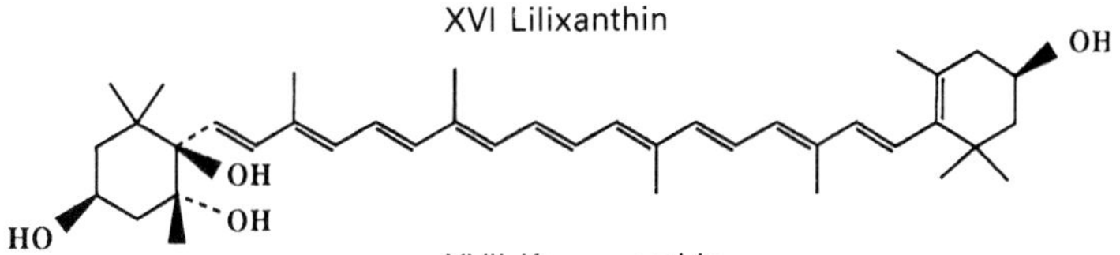

XVII Karpoxanthin

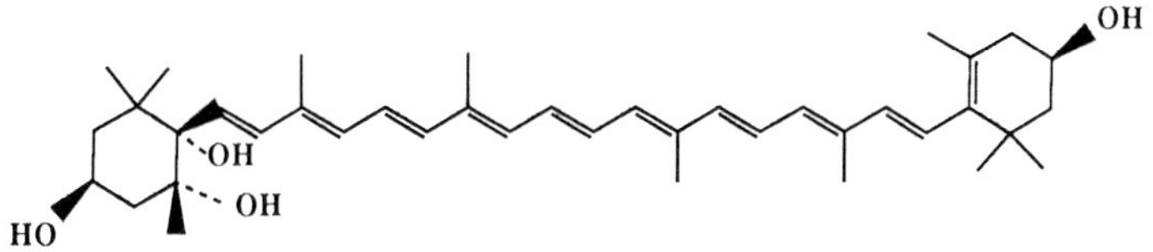

XVIII Epikarpoxanthin

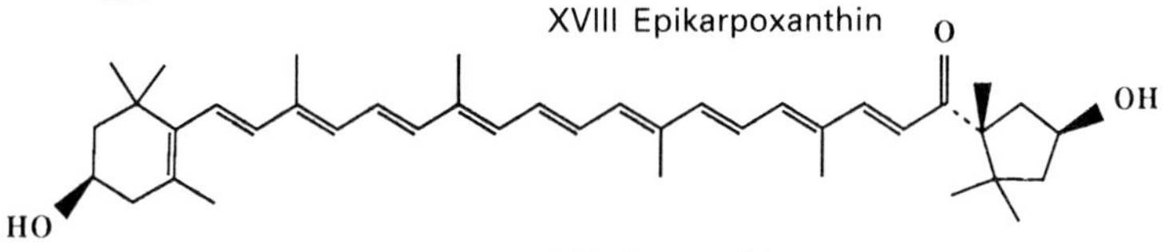

XIX Capsanthin

2 STIGMAS

The stigmas of *Crocus neopolitanus* were found to synthesize two new esters of the well known apocarotenoid acid crocetin (8,8′-diapocarotene-8,8′-dioic acid, XX). They are the mono- and diesters of a new trisaccharide neapolitanose {*O*-β-*D*-glucopyranosyl-(1→2)-*O*-[β -*D*-glucopyranosyl-(1→6)]-*D*-glucose} (Rychener *et al.*, 1984).

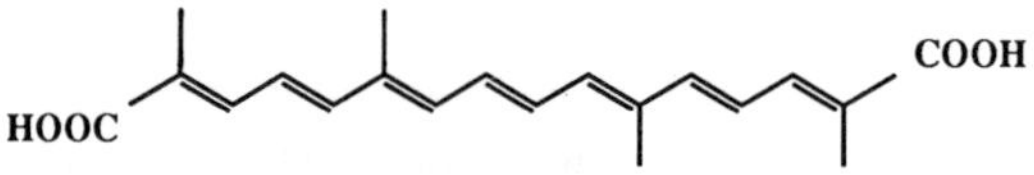

HOOC COOH

XX Crocetin

3 FLOWERS

Flower carotenoids are characterized by many unique structures and up until 1980 the isolation of some 40 carotenoids from flower petals had been reported. In contrast with chloroplast xanthophylls, flower xanthophylls are often esterified (Goodwin, 1980). Recent work has focused mainly on settling outstanding stereochemical uncertainties of known compounds and characterizing naturally occurring *cis*-isomers of known compounds.

The separating of carotenogenic flowers into three main groups (Goodwin, 1980) still holds true: those synthesizing (i) highly oxygenated xanthophylls, frequently isolated as 5,8-epoxides (furanoid carotenoids); (ii) mainly carotenes and (iii) highly species-specific carotenoids, such as eschscholtzxanthin (4′,5′-didehydro-4,5′-*retro*-β,β-carotene-3,3′-diol, XXI).

The stereochemistry (at C-8) of the diastereomeric mutatoxanthins (5,8-epoxy-5,8-dihydro-β,β-carotene-3,3′-diols, XXII) (Märki-Fischer *et al.*, 1982) and neochromes [= trollichromes (5′,8′-epoxy-6,7-didehydro-5,6,5′,8′-tetrahydro-β,β-carotene-3,5, 3′-triols, XXIII); Märki-Fischer *et al.*, 1984a] has been established. Indeed in all HPLC analyses of furanoid carotenoids from natural sources both C-8 epimers were found (Eugster,

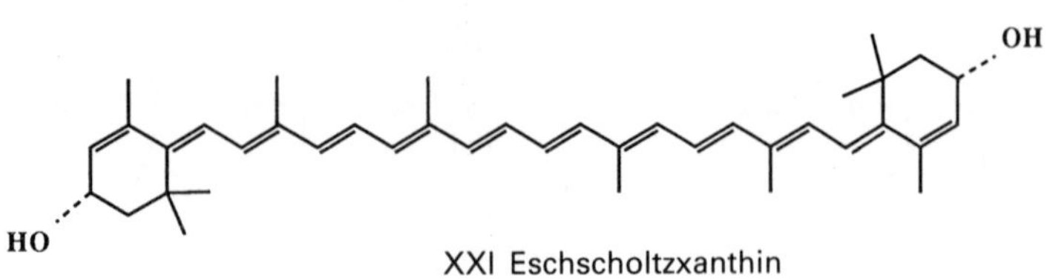

OH

HO

XXI Eschscholtzxanthin

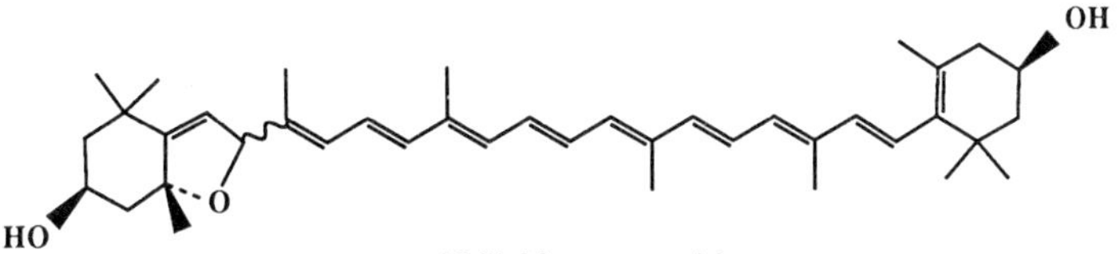

XXII Mutatoxanthin

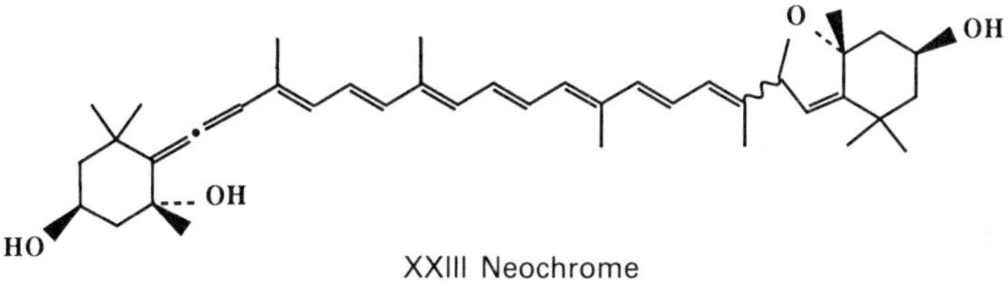

XXIII Neochrome

1985). This observation casts some doubt on whether these furanoid oxides are true natural products. If they were formed directly and enzymically, only one C-8 epimer would be expected. Even the mildly acidic conditions that may obtain in plant tissues are sufficient to cause isomerization of naturally occurring 5,6-epoxides into the 5,8-form when the tissue is disrupted during the isolation procedure. Such isomerization always produces a mixture of the 8*R*- and 8*S*-epimers.

The C-8 epimeric pairs of luteoxanthin (5,6,5′,8′-diepoxy-5,6,5′,8′-tetrahydro-β,β-carotene-3,3′-diol, XXIV) and auroxanthin (5,8,5′,8′-diepoxy-5,8,5′,8′-tetrahydro-β,β-carotene-3,3′-diols, XXV) and of their 9*Z*-isomers, that is, eight pigments in all, have been isolated from the petals of *Rosa foetida* (Eugster, 1985). New tetraols have also been obtained from the same source: latoxanthin (5′,6′-epoxy-5,6,5′,6′-tetrahydro-β,β-carotene-3,5,6,3′-tetraol (XXVI) and its 9*Z* isomer and both the (8′*R*) and (8′*S*) epimers of its furanoid isomer latochrome (5′,8′-epoxy-5,6,5′,8′-tetrahydro-β,β-carotene-3,5,6,3′ tetraol, XXVII) (Märki-Fischer

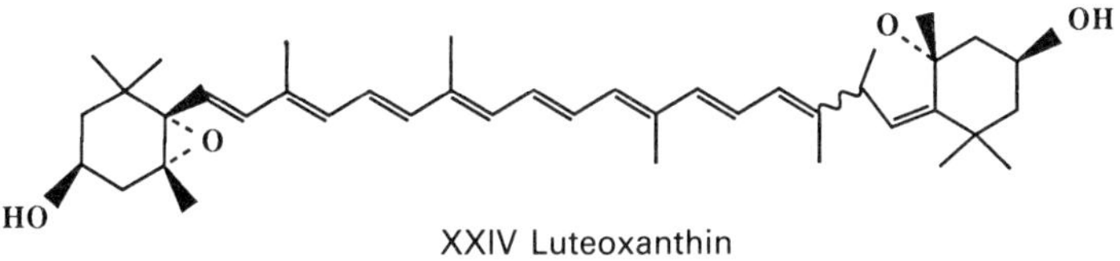

XXIV Luteoxanthin

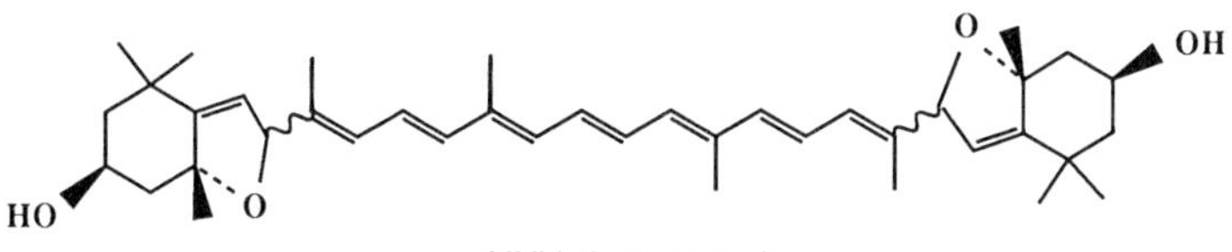

XXV Auroxanthin

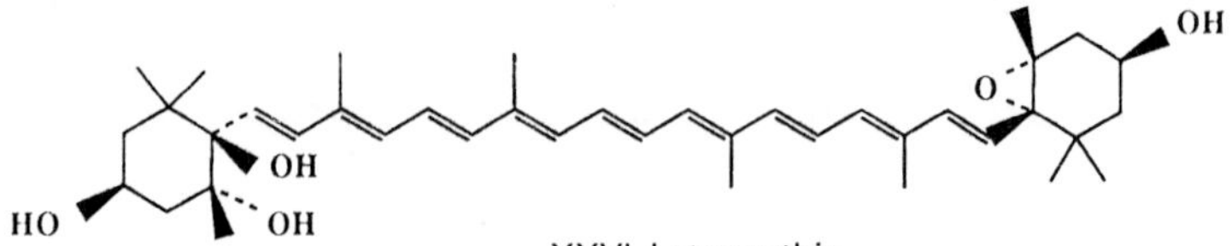

XXVI Latoxanthin

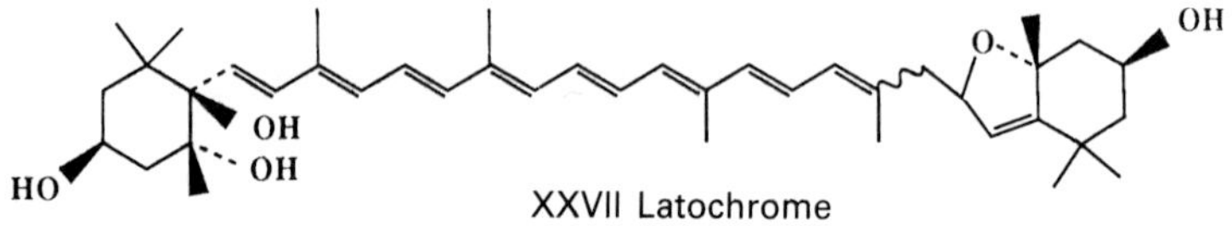

XXVII Latochrome

et al., 1984b). The 3,5,6-triol end-group present in the latoxanthin molecule (XXVI) also appears in a pentaol from the petals of *Trollius europaeus* (6′,7′-didehydro-5,6,5′,6′-tetrahydro-β,β-carotene-3,5,6,3′,5′-pentaol, XXVIII). This pigment is formally a hydration product of neoxanthin (Eugster, 1985).

The structure first suggested by Dabbagh and Egger (1974) for calthaxanthin from the flowers of *Caltha palustris* has now been firmly established (Buchecker and Eugster, 1979) as 3′-epilutein [(3*R*, 3′*S*, 6′*R*)-β,ε-carotene-3,3′-diol, XXIX]. In addition 3′-O-didehydrolutein (3-hydroxy-β,ε-caroten-3′-one, XXX) is present and it is a probable

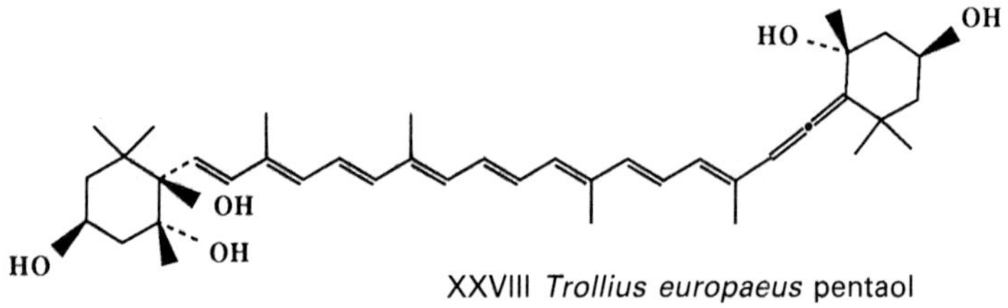

XXVIII *Trollius europaeus* pentaol

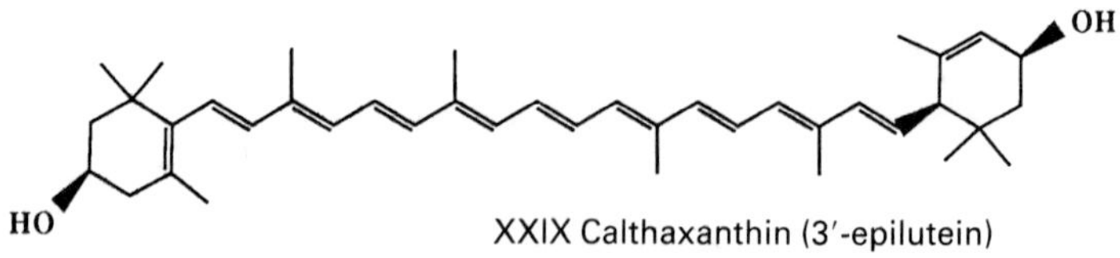

XXIX Calthaxanthin (3′-epilutein)

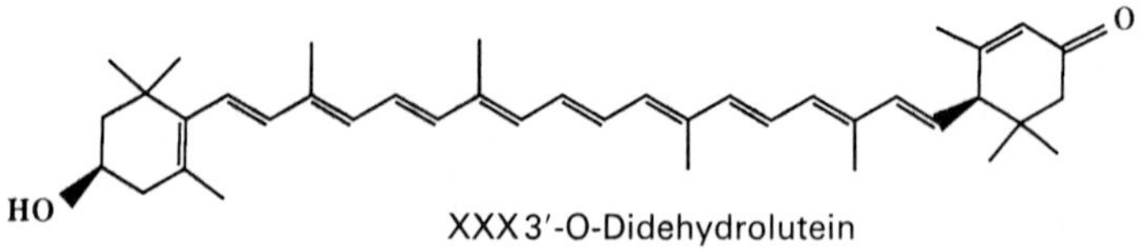

XXX 3′-O-Didehydrolutein

intermediate in the conversion of lutein into 3′-epilutein. As it is very labile to alkali, 3′-O-didehydrolutein may be more widespread in plants than is currently suspected because saponification is commonly used as the first step in purifying carotenoid extracts.

Geometrical isomers of lutein are now frequently being observed; the 9*Z* and 9′*Z* isomers have been observed in *Calendula officinalis*, *Helianthemum annuum*, and *Chrysanthemum* spp. (Toth and Szabolcs, 1981) and the 9*Z*, 9′*Z* and 13*Z* forms have been identified in daffodil flowers (Berset and Pfander, 1985). The (9*Z*) isomer of lutein-5,6-epoxide (IX) is found in considerable amounts in *Calendula officinalis*, *H. annuum* and *Chrysanthemum* spp. (Toth and Szabolcs, 1981). The antheraxanthin in *Calendula officinalis* is entirely the 9*Z* isomer (Toth and Szabolcs, 1981). Seven geometrical isomers of violaxanthin (III) from *Viola tricolor* have been characterized by ^{1}H NMR spectroscopy, *viz*. (9*Z*), (13*Z*), (15*Z*), (9*Z*,9′*Z*), (9*Z*,13′*Z*), (9*Z*,15*Z*) and (9*Z*,13*Z*) (Molnar and Szabolcs, 1980; Molnar *et al*., 1986).

The fatty acid composition of the violaxanthin esters in *V. tricolor* is complex and includes β-hydroxyacids (Hansmann and Kleinig, 1982). α-Cryptoxanthin [(3*R*, 6′*R*)-β,ε-caroten-3-ol, = physoxanthin, XXXI; Bodea *et al*., 1978] is present in *C. palustris* petals (Buchecker and Eugster, 1980). Buchecker and Eugster (1980) have also provided the full structures of mimulaxanthin (6,7,6′,7′-tetradehydro-5,6,5′,6′-tetrahydro-β,β-carotene-3,5,3′, 5′-tetraol, XXXII) first isolated as an ester from *Mimulus* spp. by Nitsche (1973, 1974a), and deepoxyneoxanthin (6,7-didehydro-5,6-dihydro-β,β-carotene-3,5,3′- triol, XXXIII).

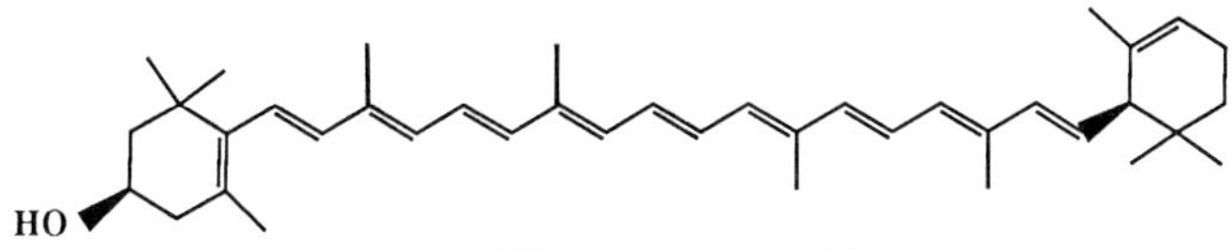

XXXI α-Cryptoxanthin

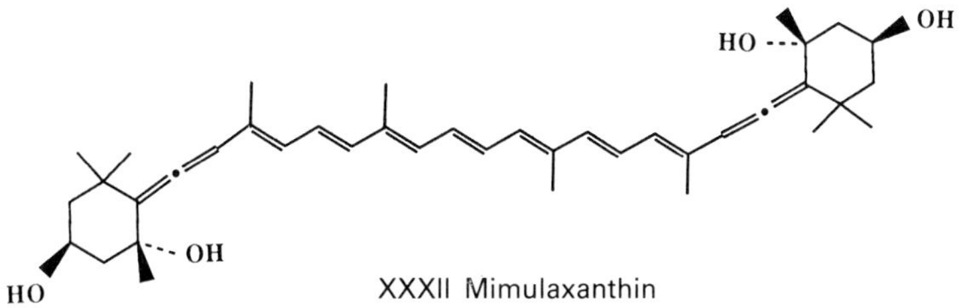

XXXII Mimulaxanthin

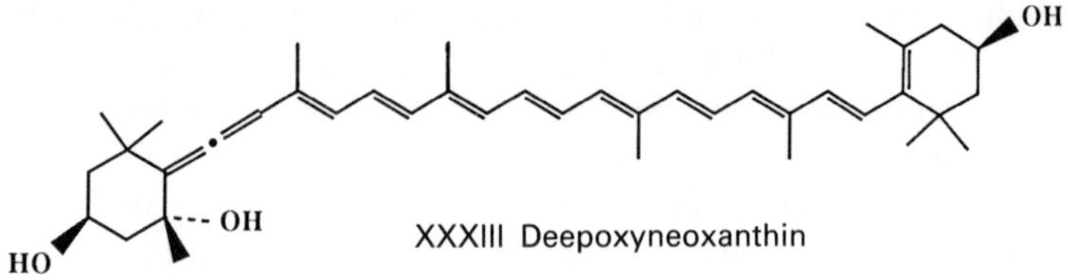

XXXIII Deepoxyneoxanthin

4 FRUIT

As is the case with flowers, fruit carotenoids are characterized by a great variety of structures and up to 1980 nearly 70 authenticated compounds had been reported (Goodwin, 1980). Eight main groups can be distinguished, although there is occasional blurring of one group into another (Table 2.1).

Table 2.1 Eight general groups into which fruit can be placed on the basis of their carotenoid compositions

Group	Pigment characteristics
I	Insignificant amounts
II	Small amounts generally of chloroplast carotenoids
III	Relatively large amounts of lycopene and its hydroxy-derivatives
IV	Relatively large amounts of β-carotene and its hydroxy-derivatives
V	Large amounts of epoxides, particularly furanoid epoxides
VI	Unusual carotenoids, e.g. capsanthin (XIX)
VII	Poly-*Z* carotenoids, e.g. prolycopene (XXXVI)
VIII	Apocarotenoids e.g. persicaxanthin (XLI)

Two new carotenoids, cucurbitaxanthin A [(3*R*,3′*S*,5′*R*,6′*R*)-3′,6′-epoxy-5′,6′-dihydro-β,β-carotene-3,5′-diol, XXXIV] and its 5,6-epoxide, cucurbitaxanthin B [(3*S*,5*R*,6*S*,3′*S*,5′*R*,6′*R*)-5,6,3′,6′-diepoxy-5,6,5′,6′-tetrahydro-β,β -carotene-3,5′-diol, XXXV] comprised 25 and 8% respectively, of the total carotenoid of the flesh of the pumpkin *Cucurbita maxima*. The chirality at C-3 and C-3′ was proved by conversion into (3*R*,3′*R*)zeaxanthin (Matsuno *et al.*, 1986). These are the first examples of carotenoids with the 3,6-epoxy end-group to be found in higher plants.

Prolycopene, the fascinating poly*cis* lycopene first isolated from tangerine tomatoes by Zechmeister in the early 1940s (see Zechmeister

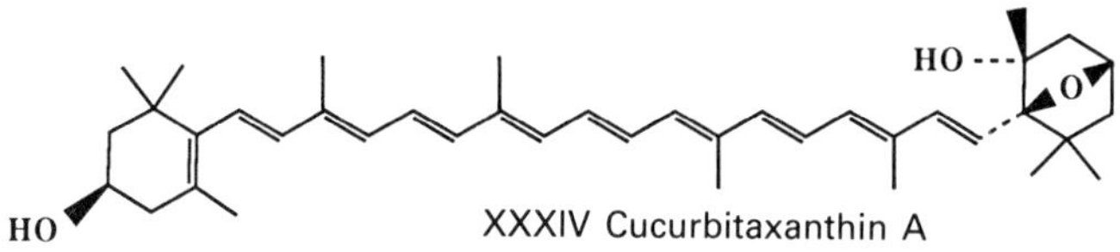

XXXIV Cucurbitaxanthin A

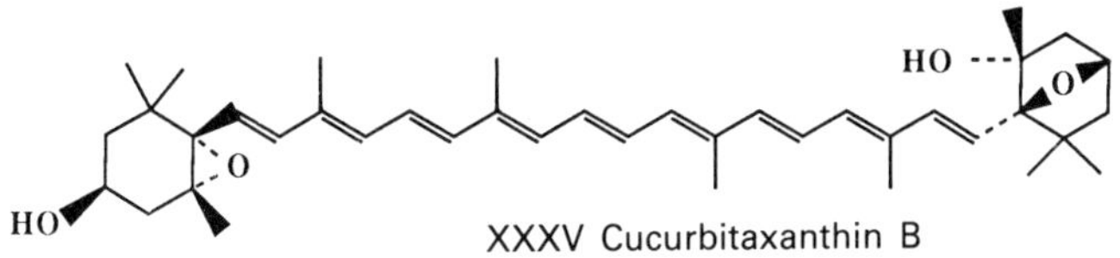

XXXV Cucurbitaxanthin B

and Pinckard, 1960), has recently been shown to be (7*Z*, 9*Z*, 7′*Z*, 9′*Z*)-lycopene [(7*Z*,9*Z*,7′*Z*,9′*Z*)-ψ,ψ -carotene, XXXVI] (Clough and Pattenden, 1979; Englert *et al.*, 1979). The partly saturated polyenes occurring with prolycopene in tangerine tomatoes are also *Z* isomers, namely (15*Z*)-phytoene, (15*Z*, 9′*Z*)-phytofluene, (9*Z*, 9′*Z*)-ζ-carotene and (9*Z*, 7′*Z*, 9′*Z*) neurosporene (Clough and Pattenden, 1979), (see Chapter 3).

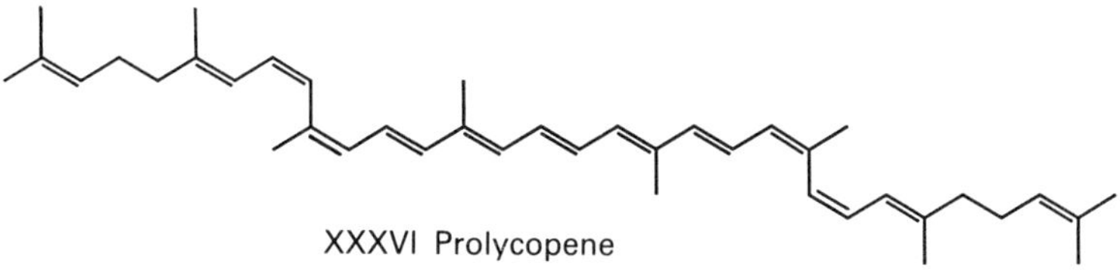

XXXVI Prolycopene

Hips of *Rosa pomifera* contain *Z* isomers of the acyclic polyenes as well as of β-cryptoxanthin (VI), zeaxanthin (VII) and rubixanthin [(3*R*)-β,ψ-caroten-3-ol, XXXVII]; the 5*Z* and 5′*Z* isomers are of considerable interest (Märki-Fischer *et al.*, 1983).

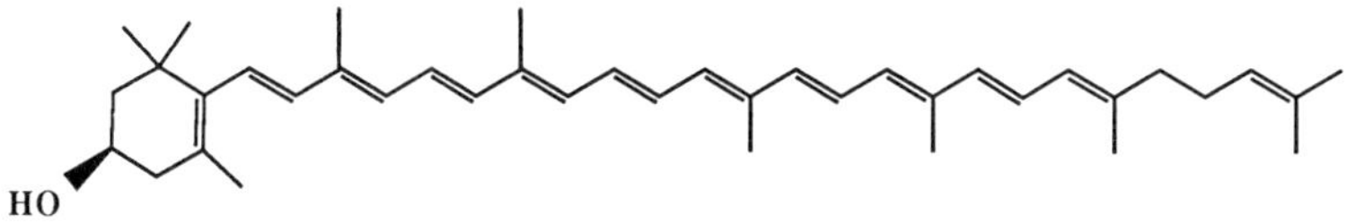

XXXVII Rubixanthin

Physoxanthin, first isolated from the berries of *Physalis alkekengi* (Bodea *et al.*, 1978) is, as indicated in the previous section, identical with α-cryptoxanthin (XXXI) (Buchecker and Eugster, 1980; Bjørnland *et al.*, 1984a), a pigment well distributed in fruit (see Goodwin, 1980).

The 1,2-epoxides of carotenes with an acyclic(ψ) end-group, e.g. 1,2-epoxylycopene (1,2-epoxy-1,2-dihydro-ψ,ψ-carotene, XXXVIII) and 1,2,1′,2′-diepoxylycopene (1,2,1′,2′-diepoxy-1,2,1′,2′-tetrahydro-ψ,ψ-carotene, XXXIX) were first isolated from tomatoes some time ago (Britton and Goodwin, 1969, 1975; Ben-Aziz *et al.*, 1973). The chirality at C-2 and C-2′ in these compounds has now been demonstrated as *S* by comparison of the circular dichroism (CD) of the natural and synthetic compounds (Berset and Pfander, 1984; Kamber *et al.*, 1984).

The absolute configurations of the well known pigments of red peppers capsanthin [(3*R*, 3′*S*, 5′*R*), XXIX] and capsorubin [(3,3′-dihydroxy-κ,κ-carotene-6,6′-dione) (3*S*, 5*R*, 3′*S*, 5′*R*), XL] (Cooper *et al.*, 1962) have been confirmed by their stereospecific synthesis from (+)-camphor (Bowden *et al.*, 1983).

Persicaxanthin, first isolated by Curl (see Curl and Bailey, 1954) from plums, is an apoviolaxanthin (5,6-epoxy-5,6-dihydro-12′-apo-β-carotene-3,12′-diol, XLI) (Gross and Eckhardt, 1981) with the stereochemistry indicated (Molnar *et al.*, 1987). The corresponding 5,8-furanoid oxide was also present in the peach extracts (Gross and Eckhardt, 1981).

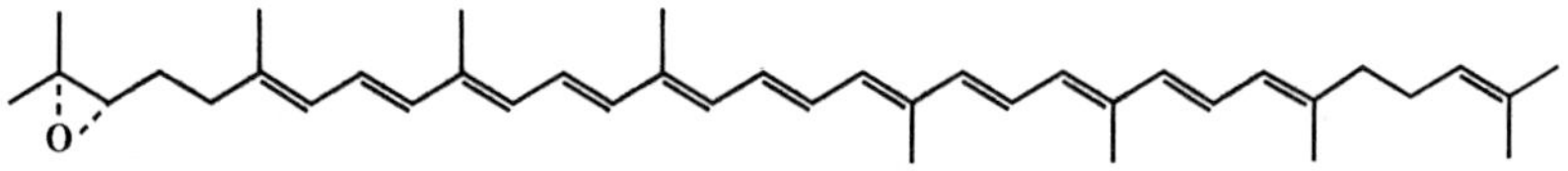

XXXVIII 1,2-Epoxylycopene

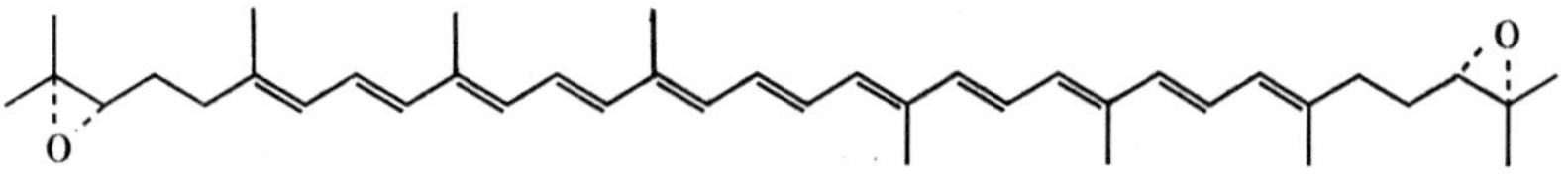

XXXIX 1,2,1′,2′-Diepoxylycopene

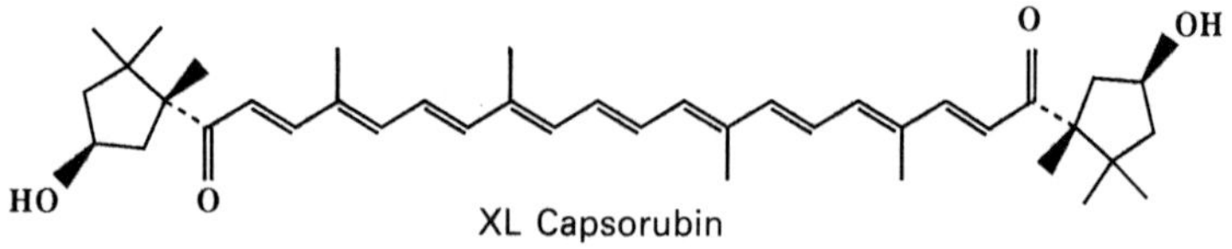

XL Capsorubin

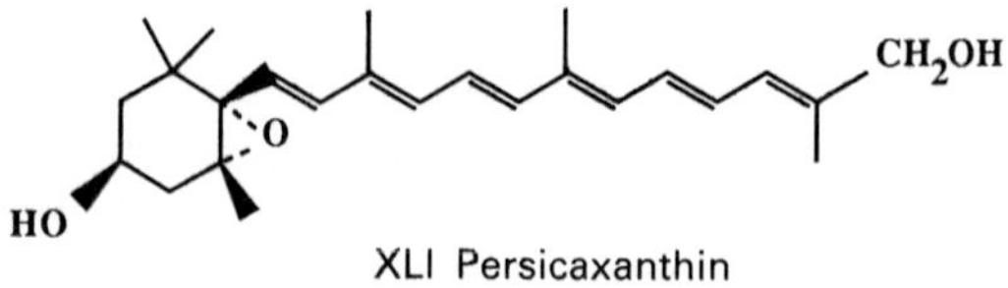

XLI Persicaxanthin

5 RHIZOMES AND ROOTS

A new apocarotenoid, cochloxanthin (3-oxo-6-hydroxy-8′-apo-ε-caroten-8′-oic acid, XLII) and its 4,5-dihydro-derivative (XLIII) have been isolated from the rhizomes of *Cochlospermum tinctorium* (Diallo and VanHaelen, 1987). The chirality at C-6 has yet to be determined.

Most varieties of sweet potato (*Ipomoea batatas*) contain β-carotene as the main carotenoid, with smaller amounts of many other components. The isolation of luteochrome [(5*R*,6*S*,5′*R*,8′*RS*)-5,6,5′,8′-diepoxy-5,6,5′,8′-tetrahydro-β,β-carotene, XLIV] from one variety has been reported (De Almeida *et al.*, 1986). This is the first example of the proven occurrence of an optically active 5,6-epoxycarotene. The chirality at C-5 and C-6 is the same as in the widespread epoxyxanthophyll, violaxanthin.

3 ALGAE

In marked contrast to higher plants the members of the various algal classes, with the exception of most of the Chlorophyceae, accumulate a variety of unusual and characteristic carotenoids in their chloroplasts; this is also evident in a few members of the Chlorophyceae.

The detailed pattern of carotenoid distribution has been frequently reviewed during the past decade or so (Goodwin, 1976, 1980; Liaaen-Jensen, 1977, 1978; Ragan and Chapman, 1978) so here only new basic information will be described in detail.

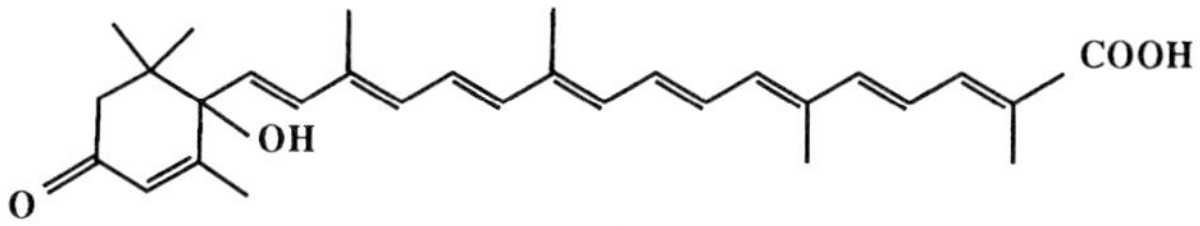

COOH
OH
O

XLII Cochloxanthin

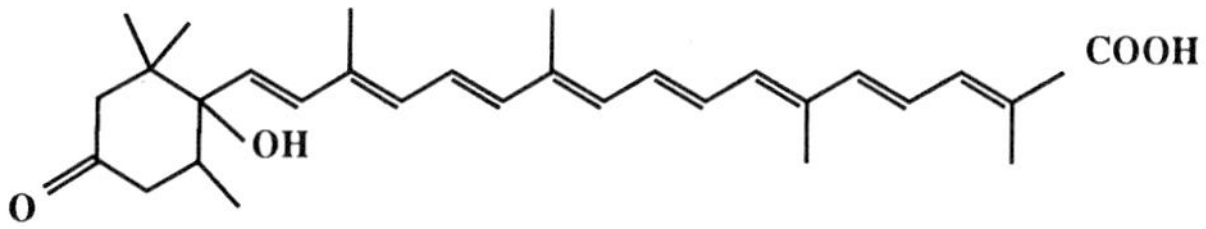

COOH
OH
O

XLIII 4,5-Dihydrocochloxanthin

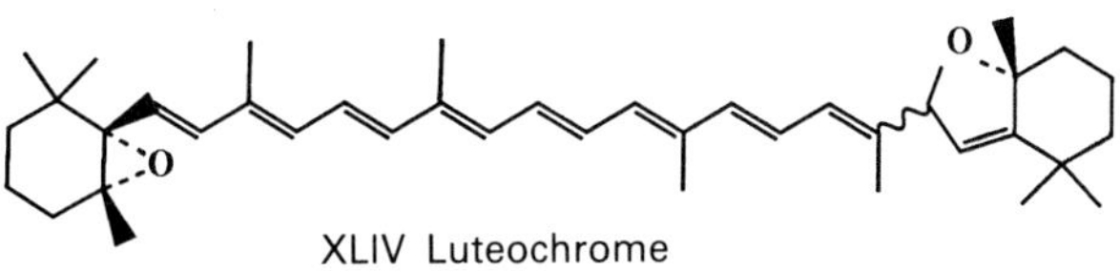

O
O

XLIV Luteochrome

A Chlorophyta

Members of the Chlorophyta generally contain the same set of major carotenoids as higher plants in their chloroplasts, namely α-carotene, β-carotene, lutein, violaxanthin and neoxanthin. Some species, however, accumulate additional, unusual xanthophylls, e.g. siphonaxanthin, which was first described by Strain (1951) and established as 3,19,3′-trihydroxy-7,8-dihydro-β,ϵ-caroten-8-one (XLV) by Walton *et al.* (1970) and which, with its ester siphonein, occurs most frequently in the order Caulerpales (Kleinig, 1969) and loroxanthin [(3R, 3′R, 6′R)-β,ϵ-carotene-3,19,3′-triol, XLVI] in the order Chlorococcales (Aitzetmüller *et al.*, 1969; Francis *et al.*, 1973). Pyrenoxanthin from *Chlorella pyrenoidosa* is now known to be identical with loroxanthin (Francis *et al.*, 1973; Nitsche, 1974b). Siphonein is in fact the generic name of a group of pigments differing in the fatty acid (most commonly C_{12}) with which siphonaxanthin is esterified (see Goodwin, 1980).

Carotenoids with hydroxy groups at C-2 and/or C-2′ have been observed in *Trentepohlia iolithus*, for example β,β-carotene-2,2′-diol (XLVII) earlier considered to be zeaxanthin, and β,ϵ-caroten-2-ol (XLVIII) (Buchecker *et al.*, 1974). All have *R* chirality at C-2.

Some members of the Prasinophyceae contain siphonein along with xanthophyll "K"(Ricketts, 1971), the structure of which has recently been elucidated as 3,6,3′-trihydroxy-7,8-dihydro-γ,ϵ-caroten-8-one (XLIX)

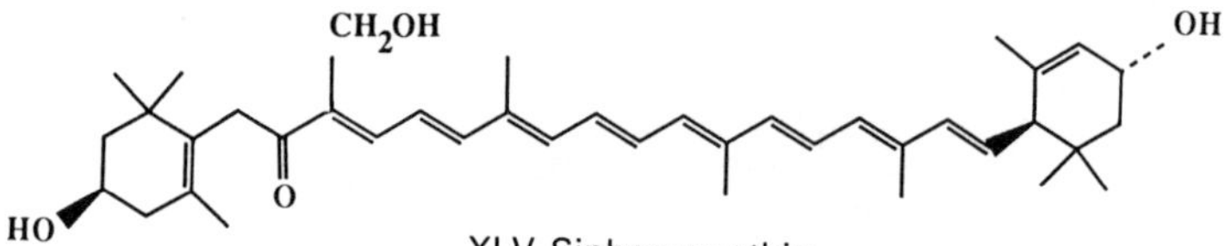

XLV Siphonaxanthin

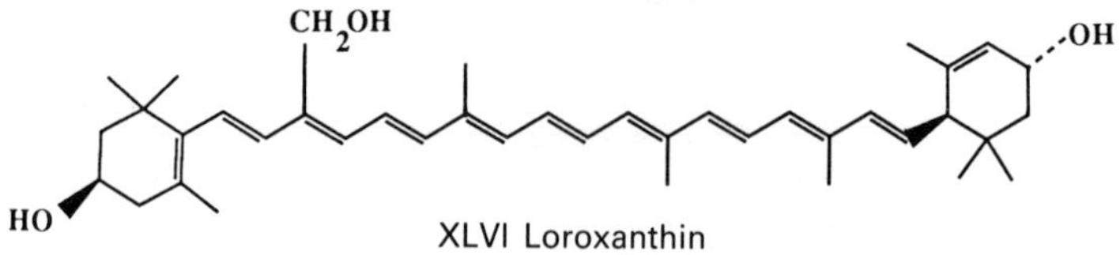

XLVI Loroxanthin

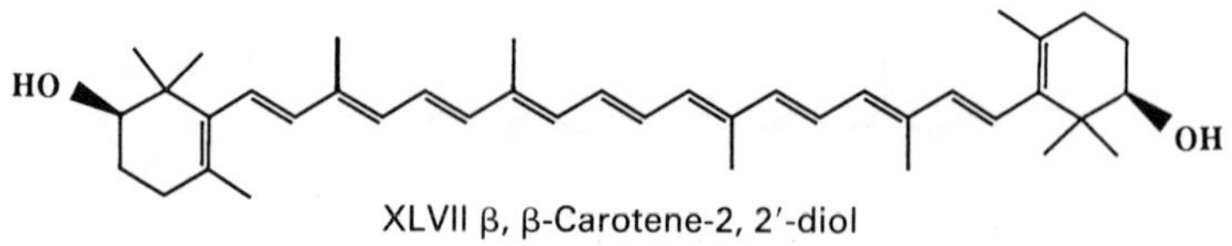

XLVII β, β-Carotene-2, 2′-diol

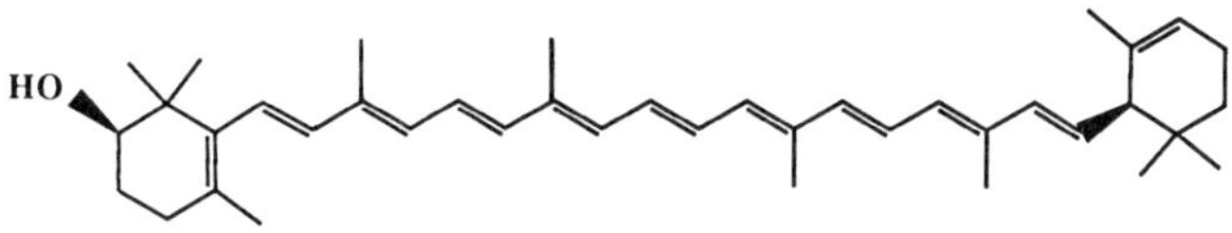

XLVIII β, ε-Caroten-2-ol

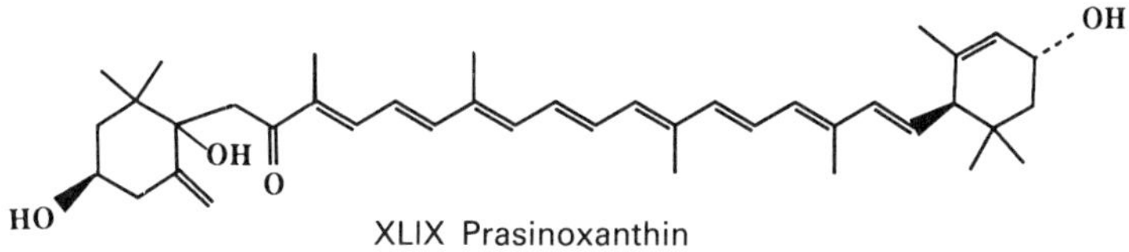

XLIX Prasinoxanthin

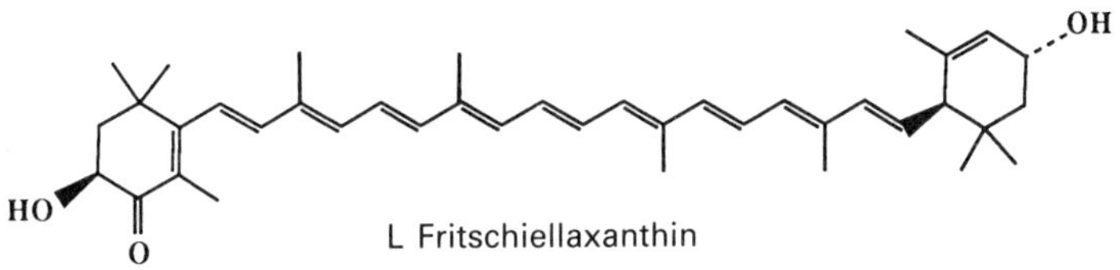

L Fritschiellaxanthin

(Liaaen-Jensen, 1985) with tentative chirality 3*R*,3′*R*,6′*R*. It has been named prasinoxanthin.

An interesting new pigment from *Fritschiella tuberosa*, fritschiellaxanthin [(3*S*,3′*R*,6′*R*)-3,3′-dihydroxy-β,ε-caroten-4-one, L] (Eugster, 1980) is the 3′-epimer of α-doradexanthin, a carotenoid which is found in many fish (see Goodwin, 1984).

Under conditions of stress, e.g. nitrogen deficiency, many species accumulate "secondary carotenoids" outside the chloroplast in oil droplets or in the cell wall (Burczyk, 1987). Most frequently it is ketocarotenoids which predominate, but one species, *Dunaliella bardawil*, accumulates enormous amounts of β-carotene (up to 9% cell dry weight) in conditions of high light intensity and high salt (NaCl) concentration (Ben-Amotz and Avron, 1983).

B Rhodophyta

These red algae are characterized by a relatively simple carotenoid pattern, the main components being α- and β-carotenes and their hydroxy-derivatives, all with the same chirality as the corresponding pigments of higher plants (Liaaen-Jensen, 1985). In contrast to the chloroplast pigments of higher plants, α-carotene frequently preponderates over β-carotene and zeaxanthin over lutein. A qualitative difference is that epoxides are rarely found in the Rhodophyta (see Goodwin, 1984).

C Pyrrophyta

The characteristic pigment of this phylum is the norcarotenoid peridinin [(3*S*,5*R*,6*R*,3′*S*,5′*R*,6′*S*)-5′,6′-epoxy-3,5,3′-trihydroxy-6,7-didehydro- 5,6, 5′,6′-tetrahydro-10,11,20-trinor-β,β-caroten-19′,11′-olide 3-acetate, LI] (Strain *et al.*, 1971; Kjøsen *et al.*, 1976). Various peridinin isomers have recently been isolated by HPLC, such as all-*trans*-6*R*, neo UI (6*R*, 11′,13′-di*cis*) and neo UIII (6*R*, 13′-mono*cis*) (Liaaen-Jensen, 1985). Apart from the presence of dinoxanthin (5′,6′-epoxy-6,7-didehydro-5,6,5′,6′-tetrahydro-β,β-carotene-3,5, 3′-triol 3-acetate, LII (= neoxanthin 3-acetate), which occurs alongside peridinin in many members of the Pyrrophyta, two C-8 epimers (*a* and *b*) of dinochrome, the furanoid oxide isomer of dinoxanthin, have been isolated and their stereochemistry clearly defined as 8′*S* and 8′*R*, respectively (Märki-Fischer *et al.*, 1984a; Eugster, 1985). Fucoxanthin [(5,6-epoxy-3,3′,5′-trihydroxy-6′,7′-didehydro-5,6,7,8,5′,6′-hexahydro-β,β-caroten-8-one 3′-acetate), LIII], which is acetylated at C-3′, occurs with its C-3 hydroxy group also acetylated in a *Gyrodinium* sp. Also present are 19′-hexanoyloxyfucoxanthin, first isolated from a chrysophyte (see section D) and a new pigment,

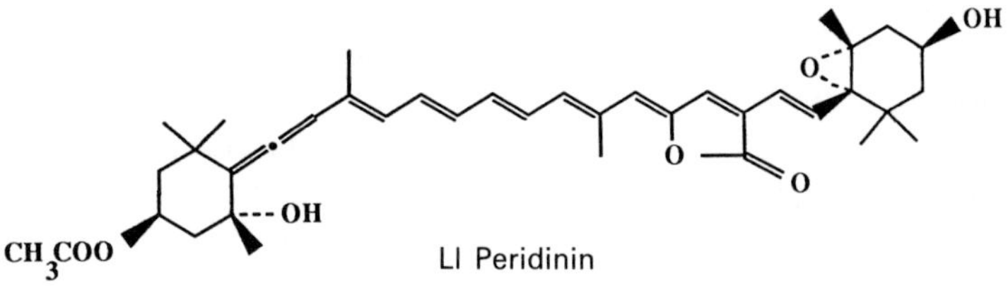

LI Peridinin

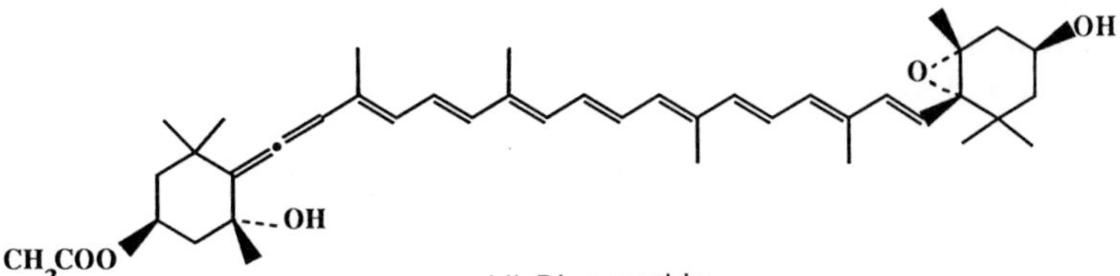

LII Dinoxanthin

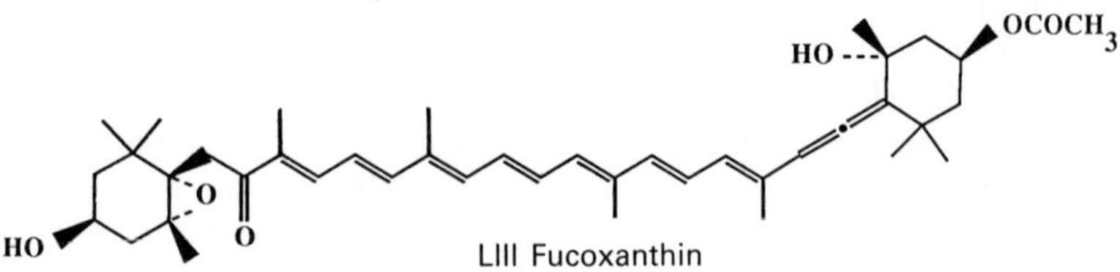

LIII Fucoxanthin

gyroxanthin, which occurs as a diester with probable structure LIV (Liaaen-Jensen, 1985). A butanoyloxy-derivative of fucoxanthin (see next section) was isolated, together with the hexanoyloxy-derivative, from *Gyrodinium breve* (Liaaen-Jensen, 1985).

A pigment, provisionally named P457 and obtained from a number of dinoflagellates (Johansen *et al.*, 1974), may be a carotenoid bonded to a disaccharide (Liaaen-Jensen, 1985).

D Chrysophyta

A number of epoxy, allenic and acetylenic carotenoids, including fucoxanthin (LIII) and diadinoxanthin [(5,6-epoxy-7′,8′-didehydro-5,6-dihydro-β,β-carotene-3,3′-diol), LV], are characteristic of the Chrysophyta.

The two fucoxanthin-derivatives mentioned in section C were first fully characterized from samples obtained from chrysophytes: 19′-hexanoyloxyfucoxanthin from *Coccolithus (Emeliana) huxleyi* (Arpin *et al.*, 1976) and 19′-butanoyloxyfucoxanthin from *Coccolithus min*. Haltenbanken (Bjørnland *et al.*, 1984b). Unexpectedly the ε,ε-carotene isolated from the latter alga probably has the 6*S*, 6′*S* chirality (LVI) (Liaaen-Jensen, 1985); this is opposite to that of other samples of algal ε,ε-carotene (see Goodwin, 1980).

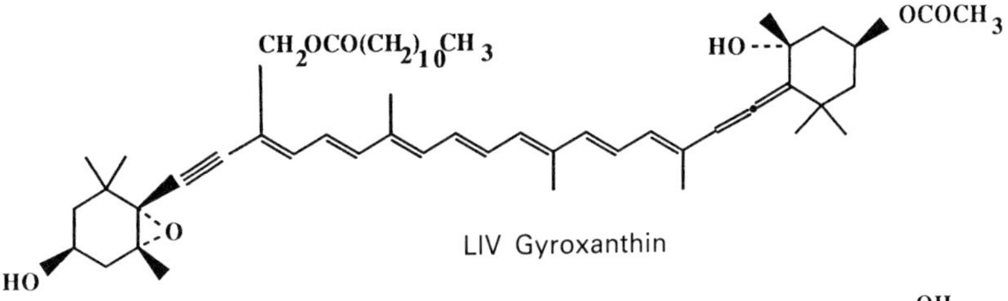

LIV Gyroxanthin

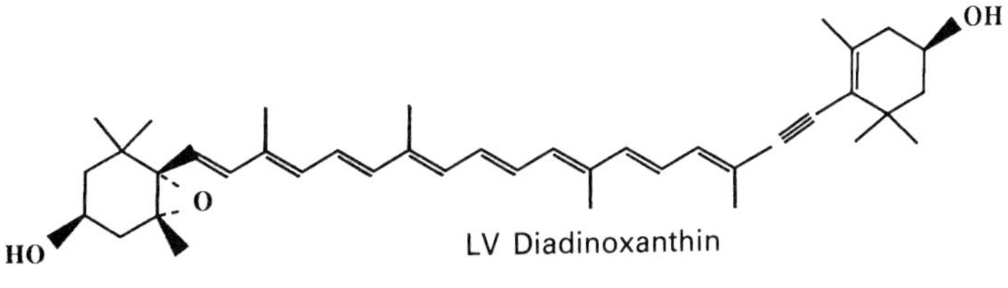

LV Diadinoxanthin

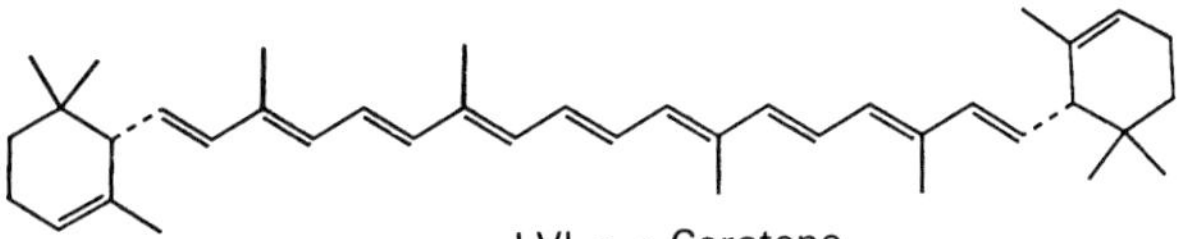

LVI ε, ε-Carotene

E Euglenophyta

New structures were encountered in the marine *Eutreptiella gymnastica*: eutreptiellanone (LVII), the first naturally occurring 3,6-epoxide or oxa-bicycloheptane carotenoid, (3,6-epoxy-3′,4′,7′,8′-tetradehydro-5,6-dihydro-β,β-caroten-4-one), and the minor carotenoids (LVIII, LIX) with the same end-group, together with anhydrodiatoxanthin [(3′, 4′, 7′, 8′-tetradehydro-β,β-caroten-3-ol), LX] have all been provisionally characterized (Fiksdahl *et al.*, 1984a; Liaaen-Jensen, 1985).

Siphonein, with *n*-2-*trans*-2-dodecenoic acid as esterifying acid, has been isolated from two marine euglenophytes (Fiksdahl *et al.*, 1984a; Liaaen-Jensen, 1985). Diatoxanthin [(7,8-didehydro-β,β-carotene-3,3′-diol, LXI] is frequently the major carotenoid in euglenophytes (Goodwin, 1980; Liaaen-Jensen, 1985).

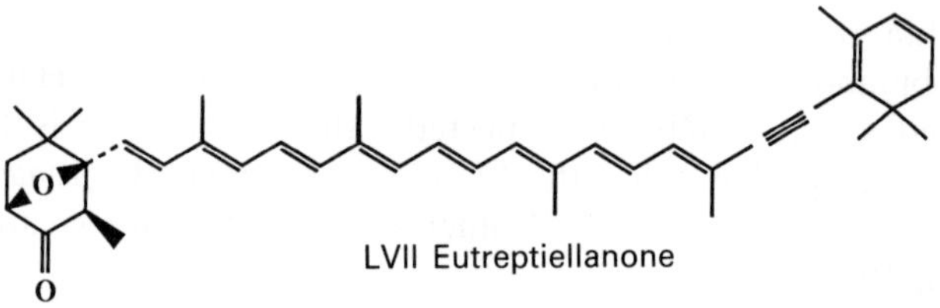

LVII Eutreptiellanone

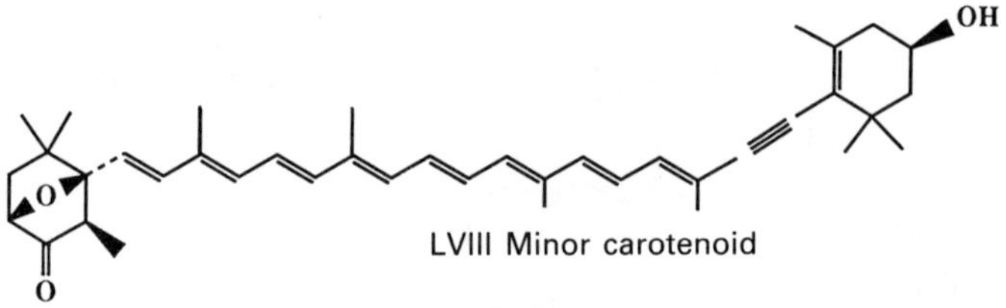

LVIII Minor carotenoid

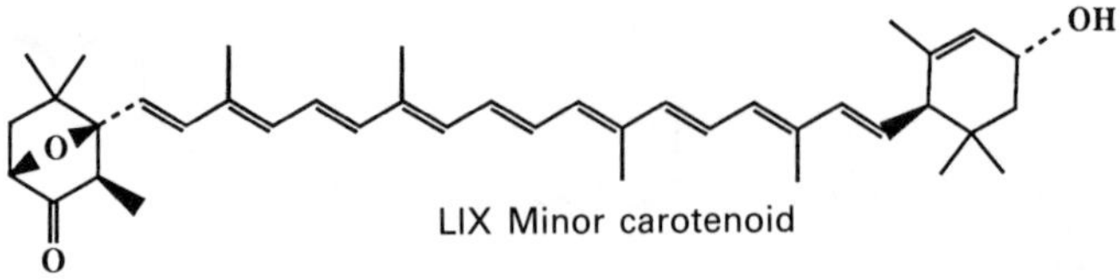

LIX Minor carotenoid

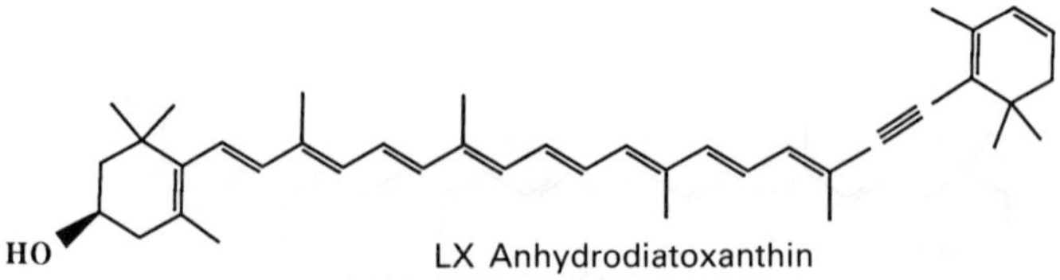

LX Anhydrodiatoxanthin

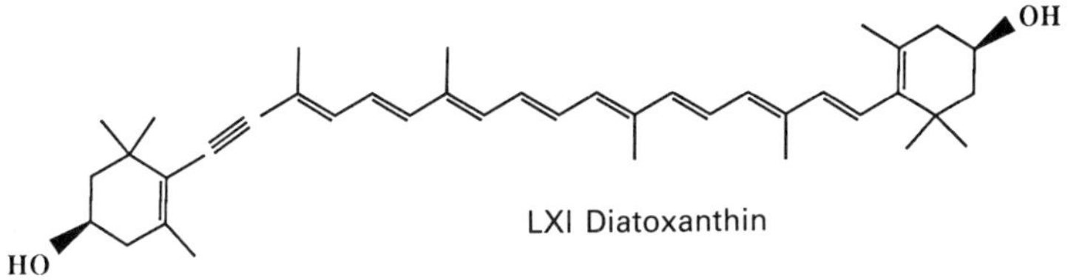

LXI Diatoxanthin

F Chloromonadophyta (Raphidophyta)

Two members, *Gonyostomum semen* and *Vacuolaria virescens* have recently been re-examined with modern analytical techniques (Fiksdahl *et al.*, 1984b). They both show a similar pigment pattern consisting mainly of acetylenic, allenic and epoxy carotenoids. The major components are diadinoxanthin (LV), heteroxanthin [(7′,8′-didehydro-5,6-dihydro-β,β-carotene-3,5,6,3′-tetraol), LXII], now assigned the (3*S*,5*R*,6*R*,3′*R*) configuration (Eugster, 1985) and vaucheriaxanthin [(5′,6′-epoxy-6,7-didehydro-5,6,5′,6′-tetradehydro-β,β-carotene- 3,5,19,3′-tetraol), LXIII]. These pigments are also characteristic of algae of the class Xanthophyceae (see section G) (Goodwin, 1980).

G Xanthophyta

The main pigments found in these algae are the same as in the Chloromonadophyta (section F), and little new information has emerged during the past few years (Goodwin, 1980).

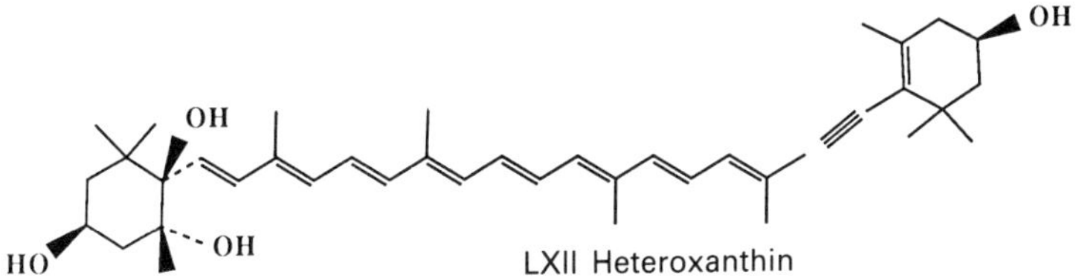

LXII Heteroxanthin

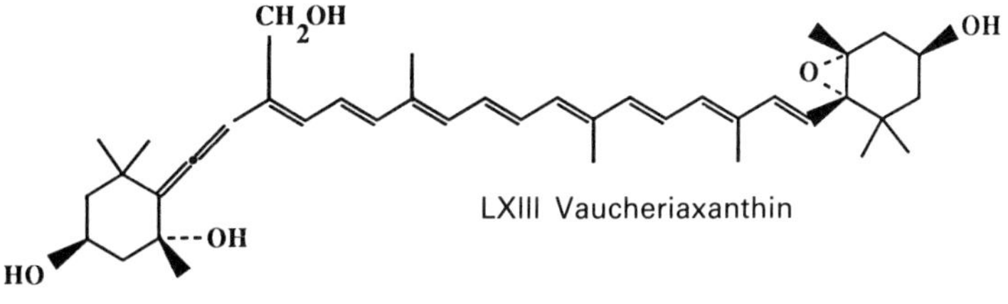

LXIII Vaucheriaxanthin

H Cryptophyta

The characteristic pigments of these algae are acetylenic carotenoids; these include alloxanthin [(3*R*, 3′*R*)-7,8,7′,8′-tetradehydro-β,β-carotene-3,3′-diol, LXIV], monadoxanthin [(3*R*, 3′*R*, 6′*R*)-7,8-didehydro-β,ε-carotene-3,3′-diol, LXV], and crocoxanthin [(3*R*, 6′*R*)-7,8-didehydro-β,ε-caroten-3-ol, LXVI]. Only recently has the chirality of each of these pigments been decided (Pennington *et al.*, 1985; Liaaen-Jensen, 1985). Allenic carotenoids reported earlier (Swift and Milborrow, 1981) were not detected in later studies (Liaaen-Jensen, 1985). An uncommon situation is that in the Cryptophytes so far examined α-carotene preponderates over β-carotene (Goodwin, 1980).

I Phaeophyta

It is well known that the main pigment in the Phaeophyta is fucoxanthin (LIII), which is thus the most abundant carotenoid on earth (Goodwin, 1980). No new pigments have been isolated from these algae in recent years.

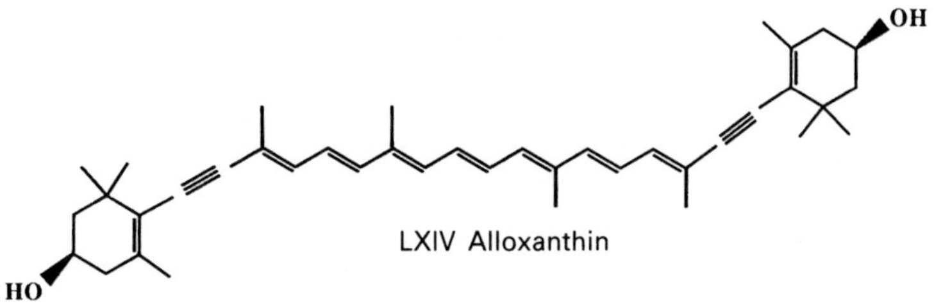

OH

HO

LXIV Alloxanthin

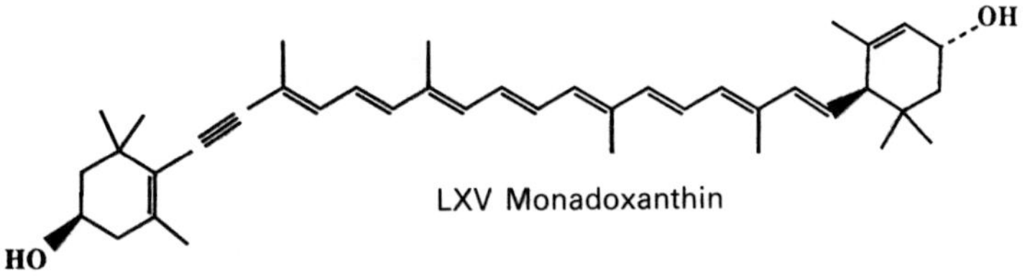

OH

HO

LXV Monadoxanthin

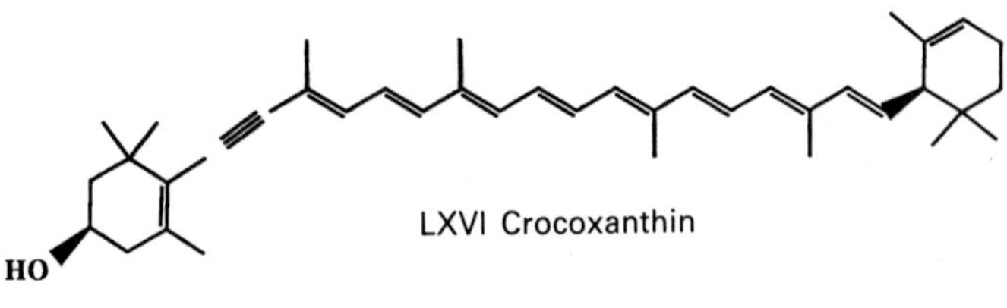

HO

LXVI Crocoxanthin

4 FUNGI

Fungi, not being photosynthetic, do not all contain carotenoids and the distribution amongst them appears to be capricious and without any major functional significance (Goodwin, 1980). The general characteristics of carotenogenic fungi are:

(i) the frequent presence of the monocyclic and bicyclic carotenoids found in higher plants;
(ii) the frequent accumulation of hydrocarbons only and
(iii) the general absence of ε-ring carotenoids and epoxides (Goodwin, 1980).

Novel pigments do, however, exist. Torulene [(3′,4′-didehydro-β,ψ-carotene), LXVII] and torularhodin [(3′,4′-didehydro-β,ψ-caroten-16′-oic acid), LXVIII] are the characteristic pigments of red yeasts (Deuteromyces), but one species *Phaffia rhodozyma* uniquely produces (3*R*, 3′*R*) astaxanthin [(3,3′-dihydroxy-β,β-carotene-4,4′-dione), LXIX] (Andrewes and Starr, 1976). The astaxanthin from algae, e.g. *Haematococcus* species, has the opposite chirality.

Characteristic pigments of Ascomycetes, the chirality of which has recently been determined, are (6′*S*)-β,γ-carotene (LXX) (Hallenstvet *et al.*, 1977) and plectaniaxanthin [(2′R)-3′,4′-didehydro-1′,2′-dihydro-β,ψ-carotene-1′,2′-diol, LXXI] (Rønneberg *et al.*, 1982).

Ketocarotenoids have been observed in some Hymenomycetes; indeed canthaxanthin (β,β-carotene-4,4′-dione, LXXII) was first isolated from *Cantharellus cinnabarinus* and astaxanthin has been reported in a number of species (see Goodwin, 1980).

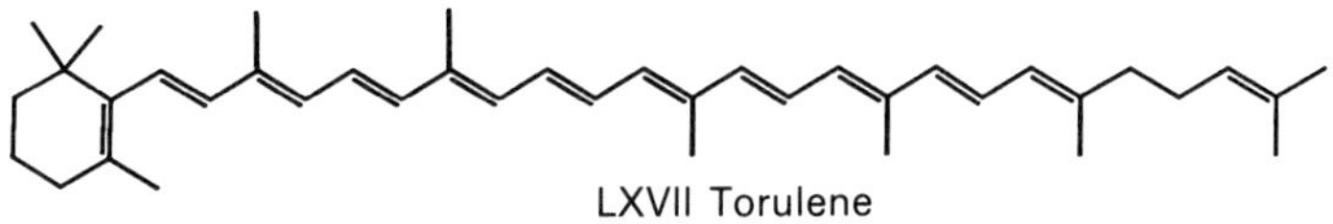

LXVII Torulene

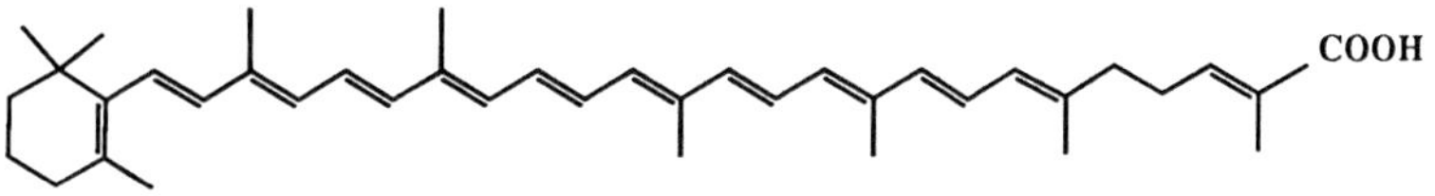

COOH

LXVIII Torularhodin

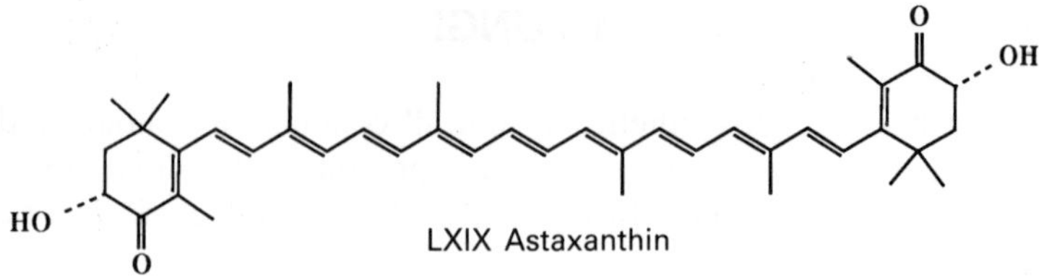

LXIX Astaxanthin

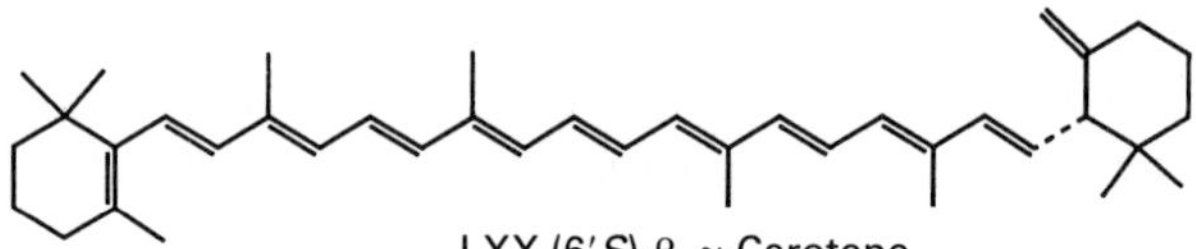

LXX (6′*S*)-β, γ-Carotene

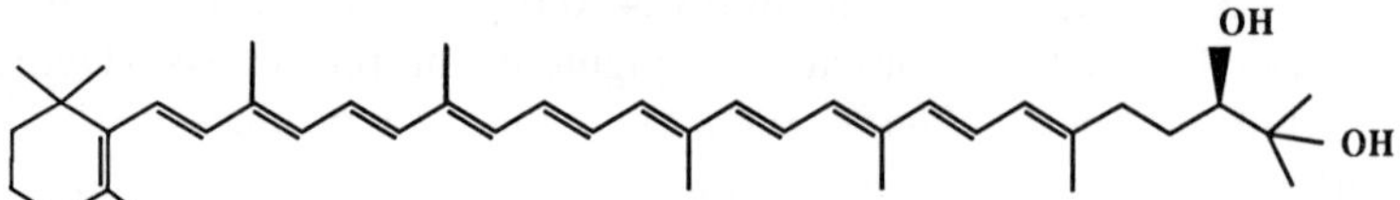

LXXI Plectaniaxanthin

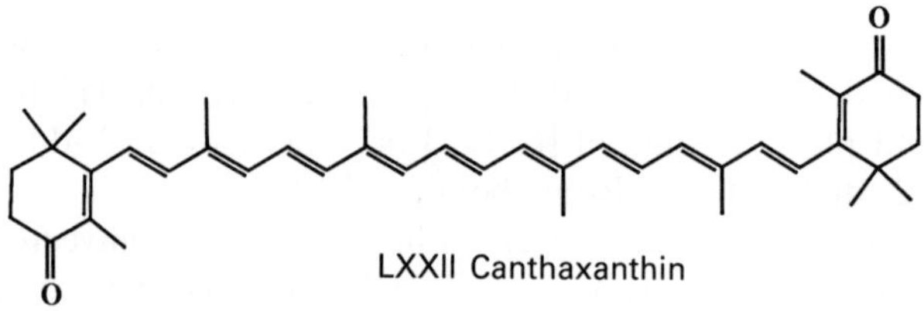

LXXII Canthaxanthin

5 BACTERIA

It is not appropriate to discuss bacterial carotenoids in this volume, with the exception of the Cyanobacteria, which have recently been remustered as bacteria from their original place as blue-green algae (Cyanophyceae). Although much work on the carotenoids of these organisms has been carried out by workers interested in algae, their new classification requires that they should now be considered under the general heading 'Bacteria' rather than 'Algae'.

A Cyanobacteria

The carotenogenic characteristics of these organisms include:

(i) synthesis of ketocarotenoids;
(ii) failure to synthesise ε-rings;

(iii) production of 2-hydroxy-derivatives;
(iv) formation of glycosides.

Echinenone (β,β-caroten-4-one, LXXIII), was the first ketocarotenoid to be discovered; both echinenone and canthaxanthin are well distributed (Goodwin, 1980). Caloxanthin [(2*R*, 3*R*, 3′*R*)-β,β-carotene-2,3,3′-triol, LXXIV)] and nostoxanthin [(2*R*, 3*R*, 2′*R*, 3′*R*)-β,β-carotene-2,3,2′,3′,-triol, LXXV], found, for example, in *Nostoc communa* and *Anacystis nidulans* (Stransky and Hager, 1970) have opposite chirality at C-2 to that of the similar 2-hydroxycarotenoids in the green alga *Trentepohlia iolithus* (see section 3A) (Buchecker *et al.*, 1976). The two well characterized glycosides are myxoxanthophyll [(3*R*, 2′*S*)-2′-(β-L-rhamnopyranosyloxy)-3′,4′-didehydro- 1′,2′-dihydro-β,ψ-carotene-3′,1′-diol, LXXVI] (Liaaen-Jensen, 1985; Rønneberg *et al.*, 1985) and oscillaxanthin [2,2′-bis(β-L-rhamnopyranosyloxy)-3,4,3′,4′-tetradehydro-1,2,1′,2′-tetrahydro-ψ,ψ-carotene-1,1′-diol, LXXVII] which probably has the 2*S*, 2′*S* configuration (Liaaen-Jensen, 1985) although the 2*R*, 2′*R* configuration has previously been suggested (Rønneberg *et al.*, 1980).

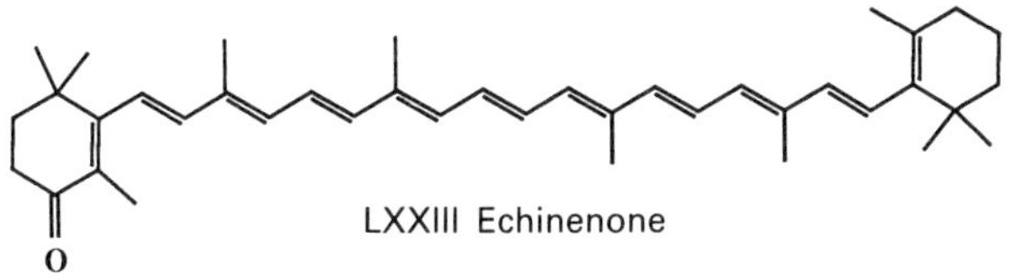

O

LXXIII Echinenone

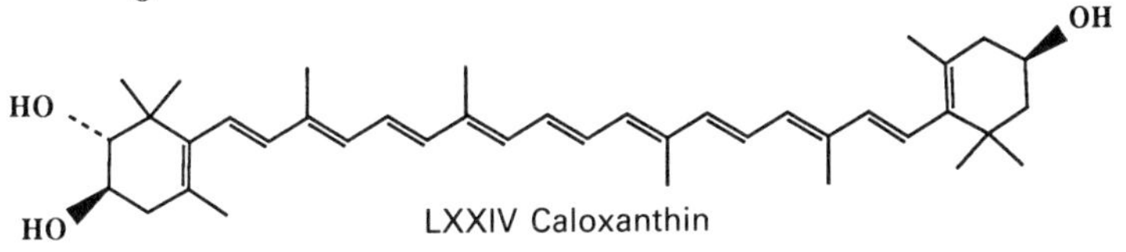

HO
HO
OH

LXXIV Caloxanthin

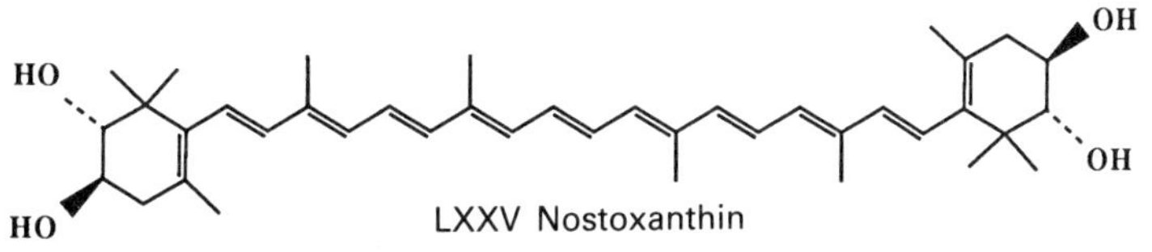

HO
HO
OH
OH

LXXV Nostoxanthin

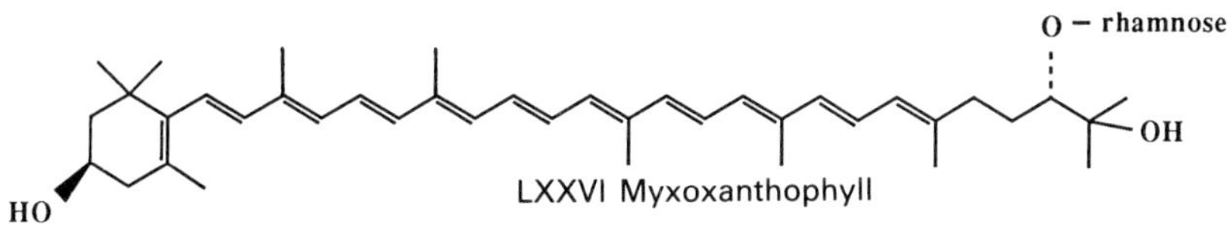

O – rhamnose
OH
HO

LXXVI Myxoxanthophyll

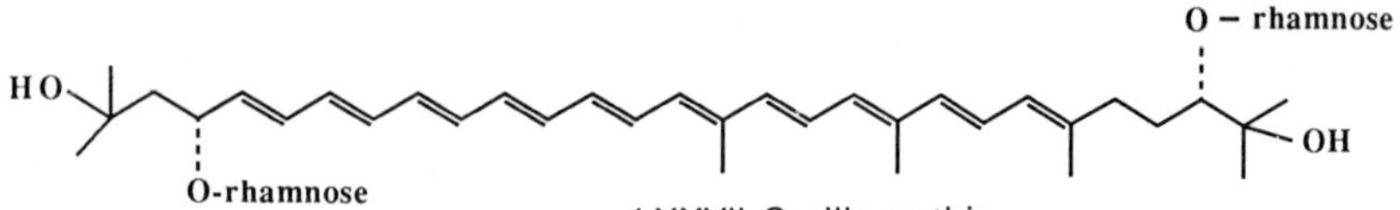

LXXVII Oscillaxanthin

PART 2: ANALYSIS

1 INTRODUCTION

In 1976, Davies presented a comprehensive account of general methods recommended for the handling of carotenoids and for their extraction from natural tissues, purification and analysis. Other articles have appeared since then which have also described these procedures (De Ritter and Purcell, 1981; Krinsky and Welankiwar, 1984; Britton, 1985a; Liaaen-Jensen and Andrewes, 1985). Although some details may be different because of personal preferences, in general the topics covered and the procedures recommended are similar in all these articles. A particular feature of Davies's (1976) survey was the inclusion of extensive tables of ultraviolet (UV)-visible light absorption maxima of plant carotenoids in several different solvents, together with lists of extinction or absorbance coefficients ($E^{1\%}_{1\,cm}$ or $A^{1\%}_{1\,cm}$). Less extensive tables of this kind, including data on some carotenoids that were not dealt with by Davies, were also given by De Ritter and Purcell (1981) and Britton (1985a).

The traditional separation methods of open column and thin-layer chromatography (TLC) remain extremely valuable for the isolation and purification of carotenoids, especially for preparative-scale work, just as the light absorption spectra are still the basis of the routine methods for carotenoid identification and quantitative analysis (see Davies, 1976, for a full description of these traditional methods). The greatest advances in the purification of carotenoids and their routine identification and analysis have come from the development of HPLC, which is now very much the method of choice for carotenoid work. Enormous advances have also been made in the development of powerful spectroscopic techniques and their application in the carotenoid field for elucidation of structures and stereochemistry.

2 GENERAL METHODS

There have been no major innovations in the general handling of carotenoids, but the development of sensitive HPLC methods increases the risk that artifacts produced during extraction and conventional purification operations will be detected. Careful working is therefore imperative. Geometrical isomers (*Z*- or *cis*-isomers) of carotenoids are being reported with increasing frequency in natural extracts, and their presence is considered to be important in relation to the physiological roles of the carotenoids in the tissues. For example, *Z-E* (*cis-trans*) isomerization may be directly involved in the functioning of a carotenoid in light absorption, energy transfer or photoprotection. Alternatively the presence of *Z*-isomers may be a consequence and therefore an indicator of the functional role that the carotenoid is playing in the tissue. It is therefore essential to avoid artificial production of *Z*-isomers and changes in geometrical isomeric composition during isolation and purification. Although exposure of carotenoids to heat and light should be avoided if at all possible, direct isomerization or photoisomerization of most carotenoids during normal manipulations—even extraction with hot solvents—occurs comparatively slowly and not to a great extent. It is most important to note, however, that with plant extracts which contain chlorophyll, chlorophyll-sensitized photoisomerization, presumably via the carotenoid triplet state, occurs very rapidly and appreciable amounts of carotenoid *Z*-isomers can be produced during even a brief exposure of a chlorophyll-containing extract to light. It is necessary to be wary also of the possible presence of other, apparently innocent, molecules which could be potentially dangerous sensitizers of carotenoid isomerization; acetone, for example, has been used deliberately as a sensitizer for producing triplet state carotenes and consequently *Z*-isomers, by UV-irradiation.

Secondly, the need to preclude oxidation of carotenoids cannot be overemphasized. The presence of even traces of oxygen in stored samples, of peroxides in solvents, especially diethyl ether, or of oxidizing agents even in crude extracts containing carotenoids, can rapidly lead to bleaching, formation of carotenoid epoxides, and polyene chain cleavage to give apocarotenoids. The presence of unsaturated lipids and metal ions can greatly enhance the oxidative breakdown. The products are readily detected by HPLC, so the exclusion of oxygen and oxidizing agents from samples containing carotenoids must be rigorous. It has often been assumed that, although extracted carotenoids are at risk of oxidation, the carotenoid molecules are stabilized *in vivo* by protein etc. However, it is now clear that when plant tissues are disrupted, and especially during the isolation or incubation of chloroplasts, thylakoids and pigment–protein complexes, rapid oxidation of the carotenoids, particularly β-carotene, occurs and β-carotene-5,6-epoxide (LXXVIII) can be produced. The well established procedure for separating chloroplast pigment–protein complexes by polyacrylamide gel electrophoresis can also cause destruction of the carotenoids, particularly β-carotene, in the complexes. This is a consequence of the use of the powerful oxidizing agent, ammonium persulphate, as polymerizing agent during preparation of the gels. Attempts to rectify this problem by including antioxidants in the sample are usually not successful. The inclusion of ascorbate, for example, can increase rather than decrease the destruction and lead to other changes in the carotenoid compositions.

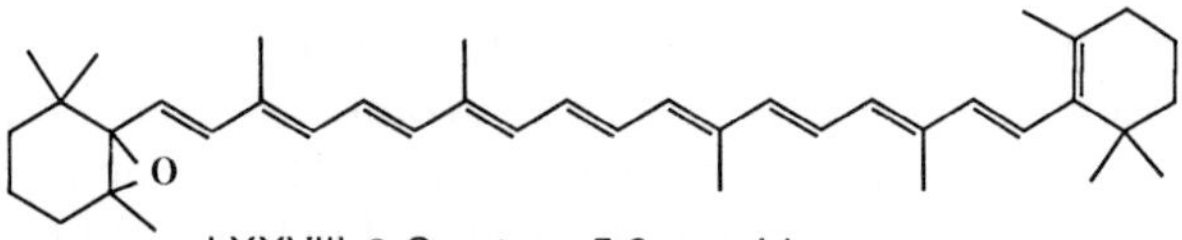

LXXVIII β-Carotene-5,6-epoxide

The vastly increased sensitivity of the spectroscopic methods now available places great demands on any purification procedure. When very small samples (e.g. a few μg) are being prepared for mass spectrometric or NMR analysis, it is very easy for the purification procedure to introduce much larger amounts of impurities which only become apparent from their deleterious effect on the spectra obtained. Plasticizers (especially phthalates) and trace impurities in solvents present the biggest problem, and it can be very difficult to identify their source. All possible precautions must be taken to ensure that solvents are absolutely pure; double distillation and filtration through an activated material such as alumina are recommended. TLC plates should be prewashed with a solvent at least as strong in polarity as that which will eventually be used for elution. All possible contact of samples, solvents, reagents etc. with plastic materials must be avoided and glassware should be specially and rigorously cleaned and rinsed; the use of distilled water stored in polythene containers should be avoided. Solvents themselves must be regarded as impurities; samples for NMR analysis, for example, should be dried *in vacuo* to remove all traces of residual solvents.

3 HIGH-PERFORMANCE LIQUID CHROMATOGRAPHY (HPLC)

A Principles and instrumentation

It is now usually understood that the initials HPLC indicate 'high-performance liquid chromatography', although the alternative 'high-pressure' is still sometimes encountered. There are no fundamental new principles underlying separation by HPLC, simply vast improvements in technology. HPLC is a column chromatography method and the key factor is the nature of the stationary phase. Column packing materials have now been developed which consist of very small particles (typically 3 or 5 μm) of uniform size, shape and porosity. These materials permit very efficient resolution, but considerable pressure may be needed to force the solvent through the column. An HPLC system therefore requires at least one high-pressure metering pump which provides uniform solvent flow at a selected flow rate. For many applications, solvent gradients have been devised. These may be generated by using more than one pump, changing the relative rates of delivery of solvent from each and then mixing the solvents under pressure just before they pass to the column. Alternatively, a gradient may be generated by controlled mixing of monitored amounts of different solvents at atmospheric pressure and then using a single pump to drive the solvent mixture through the column. Sophisticated control systems are available to programme and control the solvent gradient, flow rate, etc. Another essential is an injection port through which an accurately known volume of sample is applied to the high-pressure end of the column.

The components of the sample are then separated during passage through the column. They are eluted successively and pass to the detector. For the carotenoids, detection by means of UV-visible light absorption is used. An ordinary UV-visible absorption spectrophotometer can be adapted for this purpose by the incorporation of a suitable low-volume flow cell, but it is more usual to use a purpose-built UV-visible detector with variable wavelength monitoring capacity. Light absorbance at the chosen wavelength or wavelengths is monitored continuously, so that chromatograms can be obtained as a plot of absorbance versus time, by use of a suitable chart recorder. The inclusion of an automatic integrator for determination of peak areas allows accurate quantitative analysis.

HPLC has many advantages over traditional column and TLC methods, e.g.:

(i) sensitivity: with a suitable detector, analysis of nanogram quantities of carotenoids is possible.
(ii) resolution: with the specially prepared uniform stationary phases very efficient resolution of carotenoids is achieved on a relatively short column (usually 25 cm).
(iii) reproducibility: identical separations can be obtained in repeated runs, with accurately reproducible retention times. (When solvent gradients are used, sufficient equilibration time must be allowed between runs if satisfactory reproducibility is to be achieved.)

(iv) inert conditions: separations are achieved with the virtual exclusion of light and oxygen, and the carotenoids are generally stable on the commonly used column materials, so the risk of carotenoid degradation is minimal (but see below).
(v) speed: for most purposes a run time of no more than 30 min is sufficient for full separation and analysis.

1 SPECTRAL SCANNING DETECTORS

The first evidence for the identification of components of a chromatogram is provided by their retention times, especially when compared directly with those of authentic standard compounds, and by their UV-visible light absorption spectra which, with suitable instrumentation, can be determined for each component during the course of the chromatography. The simplest way to achieve this is to employ a stop-flow procedure in which the solvent flow is discontinued and the absorption spectrum of the material in the flow cell is then determined by scanning. (If this is not possible, fractions of the column eluate may be collected and the spectra determined in a scanning spectrophotometer in the normal way.) More recent developments have provided instruments which will rapidly scan spectra "on the fly" without the need for interrupting solvent flow. The most powerful and ideal detectors for carotenoid work, however, are those which employ photodiode array detection. These contain an array of photodetectors, each positioned to receive light of a different fixed 2 nm wavelength band so that they function effectively as an array of fixed-wavelength detectors, monitoring simultaneously 200 different 2 nm light wavelength bands. This detection system, together with an associated powerful computer, permits simultaneous detection and monitoring at several selected wavelengths and instantaneous and continuous determination and memorizing of spectra during the progress of the chromatography. It also offers convenient and flexible raw data-handling and evaluation after the chromatographic run, in the form of multi-signal plots, signal and spectra plots, 3D-plots etc., so that optimal qualitative and quantitative information for a sample can be obtained from a single run.

B HPLC of carotenoids—historical development

The earliest attempts to develop carotenoid HPLC adapted conventional open column and TLC separations to the HPLC technology then available.

As early as 1971, Stewart and Wheaton reported the separation of complex mixtures on enclosed columns of $ZnCO_3$ and MgO, the eluent being continuously monitored at 440 nm with a spectrophotometer. Cadosch and Eugster (1974) then used a specially prepared MgO to separate natural and semi-synthetic lutein-5,6-epoxide (IX), and the C-8 epimers of the corresponding 5,8-epoxide, flavoxanthin (8*R*, LXXIX) and chrysanthemaxanthin (8*S*, LXXX). In spite of much work to prepare and evaluate improved basic materials derived from the traditional adsorbents that had been used so successfully in open column and TLC, significant progress in carotenoid HPLC only began when stable and reproducible microparticles specially manufactured for HPLC became available. With their small and uniform particle size and controlled porosity, these materials, most commonly silica either unmodified or with suitable bonded groups attached, such as C_{18} hydrocarbon side chains, have become the standard materials used for carotenoid analysis (with the exception of some specialized separations of geometrical and optical isomers, discussed below).

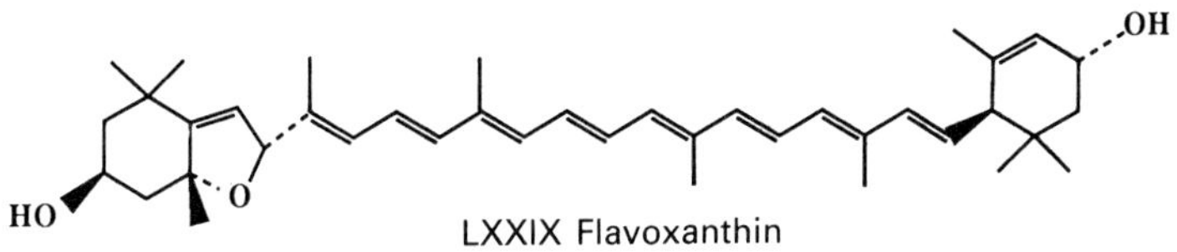

LXXIX Flavoxanthin

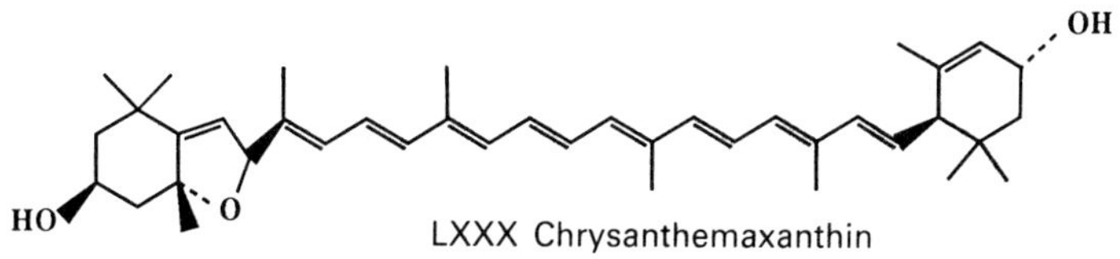

LXXX Chrysanthemaxanthin

The behaviour of carotenoids on these standard materials is now understood satisfactorily, so that modification of solvent systems in order to achieve optimal resolution of particular mixtures can usually be achieved relatively easily.

Rather than simply presenting a catalogue of the HPLC procedures that have been reported for carotenoid analysis in recent years, this article will describe the main features of carotenoid separations on the most widely used HPLC column materials. For further details and references to the wide variety of procedures that have been published, the reader should consult reviews by Lambert *et al.* (1985), Rüedi (1985), Ruddat and Will (1985), Taylor (1983) and Taylor and Ikawa (1980), and the series of surveys of the literature on carotenoids published each year or so in the Specialist Periodical Reports and now Natural Product Reports series of the Royal Society of Chemistry (Britton, 1976, 1977, 1978, 1979, 1981, 1982, 1983, 1984, 1985b, 1986).

C Normal (adsorption) phase HPLC

1 SILICA COLUMNS

Since efficient separation of carotenoids can be achieved by TLC on silica, it is not surprising that silica has been widely used, and is extremely useful for HPLC separations also. The early papers of Fiksdahl *et al.* (1978) first revealed the potential of this procedure, and described the resolution of xanthophylls (lutein, II, and zeaxanthin, VII), acyclic, monocyclic and bicyclic carotenes, geometrical isomers of, for example, lutein, and diastereoisomers, such as C-8 epimers of auroxanthin (XXV) and neochrome (XXIII) on a silica column with a solvent consisting of mixtures of hexane and acetone. Since then, many applications of adsorption phase HPLC on silica for carotenoid analysis have been reported. It is outside the scope of this article to list all of these; the reader should consult the reviews listed in the previous section.

The solvent mixtures used for HPLC of carotenoids on silica are usually translated from ones that have been used successfully for TLC. Most commonly they employ hexane with appropriate amounts of a more polar solvent, such as propan-2-ol, acetone or an ether (methyl-*t*-butyl ether and di-isopropyl ether are more suitable than diethyl ether for HPLC). Gradient elution (continuous or stepwise) with increasing amounts of the more polar solvent is frequently employed for complex extracts that contain carotenoids of widely differing polarity. Obviously, the best resolution is achieved with silica of the smallest particle size (3 μm).

Unfortunately the adsorption of carotene hydrocarbons on silica is so weak that they are eluted very rapidly and are poorly resolved, even when the apolar solvent hexane alone is used as the eluting solvent. The resolution of carotene diols and more polar xanthophylls, however, is impressive and very efficient separation of geometrical isomers, diastereoisomers and epoxides can be achieved (see Fig. 2.1 below, which illustrates the separation of leaf carotenoids on a silica column with a hexane—propan-2-ol solvent system). The best resolution of specific mixtures of closely related compounds is obtained by isocratic elution, but programmed continuous or stepwise gradient elution procedures are most useful for obtaining overall pigment profiles—both carotenes and xanthophylls—of any extract.

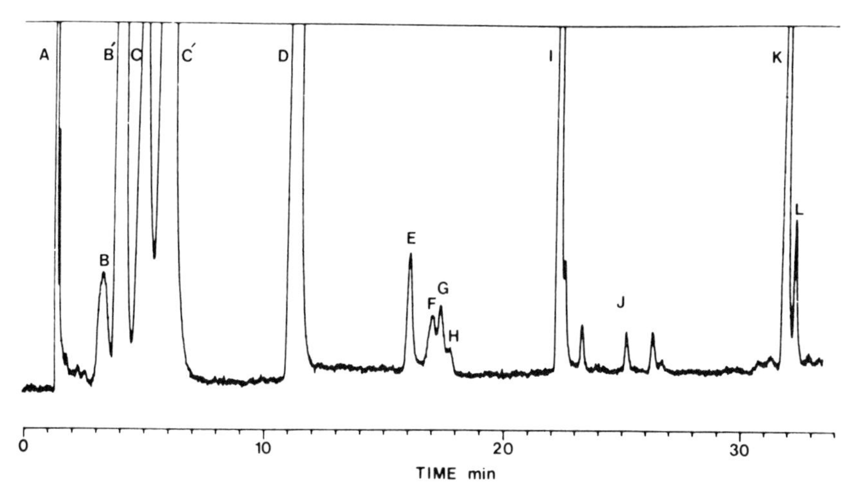

Fig. 2.1 H.p.l.c. separation of chlorophylls and carotenoids from 6-day old barley cotyledons on a normal adsorption phase column (3μ silica, Thames Chromatography). Solvent system: solvent A – hexane/propan-2-ol (8:2), solvent B – hexane. 0–8 min (10% A), 8–20 min (10–25% A), 20–30 min (40% A), 30–40 min (70% A). Flow rate = 2.0 ml min^{-1}.

Peak identification. A. β-carotene; B, B′. chlorophyll *a*; C, and C′. chlorophyll *b*; D. all-*E* lutein; E. 9 and 9′-*Z* lutein; F. 13-*Z* lutein; G. 13′-*Z* lutein; H. 15*Z* lutein; I. all-*E* violaxanthin; J. *Z* isomers of violaxanthin; K. 9′-*Z* neoxanthin, L. all-*E* neoxanthin.

There can be some disadvantages to the use of silica columns. In particular, some silica preparations may be slightly acidic and this can be sufficient to cause some isomerization of carotenoid 5,6-epoxides to the 5,8-(furanoid) epoxides. Also, if analysis of total plant or algal pigment extracts is required, silica may not be suitable, because substantial degradation of chlorophylls can occur, producing chlorophyll-derived artifacts which can mask the presence of minor carotenoids, and also adversely affect quantitative analysis.

2 BONDED NITRILE COLUMNS

As an alternative to silica, the use of a bonded nitrile phase material (Spherisorb S5-CN) for carotenoid HPLC was suggested by Vecchi in 1978, and its use has now been investigated intensively by Vecchi and coworkers and by the Zurich group of Eugster and Rüedi. Rüedi (1985) has briefly summarized some of these developments. The great advantage of the nitrile column is the efficient resolution of mixtures of similar carotenoids. The results obtained have been impressive, and separations have been described of structural isomers, e.g. lutein and zeaxanthin, and of geometrical (*Z*/*E*) isomers of xanthophylls. The system is particularly powerful for resolving isomeric mixtures of carotene or xanthophyll epoxides. Not only are the 5,6-epoxides well resolved from the corresponding 5,8 (furanoid) epoxides but the 8*R* and 8*S* epimers of the furanoid oxides are clearly resolved. The separations that can be achieved are so effective that the chromatograms can be very complex, and many minor compounds can be detected. Gradient systems can be used but the best resolutions have been achieved by isocratic separation of fractions (e.g. "dihydroxy-fraction") obtained from natural extracts by classical column or TLC. Rüedi presents a useful table which lists the conditions and solvent compositions recommended for separating the carotenoids within the various polarity groups on a 25 cm column of Spherisorb S5-CN. Two solvent mixtures are used: solvent A is of constant composition, namely hexane containing 0.1% N-ethyldiisopropylamine, whereas solvent B comprises dichloromethane containing variable proportions of methanol. A summary of Rüedi's recommendations is given in Table 2.2.

Table 2.2 Solvent compositions recommended for HPLC of different polarity groups of carotenoids on Spherisorb S5-CN (250 × 4.6 mm) (data from Rüedi 1985)

Carotenoid group	% Solvent B	% Methanol in solvent B
Carotenes	0	—
Carotene mono- and diepoxides	2–6	0.5
Monohydroxycarotenoids	10–20	0.5
Dihydroxycarotenoids	35–40	1–2
Dihydroxycarotenoid epoxides	40–45	2
Trihydroxycarotenoids	50	3
Tetrahydroxycarotenoids and more polar	50–60	3–5

Solvent A = hexane/0.1% N-ethyldiisopropylamine; solvent B = dichloromethane-methanol.

D Reversed-phase HPLC

1 GENERAL FEATURES

Reversed-phase partition chromatography is now most widely used for the routine analysis of carotenoids in natural extracts. The column materials used are very inert and there is virtually no risk of decomposition or structural modification, even of the chlorophylls and the most labile carotenoids. The stationary phases most commonly employed consist of silica particles, to which C_{18} hydrocarbon chains have been chemically bonded (ODS = octadecylsilyl). In some cases materials with shorter bonded side chains (C_6 or C_8) have been used.

Separation depends on partition between the eluting solvent and the hydrophobic environment presented by the bonded hydrocarbon chains on the stationary phase particles. Classically, the solvents used have contained water, e.g. methanol–water or acetonitrile–water mixtures, but more recently Nelis and de Leenheer (1983) have explored in detail the merits of non-aqueous reversed-phase chromatography for carotenoids.

2 AQUEOUS SYSTEMS

With the aqueous systems, simple solvent mixtures consisting of appropriate concentrations of water (usually 0-10%) in methanol or acetonitrile (more rarely acetone or isopropanol) have traditionally been used. Braumann and Grimme (1981) described the development and evaluation of a procedure which employed a C_{18} reversed-phase column and a linear gradient running from 75% solvent A (methanol-acetonitrile, 1:3) with 25% water to 100% solvent A. This method is useful for analysing the chloroplast pigments of plants and green algae.

In the author's (G.B.) laboratory a gradient procedure derived from a method previously used successfully by Wright and Shearer (1984) for analysing algal pigments is currently in favour as a useful method for screening natural plant extracts. This procedure employs a linear gradient of ethyl acetate (0-100%) in acetonitrile–water (9:1) over 25 min and gives good resolution both of polar xanthophylls and of non-polar carotenes, carotene epoxides and xanthophyll acyl esters in the same run. The gradient is easily modified to give improved resolution of selected parts of the chromatogram, e.g. the hydrocarbon area. Almost any C_{18} reversed-phase column is suitable, though there is considerable variation in the resolution achieved. Some column materials, e.g. Spherisorb 2, will resolve geometrical isomers, especially of xanthophylls, reasonably well. A typical reversed-phase chromatogram of a leaf extract is illustrated in Fig. 2.2. The major leaf pigments, both carotenoids and chlorophylls, are well resolved, as are quantitatively minor though physiologically important compounds such as zeaxanthin (VII), antheraxanthin (VIII), lutein-5,6-epoxide (IX) and β-carotene-5,6-epoxide (LXXVIII). This procedure has proved extremely valuable for the accurate and reproducible quantitative analysis of pigment compositions of leaves, chloroplasts and especially of thylakoid pigment–protein complexes which have been isolated by polyacrylamide gel electrophoresis. As outlined below, this system is also very efficient for separating xanthophyll acyl esters.

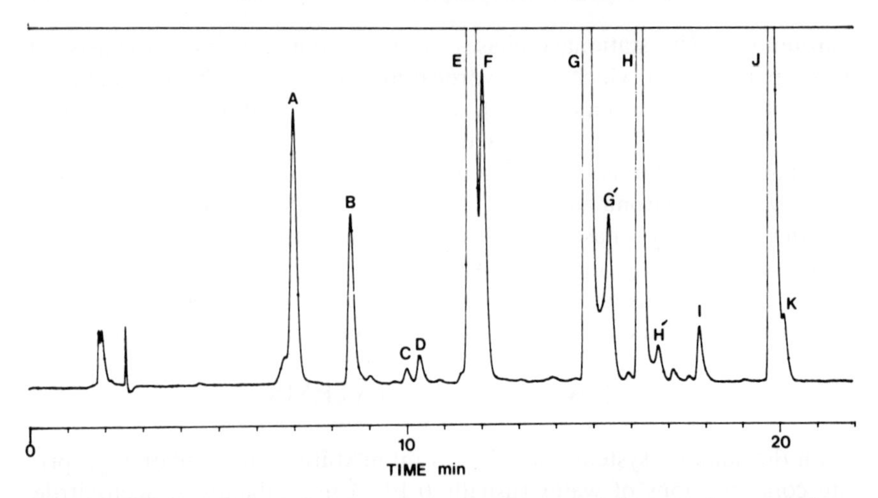

Fig. 2.2 Reversed-phase h.p.l.c. of pigments extracted from barley cotyledons (Spherisorb ODS2 5μ). Solvent system: 0–100% ethyl acetate in acetonitrile/water (9:1), containing 0.1% (v/v) triethylamine, over 25 min. Flow rate = 1.0 ml min^{-1}.

Peak identifications: A. neoxanthin; B. violaxanthin; C. lutein-5,6-epoxide; D. antheraxanthin; E. lutein; F. zeaxanthin; G , G′ chlorophyll *b*; H, H′ chlorophyll *a*; I. β-carotene-5,6-epoxide; J. β-carotene; K. *Z*-isomers of β-carotene.

3 NON-AQUEOUS SYSTEMS

Nelis and de Leenheer (1983), considering the reported (Parris, 1978) general advantages of non-aqueous reversed-phase chromatography, i.e. optimal sample solubility and hence minimal risk of precipitation or crystallization, increased sample capacity, high chromatographic efficiency and prolonged column lifetime, undertook a systematic investigation of non-aqueous reversed-phase HPLC of carotenoids (and the related retinoids). They found that Zorbax-ODS, in which end-capping with the C_{18} chains is not complete, had advantages over other reversed-phase column materials for this application. Many solvent mixtures were surveyed, and the most satisfactory results were obtained with ternary mixtures of acetonitrile as the polar basis, dichloromethane as a non-polar modifier to act as a solubilizer and adjust the solvent strength, and methanol as a further modifier to optimize selectivity. Efficient separation of a range of carotenoids of different polarities was achieved with an isocratic mobile phase consisting of acetonitrile–dichloromethane–methanol (7:2:1). This procedure resolved difficult mixtures such as lutein and zeaxanthin and gave wide separation of the hydrocarbons lycopene (ψ,ψ-carotene, LXXXI), torulene (LXVII), α-carotene (V) and β-carotene (I). Again, by modifying the solvent compositions, optimal resolution of particular mixtures can be achieved, and gradient elution procedures have been developed which permit improved separation of polar xanthophylls, thereby giving advantages when total natural extracts are being screened. A somewhat similar non-aqueous reversed-phase procedure employing a Supelcosil C_{18} column and a gradient of 1,2-dichloroethane–propan-2-ol (2:1)–acetonitrile recently reported (Zonta *et al.*, 1987) extended the achievements of non-aqueous reversed-phase HPLC to the resolution of xanthophyll acyl esters.

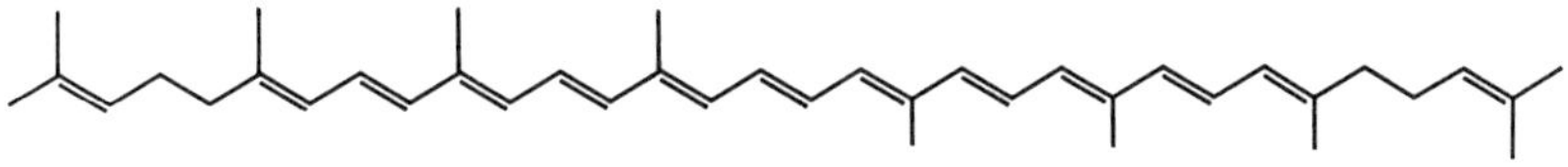

LXXXI Lycopene

E Quantitative analysis by HPLC

One of the major benefits of HPLC applied to carotenoids is that it provides the most sensitive, accurate and reproducible method yet available for quantitative analysis, especially when the instrumentation includes automatic integration facilities for determination of peak areas. The relationship between the peak areas for different components in a chromatogram allows the quantitative composition and ratios of different pigments to be determined accurately and reproducibly, provided the peak area can be calculated for each component at its absorption maximum (λ_{max}) by use of a multi-wavelength detector. Alternatively, if monitoring at only a single wavelength is possible, corrections must be made for the difference between the absorbance coefficient at the monitoring wavelength and that at the λ_{max} of each pigment.

This procedure allows the quantitative relationship between different components in a chromatogram to be determined, but for the estimation of absolute amounts or concentrations calibration is necessary. This can be achieved by means of a calibration curve or by inclusion of an internal standard. Braumann and Grimme (1981), using a reversed-phase method to separate the carotenoids and chlorophylls of plant chloroplasts and green algae (see above), prepared calibration curves by injecting known (spectroscopically determined) amounts of pure pigments, in the 20-2000 pmol range, and determining the peak areas. By use of these calibration curves, which were linear over the whole concentration range, the amount of each pigment present in a natural extract could be determined with great accuracy and precision at the nanogram level from peak areas in the chromatogram.

As an alternative to the use of calibration curves, or indeed as a check on the validity of that approach, internal standards may be used. The usual procedure is to add to the sample an accurately known amount of a pure standard compound. The amount of each component in the chromatogram is then estimated by comparing its peak area with that of the standard, again correcting for the difference between the absorbance coefficients at λ_{max} and the monitoring wavelength. Amongst internal standards used are purely synthetic compounds such as 'decapreno-β-carotene, (LXXXII), the phenylcarotene analogue (LXXXIII), and ethyl 8′-apo-β-caroten-8′-oate (LXXXIV), as well as selected natural xanthophylls. The main criteria for selection of a substance for use as an internal standard are that it should not be present in the extract under investigation, that it should clearly separate from all components of that extract, that its stability and λ_{max} should be similar to those of the components of the extract and that its absorbance coefficient should be accurately known. An alternative strategy which is often acceptable, but more time-consuming, is to use the substance or one of the substances under investigation as the internal standard. The experimental sample is analysed by HPLC and the peak areas are determined in the normal way. Then a known amount of the pure standard compound is added to the

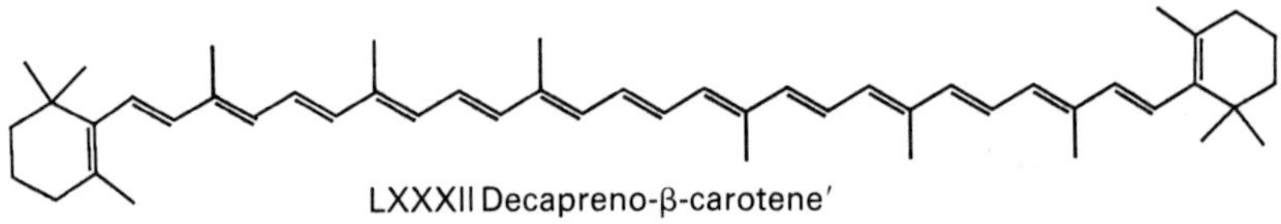

LXXXII Decapreno-β-carotene′

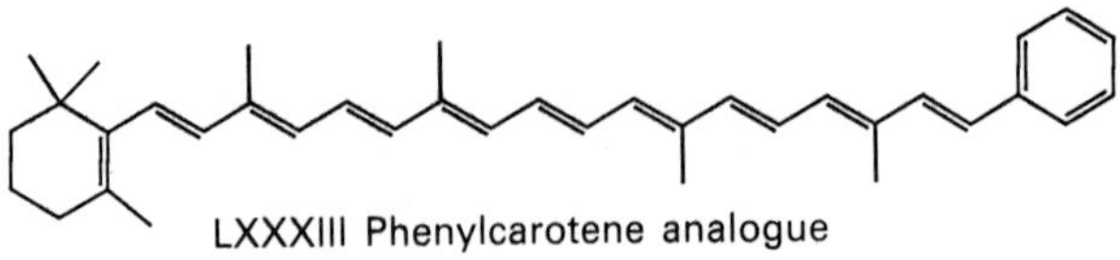

LXXXIII Phenylcarotene analogue

COOEt

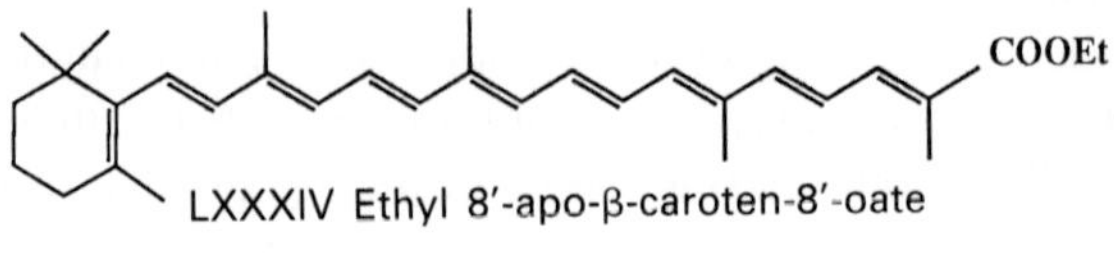

LXXXIV Ethyl 8′-apo-β-caroten-8′-oate

sample and a second HPLC analysis is performed. The increment in peak area for the standard compound is then related to the known amount that was added to give a calibration that can be used to determine the amounts of substances present in the extract.

With simpler instrumentation which does not have the automatic integration capability, compounds separated by chromatography may be collected, and the pigment concentrations determined accurately in a spectrophotometer. The virtually quantitative recovery of carotenoids from HPLC, especially with reversed-phase procedures, allows much greater accuracy than is possible when other separation methods are used; pigment losses on TLC, for example, can be significant.

F Some practical points on HPLC

Since HPLC is a relatively new development it is important to draw attention to some practical points about HPLC procedure which are essential for good and reproducible chromatography and just as important as the correct choice of column material and solvents. Only the briefest outline of a few of these points can be included here. Any worker planning to use HPLC for carotenoid analysis should make sure that he or she is familiar with the basic principles and practice by consulting some of the books and manuals on HPLC technique now available.

1 CARE OF THE COLUMN

If a good column is treated properly, it should be able to withstand repeated use for a year or more and still give good resolution and reproducibility. The use of a precolumn helps to prolong column life without appreciably affecting the separations. A precolumn consists of a short column or cartridge packed with a material similar to the column material but usually of rather larger particle size. Its position between injector and column allows it to remove any particulate, insoluble or irreversibly adsorbed material in the sample so that such extraneous material never reaches and contaminates the column. The precolumn is easily and much more cheaply replaced than the column itself.

From time to time the performance of a column may deteriorate, but it can usually be restored by a regeneration routine. The filter frits at the ends of the column should be renewed, and contaminating materials that have accumulated stripped from the column by treatment with a succession of solvents. If instructions for column regeneration are provided by the supplier/manufacturer these should be followed. Otherwise, for a normal adsorption phase silica column a recommended procedure is to pump through the column successively hexane, dimethoxypropane and again hexane. For a reversed-phase (ODS) column, successive washing with methanol, methanol–dichloromethane (1:1), dichloromethane, methanol–dichloromethane (1:1) and methanol is usually satisfactory (acetonitrile can replace methanol in this sequence). After regeneration, a column must always be equilibrated thoroughly with the solvent that is to be used at the beginning of the HPLC run.

2 PREPARATION AND USE OF SOLVENTS

Solvents for HPLC should be of high purity; HPLC-grade solvents are available from many manufacturers. For most work on carotenoid analysis, monitoring in the visible region is used, so UV absorption by the solvent does not present a major problem. All solvents must, however, be filtered and degassed, and high room temperatures must be avoided when solvents such as dichloromethane, which have very low boiling points, are used. Care must also be taken to ensure that when changes of solvent are made, the two solvents involved are completely miscible. The boundary between two immiscible solvents can form a barrier which greatly restricts flow through the column, thereby causing alarming increases in pressure and possible damage to the column.

The need for adequate equilibration must also be stressed. Either when a change needs to be made in eluting solvent or between analyses by a gradient method, the column needs to be equilibrated thoroughly with the mobile phase that is to be used at the beginning of the next chromatography run. If the equilibration time is not adequate, resolution efficiency will be reduced and retention times will not be reproducible.

3 IDENTIFICATION OF CAROTENOIDS

For anything other than the routine repeated analyses of known compounds, retention time alone is not a satisfactory criterion for identifying components in a chromatogram, though co-chromatography can give a good first indication. A combination of retention time and absorption spectrum is much more reliable, but it must be remembered that for any compound the λ_{max} value depends on the solvent. The values recorded in the HPLC solvent are not necessarily comparable with those that have been reported for the same carotenoid in selected pure solvents (e.g. hexane, ethanol) and which are given in the extensive tables published elsewhere. Also, with gradient HPLC, the solvent composition is changing throughout the chromatography so compounds which in the tables are given the same λ_{max} values may not give the same λ_{max} during chromatography. The differences are especially marked if the gradient includes a changing proportion of a chlorinated solvent, such as dichloromethane: differences of as much as 8–10 nm can be seen between the λ_{max} of zeaxanthin and β-carotene, even though these compounds have identical spectra in the same pure solvent.

The overall shape or fine structure of a spectrum also provides valuable information, but the shape of a spectrum determined on-line can be distorted if the concentration is very low or very high. Much restrained common sense must be exercised when attempting to identify components of HPLC chromatograms from retention times and absorption spectra, and other data such as mass spectrometry (MS) and NMR should be applied whenever possible.

G Some special applications

1 SEPARATION OF GEOMETRICAL ISOMERS

The importance of separating geometrical isomers of carotenoids is now being realised as more and more evidence is accumulated to show that *Z*-isomers occur widely as natural products and may have considerable physiological significance. Isomeric purity is also essential in samples prepared for examination by NMR, resonance Raman and especially CD. In the classic work of Zechmeister (1962) and others, sets of carotenoid geometrical isomers were separated by open column chromatography on basic adsorbents such as $Ca(OH)_2$, $ZnCO_3$ and many others. These materials, however, are not easily adapted for use in HPLC, and no uniform particles specially prepared for HPLC are available.

The separation of geometrical isomers of the apolar carotenes on commercially available HPLC stationary phases still poses a problem. In contrast, *Z*/*E* isomers of many xanthophylls can be separated satisfactorily by normal adsorption phase HPLC on silica or nitrile columns. Generally the most that can be hoped for with reversed-phase methods is some separation of a mixture of *Z*-isomers from the all-*E*-compound, but no resolution of the individual *Z*-isomers. Some separation of the (15*Z*) and (all-*E*)-isomers of phytoene, phytofluene and ζ-carotene (see Chapter 3 for structures) can be achieved, though the resolution is not good. With normal adsorption phase columns, however, impressive separations of xanthophyll isomers can be achieved. Lutein provides a good illustration, but similar success can be achieved with other xanthophylls, e.g. zeaxanthin, violaxanthin and neoxanthin. Berset and Pfander (1985) reported the separation of the (all-*E*)-, (9*Z*) (9′*Z*) and 13*Z* + 13′*Z*)-isomers of lutein from daffodil flowers by HPLC on a nitrile column with hexane–dichloromethane–methanol–base (60:40:0.1:0.1) as solvent. An ordinary silica column on the other hand resolves the (13*Z*), (13′*Z*) [and (15*Z*)] isomers that co-chromatograph on the nitrile column but does not separate the (9*Z*) and (9′*Z*) isomers. In both cases the Z-isomers are eluted well after the all-*E*-components. The (9*Z*) [and (9′*Z*)] are the first of the *Z*-isomers to be eluted, and can be identified by their relatively small spectral shift and minimal "*cis*-peak" in the UV region of the spectrum. The (13*Z*) [and (13′*Z*)] and (15*Z*)-isomers which are eluted later are characterized by a large "*cis*-peak" and a somewhat greater reduction in λ_{max} and spectral fine structure.

For separation of geometrical isomers of carotenes, silica and nitrile columns have not yet proved satisfactory. The interactions are too weak and the carotenes are eluted too rapidly, even with apolar solvents like hexane. The separation of a large number of geometrical isomers of β-carotene by two procedures which use other adsorbents has been reported. Vecchi *et al.* (1981) used alumina as stationary phase and hexane with a controlled water content as the mobile phase and were able to separate, on a semi-preparative scale, the all-*E* and 11 different mono-*Z*, di-*Z* and tri-*Z* isomers of β-carotene. Although the separations achieved were excellent, the procedure is difficult, requiring rigorous control of temperature and of water content in the solvent. Ben-Amotz *et al.* (1982) obtained acceptable separation of (all-*E*), (9*Z*)- and (15*Z*)-β-carotene on a column of Alox-T which they packed themselves, and used without any special precautions to control the water content, and with hexane–dichloromethane (13:7) as developing solvent. Tsukida *et al.* (1982) adapted the classical separation on a "home-packed" column of $Ca(OH)_2$ operating at ambient temperature. This procedure is reproducible and much simpler, though the separation depends critically on the origin and physical form of the Ca $(OH)_2$ used. With 0.5% acetone–hexane as the mobile phase, up to 17 peaks were resolved from the mixture of geometrical isomers obtained by thermal isomerization and photoisomerization of β-carotene, including the "hindered" (7Z) isomer that was not detected by Vecchi *et al.* Of the 16 mono-Z and di-Z isomers that were resolved, only two, namely (7Z) and (9Z), were eluted after (all-*E*)-β-carotene. With improvements in procedures and materials, some progress has now been reported in the separation of carotene geometrical isomers by reversed-phase HPLC. Bushway (1985) evaluated several normal-phase and reversed-phase procedures, and reported the separation of five geometrical isomers of β-carotene on C_{18} Vydac columns with isocratic, non-aqueous solvent mixtures (acetonitrile–methanol–tetrahydrofuran).

The problem of the separation of geometrical isomers of carotenes has also been addressed by Rüedi (1985). He also found that adsorption chromatography on silica or bonded nitrile columns was not satisfactory but useful separations could be achieved on a stationary phase with bonded amine groups, Spherisorb S5-NH_2, also an adsorption phase material. Although β- and ϵ-ring carotenes are not separated by this column, appreciable resolution of *Z*/*E* isomers has been obtained with hexane containing 0.1% N-ethyldiisopropylamine as mobile phase. The all-*E* compound is always eluted last. The reversed-phase procedure which is recommended by Rüedi for separating carotene structural isomers such as α-carotene (V), β-carotene (I), γ-carotene (β,ψ-carotene, LXXXV) and δ-carotene (ϵ,ψ-carotene, LXXXVI) (Spherisorb S5-ODS column, with mixtures of acetonitrile–tetrahydrofuran or acetonitrile–methanol–-propan-2-ol as mobile phase) also gives good separation of geometrical isomers in which the terminal double bond of an acyclic end-group is in the 5*Z* configuration, e.g. (5*Z*)- and (5*Z*, 5′*Z*)-lycopene from the all-*E* isomer. In this case the all-*E* isomers are eluted first. Some separations achieved with this procedure are illustrated by Märki-Fischer *et al.* (1983).

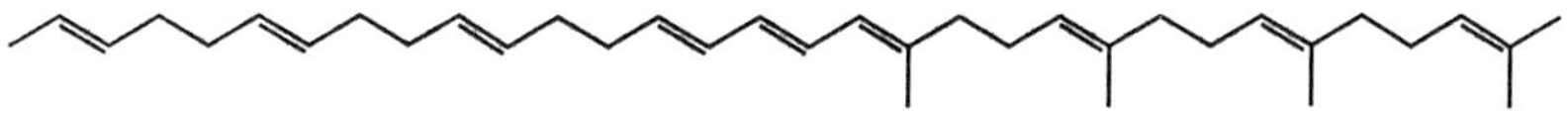

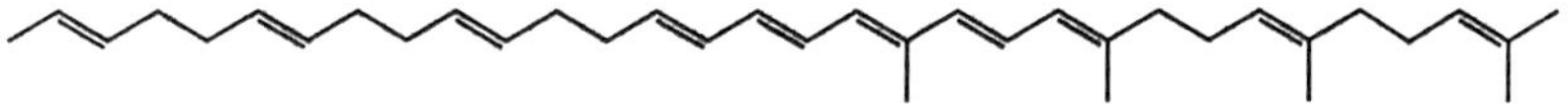

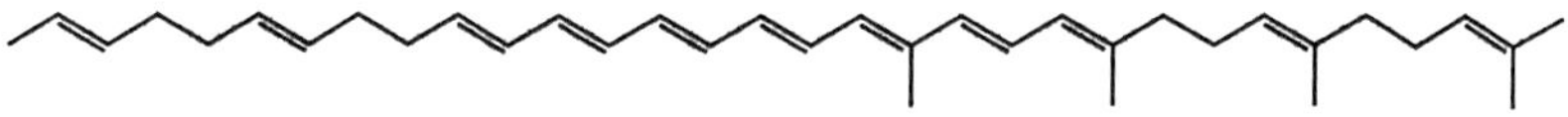

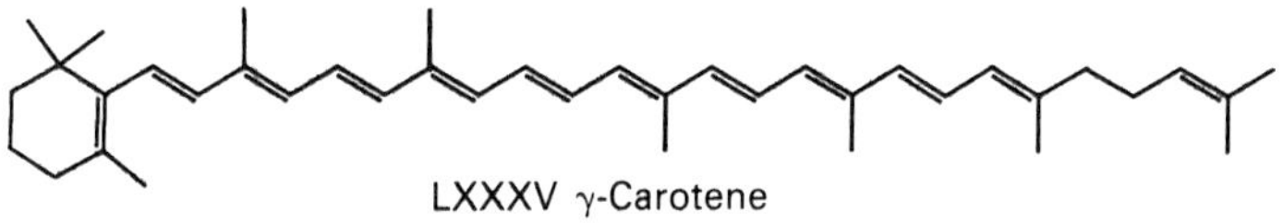

LXXXV γ-Carotene

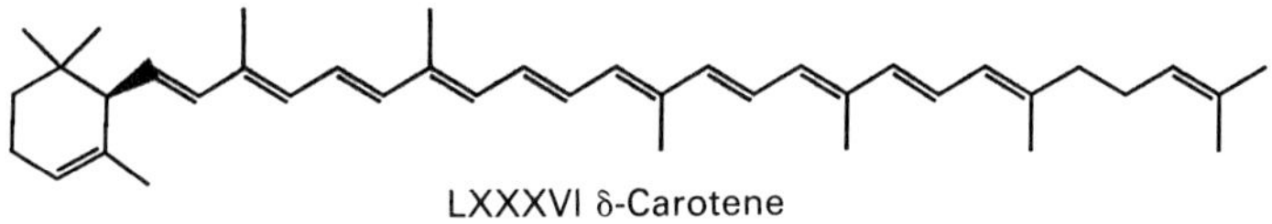

LXXXVI δ-Carotene

Joint application of these two systems has allowed a large number of *Z*-isomers of carotenes to be detected in the hydrocarbon fraction from roses, but Rüedi stressed that preseparation on classical adsorbents and application of combinations of HPLC procedures are essential to generate the maximum benefit and information. Geometrical isomer compositions of carotenes cannot yet be determined as part of a full analysis of a natural extract in a single HPLC run.

2 RESOLUTION OF OPTICAL ISOMERS

The structure of an optically active natural carotenoid cannot be considered to be fully elucidated unless the chirality has been established. Also any carotenoid produced by synthesis must have the correct chirality if it is to be considered identical to the natural product. The identification and resolution of optical isomers are therefore important and HPLC now provides the basis for direct and indirect means of doing so. The first chromatographic resolution of carotenoid optical isomers was devised by Vecchi and Müller (1979). The method involved the preparation of diastereoisomeric esters of astaxanthin by esterification with optically active (−)-camphanoyl chloride. The products were readily separated by HPLC on silica. This has now become a routine method for analysing the enantiomeric composition of astaxanthin (LXXXVII) and related carotenoids, and has shown that many samples of astaxanthin extracted from animals contain a mixture of (3*R*, 3*R*), (3*R*, 3′*S*) and (3*S*, 3′*S*) isomers, whereas samples of plant origin are optically pure. The camphanate ester method is effective only with carotenoids which have one or two 3-hydroxy-4-oxo-β-rings, and not with simple hydroxycarotenoids. A method has, however, been developed which is similar in principle and will allow the resolution of optical isomers of lutein or zeaxanthin after derivatization with (*S*)-(+)-α-(1-naphthyl) ethyl isocyanate (Rüttimann *et al.*, 1983).

Now that HPLC columns with optically active stationary phases have become available, increasing efforts are likely to be made to resolve enantiomeric mixtures of carotenoids directly. The first report of the

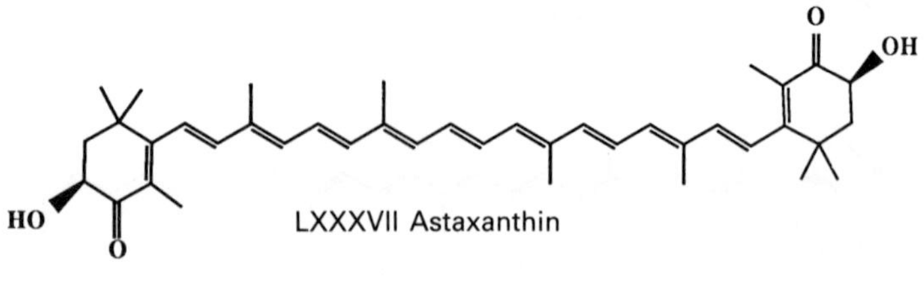

LXXXVII Astaxanthin

successful use of this approach described the direct resolution of all 10 stereoisomers of ε,ε-carotene-3,3′-diol (LXXXVIII), a symmetrical carotenoid with four chiral centres at C-3, C-6, C-3′ and C-6′ (Ikuno *et al.*, 1985). Previously adsorption phase chromatography on a nitrile column had separated the mixture only into four fractions (Vecchi *et al.*, 1982). With the exception of one pair, all the isomers were resolved, without the need for prior derivatization, on a Sumipax OA 2000 column. The overlapping enantiomeric pair was resolved by separation of the acetate or benzoate esters, i.e. without the need for preparation of diastereoisomeric derivatives with an optically active reagent. Further developments of this kind to provide methods for directly resolving enantiomeric carotenoids can be expected as improved optically active column materials become available.

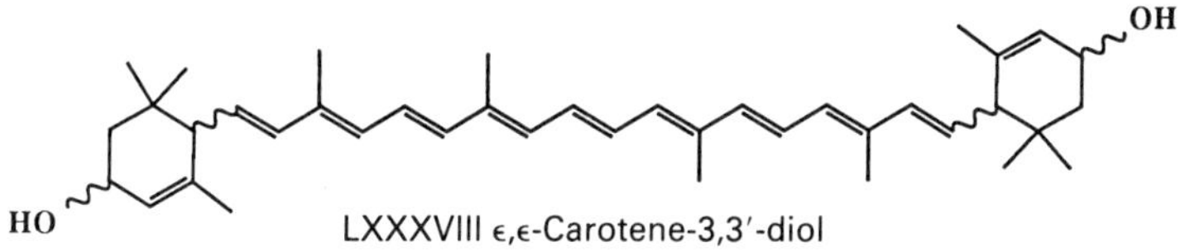

OH

HO

LXXXVIII ε,ε-Carotene-3,3′-diol

3 ANALYSIS OF CAROTENOID ACYL ESTERS

It is still common practice to saponify natural extracts to remove chlorophyll and troublesome colourless neutral lipids and also to hydrolyse any xanthophyll esters that may be present, in order to simplify the separation and analysis of carotenoids. This, of course, also destroys what may be valuable information, since it is increasingly becoming apparent that the presence and nature (e.g. fatty acid compositions) of carotenoid esters may have functional or physiological significance. The isolation and identification of esterified carotenoids by conventional chromatographic methods can be difficult and time-consuming, but reversed-phase HPLC now provides a means for the routine separation and analysis of naturally occurring mixtures of xanthophyll esters. When combined with UV-visible absorption spectroscopy and mass spectrometric analysis, this method becomes very powerful.

A reversed-phase procedure such as that illustrated in Fig. 2.2 for the analysis of leaf extracts is also very good for detecting the presence of esters. Their presence is usually revealed by a pattern of unusual peaks in the chromatogram in the vicinity of β-carotene and their resolution can be greatly improved by modifying the solvent gradient. The absorption spectra of xanthophylls are not altered by esterification, so the spectra of the individual peaks will show how many different carotenoids are present in esterified form and give a first indication of the xanthophylls concerned. It is unusual for a natural sample of esterified carotenoid to contain only a single esterifying fatty acid. Normally a collection of peaks will be seen in the chromatogram (Fig. 2.3) for each esterified carotenoid, each individual peak containing a molecular species with a particular esterifying fatty acid (or acids if more than one carotenoid hydroxy group is esterified). The separation of the different molecular species depends on several factors. Obviously the nature of the carotenoid itself plays a part but the most important features seem to be the chain length and degree of unsaturation of the esterifying fatty acid. This is nicely illustrated by considering diacyl esters of lutein. If only saturated acyl groups are present, the resolution depends only on their chain lengths. Esters with shorter-chain fatty acids are eluted first, e.g. lutein bis-palmitate (16:0) before the bis-stearate (18:0). Unsaturation in the acyl group substantially reduces the retention time, so that lutein bis-oleate (18:1) is eluted before the bis-palmitate (16:0), and the bis-linoleate (18:2) and bis-linolenate (18:3) earlier still. Provided some standards of known carotenoid and fatty acid constitution are available, carotenoid esters can be identified with reasonable confidence by HPLC and their UV-visible spectra; for

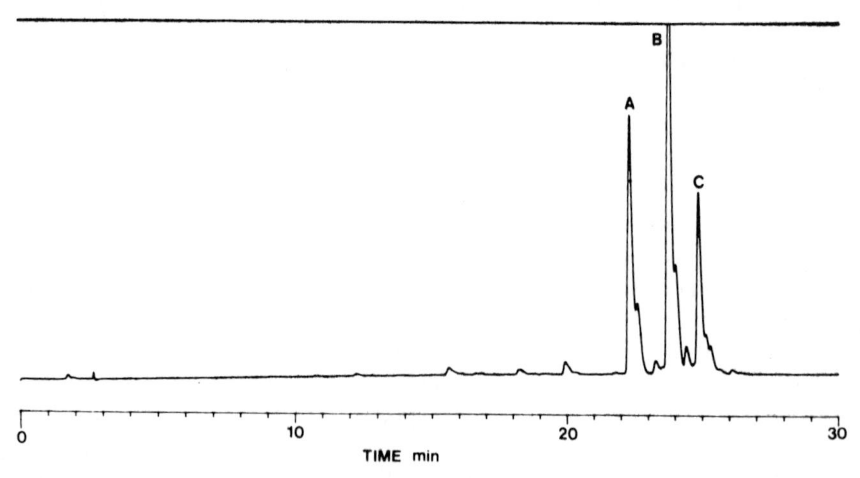

Fig. 2.3 Reversed-phase h.p.l.c. separation of C_{12} and C_{18} acyl esters of lutein. Chromatographic conditions as in Fig. 2.

Peak identifications: A. lutein *bis*-laurate; B. mixture of 3-lauryl, 3′-stearyl lutein and 3-stearyl, 3′-lauryl lutein; C. lutein *bis*-stearate.

unequivocal characterization, however, both the carotenoid and fatty acid must be identified rigorously. For identification of the fatty acid the classical method of hydrolysis followed by gas-liquid chromatography (GLC) (or, better, GC–MS) of the methyl ester is still applicable, but it is now possible to identify both the carotenoid and the esterifying fatty acid(s) directly by mass spectrometry of the esters themselves.

4 APOCAROTENOIDS

If apocarotenoids are present in a natural extract but their presence is not suspected, their chromatographic behaviour on a reversed-phase column can cause some confusion. Unless different functional groups are present, a mixture of apocarotenals of different chain lengths, e.g. the C_{25} 12′-apo-β-caroten-12′-al (LXXXIX), the C_{27} 10′-apo-β-caroten-10′-al (XC) and the C_{30} 8′-apo-β-caroten-8′-al (XCI), is virtually unresolved by normal adsorption phase chromatography on silica. In most cases only an "apocarotenal fraction" can be obtained. The components of such a mixture are, however, widely separated on reversed-phase systems, eluting in order of increasing chain length. Thus in the ethyl acetate gradient procedure outlined above, apo-β-carotenals may be found in the same area of the chromatogram as the very polar xanthophylls such as neoxanthin. The retention times of the corresponding apocarotenals and apocarotenols on reversed-phase HPLC are not greatly different; the length of the hydrocarbon chain is the dominant feature determining the

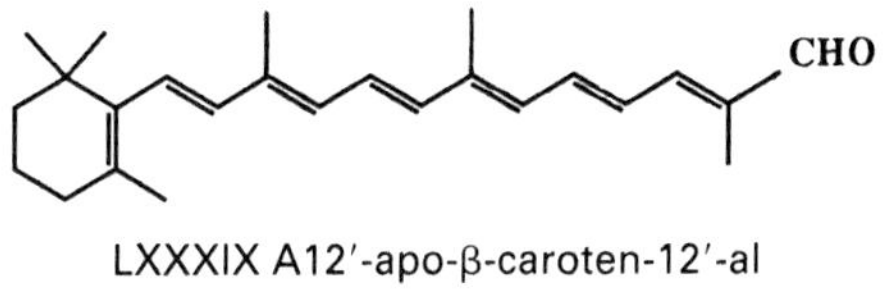

CHO

LXXXIX A12′-apo-β-caroten-12′-al

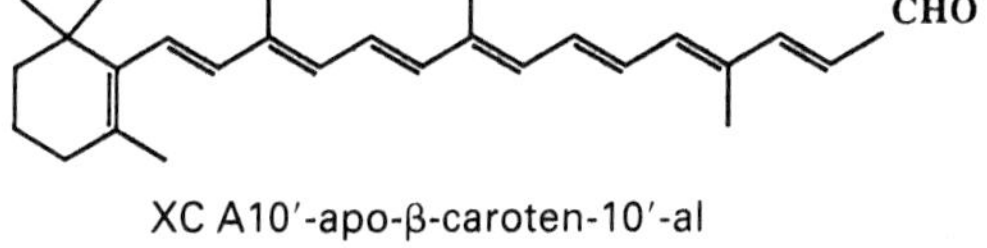

CHO

XC A10′-apo-β-caroten-10′-al

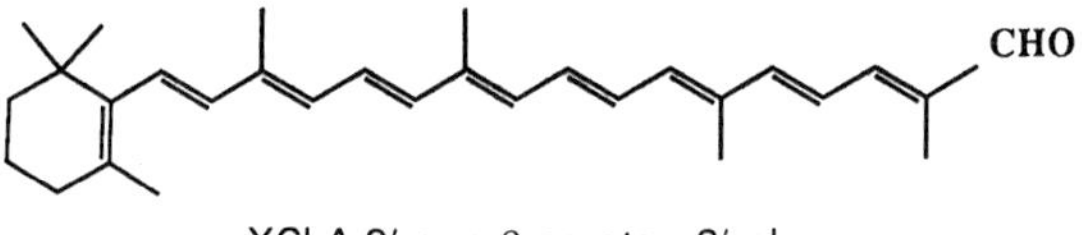

CHO

XCI A 8′-apo-β-caroten-8′-al

chromatographic behaviour. Apocarotenoids with the same chain length but different functional groups are well separated by normal adsorption phase HPLC. Components of a complex mixture of apocarotenoids can be separated by judicious application of a combination of normal and reversed-phase procedures.

4 SPECTROSCOPIC METHODS

A Introduction

There have been many refinements and developments in the use of spectroscopic and other physicochemical techniques for determining the structure and stereochemistry of novel carotenoids. A very useful and extensive survey of the use of these methods in the carotenoid field (Moss and Weedon, 1976) gave tabulated NMR, infrared and MS data for a wide range of carotenoid end-groups. All these techniques are now in routine use and any publication on carotenoid chemistry and the elucidation of new structures is expected to give MS, NMR and, when appropriate, CD data. Because of the limitations of space no attempt will be made to give exhaustive spectroscopic data for all natural plant carotenoids. However, some developments relevant to the general use of these techniques will be discussed briefly, and attention will be drawn to some particularly useful review articles.

B Nuclear magnetic resonance (NMR)

1 INTRODUCTION

NMR spectroscopy is undoubtedly the most powerful technique for investigating carotenoid structures. The determination of ^{1}H NMR spectra is a routine part of any chemical investigation and the use of ^{13}C NMR, when sufficient sample is available, is also becoming widespread. Moss and Weedon (1976) discussed the general features of NMR spectra of carotenoids and presented in tabulated form a large amount of data on the ^{1}H NMR chemical shifts of many carotenoid end-groups, especially the methyl group protons that were most easily identified with the low-resolution instruments then available. Some early ^{13}C NMR data were also given. Since then, spectacular advances have been made in instrumentation and data handling, and many new and sophisticated experimental procedures have been devised to facilitate the assignment of complex spectra. To describe all these developments in detail would be far beyond the scope of this volume, as would an exhaustive description and discussion of the enormous amount of NMR data that has accumulated in recent years in published reports of the characterization of synthetic and naturally occurring carotenoids. However, two authoritative articles by Englert (1982, 1985) describe some of the complicated procedures now available and give examples of their application in the carotenoid field. Englert also gives an extremely valuable collection of figures which illustrate the ^{1}H and ^{13}C NMR assignments for a wide range of natural and synthetic carotenoid end-groups. Later in this article, tables are given (Tables 2.3, 2.4) which summarize the data for some of the end-groups most likely to be encountered in plant and algal carotenoids. The reader should consult the primary literature for full details of the NMR investigations and assignments for invididual carotenoids.

2 ADVANCES IN INSTRUMENTATION

In recent years the development of stable, uniform, superconducting magnets has made available instruments which use very high magnetic fields corresponding to ^{1}H NMR frequencies of 400 and even 600 MHz. With these instruments, signals which are strongly overlapping in a lower field (60-90 MHz) instrument are clearly resolved. In particular the signals for single proton substituents on a ring end-group and for the individual olefinic protons of the polyene chromophore can be identified and assigned. The increases in the magnetic field strength, together with the advent of Fourier-transform techniques, have also led to a significant increase in sensitivity. Good ^{1}H NMR spectra can now be obtained from about 100 μg and ^{13}C spectra from about 5 mg of carotenoid with accumulation times of only a few minutes. Englert (1982) illustrates 400 MHz ^{1}H NMR spectra obtained from samples of 50 μg in 11 min and 1.5 μg in 13 h, and even as little as 0.25 μg of canthaxanthin (LXXII) gave an interpretable spectrum after 116 h. In the case of natural abundance ^{13}C spectra, for an instrument operating at the equivalent frequency (100.6 mHz), 3 mg of carotenoid gave a good spectrum in 3.2 h, whereas a 1 mg sample required 12 h, and 48 h accumulation time was needed when the sample size was reduced to 0.4 mg.

3 NEW EXPERIMENTAL METHODS

Many new and powerful experimental methods have been devised in recent years which greatly extend the amount of information that can be obtained from an NMR study, and facilitate the solution of complex structural problems. Englert (1982) outlined the principles and described the application in the carotenoid field of two of these, the nuclear Overhauser technique and two-dimensional (2D)-NMR spectroscopy. His review should be consulted for details. Nuclear Overhauser experiments have proved extremely useful for elucidating the structures of stereoisomers. The nuclear Overhauser effect arises from the fact that two protons or groups of protons in a molecule, which are sufficiently close to each other spatially, are the main influences on their mutual relaxation. Any change in the population of the spin state of one proton is therefore partly transferred to the other via relaxation processes. Thus if the NMR signal of one proton is saturated by the application of appropriate radiofrequency, the nuclear Overhauser effect would lead to a change in the population of the spin states of the neighbouring proton(s) and this would be detected as a change, normally an enhancement, of their signal intensity. The enhancements are most easily and accurately observed when nuclear Overhauser difference spectra are determined, and give valuable information about the spatial relationships between different protons in a molecule, i.e. which protons are close together in space and which are distant. The technique has been used not only to determine the geometry of carotenoid polyene chains but also to assign the axial/equatorial or α/β orientation of protons or methyl groups in cyclic carotenoid end-groups, e.g. in zeaxanthin (VII) bis-acetate.

Several different 2D-NMR methods have been developed. Englert (1982) describes one of these, the so-called 2DJ-resolved spectroscopy (J = spin–spin coupling). This method provides an effective and elegant means of separating and identifying the different multiplets of a strongly overlapping spectrum and can also reveal additional long-range couplings which may not be observed in a basic 1D-experiment. In a normal 1D-spectrum the pattern of signals is determined by the chemical shifts (δ) and coupling constants (J) which are displayed along the same frequency axis and which therefore together give a strongly overlapping pattern. The 2D-spectrum allows the chemical shift and spin-coupling effects to be separated and the information to be displayed along two different frequency axes, the δ and J axes. This obviously greatly facilitates assignment, and is especially valuable for sorting out the very complex olefinic proton region of the spectrum. It is therefore a very powerful, though time-consuming method for distinguishing between and identifying carotenoid geometrical isomers.

Developments in NMR methods have been so rapid that in his 1985 article Englert was able to introduce several new experimental techniques. These include DID (double INDOR difference) which greatly simplifies the assignment of protons in the crowded spectra of carotenoids, DEPT (distortion enhancement by polarization transfer) for investigating the multiplicity of ^{13}C signals, and further 2D methods such as homonuclear (COSY) and heteronuclear chemical shift-correlated 2D-spectra for elucidating ^{1}H–^{1}H and ^{1}H–^{13}C coupling relationships respectively, and INADEQUATE or carbon–carbon connectivity which allows the measurement of ^{13}C–^{13}C coupling in natural abundance spectra. Much of the discussion given by Englert is highly technical and specialized and further elaboration in this article would not be appropriate. Also the amount of sample required for some of these procedures can be substantial (e.g. 10 mg–1 g) so they are not likely to be used routinely in the characterization of small amounts of newly discovered carotenoids. However, they have been used in some of the work which led to the assignments for carotenoid geometrical isomers and end-groups summarized below.

Finally, although carotenoid NMR spectra are usually determined conventionally with samples in solution, techniques are now available for determining spectra of samples in the solid state. The MASS (magic angle sample spinning) technique has not yet been applied in the carotenoid field, but has been used successfully to study the retinaldehyde chromophore of bacteriorhodopsin in the purple membrane of *Halobacterium* species (Lugtenburg, 1985) and should be applicable to similar studies of carotenoid molecules *in situ*, provided suitable samples can be prepared (i.e. with the carotenoid highly and specifically enriched with ^{13}C).

4 CAROTENOID END-GROUP ASSIGNMENTS

In their article Moss and Weedon (1976) included tabulated ^{1}H NMR data for a large number of carotenoid end-groups; they concentrated on the methyl proton signals since the data were obtained mainly with comparatively low-field instruments. In his recent reviews Englert (1982, 1985) has given extensive figures which also illustrate the assignments of proton chemical shifts for many end-groups, but these data were taken from high magnetic field spectra, mostly at 400 MHz, and in some cases include the assignment of α/β or axial/equatorial orientations of protons and methyl groups. The chemical shifts are considered to be reliable (within $\pm$ 0.02 ppm) for that end-group in any carotenoid, provided that the chain attached to the end-group is at least a tetraene with no strongly anisotropic group at the end of the chain. Englert (1985) has also presented extensive figures showing the ^{13}C NMR assignments for a wide range of carotenoid end-groups. Tables 2.3 and 2.4 list the ^{1}H and ^{13}C chemical shift values recorded for some of the end-groups most likely to be encountered in plant and algal carotenoids. The nature of the end-

Table 2.3 ^{1}H n.m.r. chemical shift assignments for some common carotenoid end-groups (solvent $CDCl_3$). Data from Englert (1982, 1985)

Carotenoid end-group	16	17	18	19	2		3		4		6	7	8	10
β (all-*E*)	1.03		1.72	1.98	1.46		1.62		2.02		—	6.16	6.14	6.16
(2*R*)-2-Hydroxy-β	1.04 1.09	(β, ax) (α, eq)	1.71	1.97	3.55 1.44	(Hax) (OH)	1.75 1.85	(β, ax) (α, ax)	2.15		—	6.11	6.12	6.16
(3*R*)-3-Hydroxy-β	1.07		1.74	1.97	1.48 1.77	(β, ax) (α, eq)	4.00 1.34	(Hax) (OH)	2.04 2.39	(β, ax) (α, eq)	—	6.10	6.16	6.16
(3*R*)-3-Hydroxy-7,8-didehydro-β	1.20 1.15	(β, eq) (α, ax)	1.92	2.01	1.45 1.84	(β, ax) (α, eq)	3.99 1.41	(Hax) (OH)	2.07 2.43	(β, ax) (α, eq)	—	—	—	6.45
(3*R*)-3-Hydroxy-7,8-didehydro-β (9*Z*)	1.25 1.19	(β, eq) (α, ax)	1.97	2.00	1.48 1.85	(β, ax) (α, eq)	4.01 1.41	(Hax) (HO)	2.09 2.46	(β, ax) (α, eq)	—	—	—	6.29
4-Hydroxy-β	1.02 1.05		1.83	1.98	1.43 1.65	(eq) (ax)	1.72 1.91	(eq) (ax)	4.01	(Hax)	—	6.12	6.16	6.17
(2*R*,3*R*)-2,3-Dihydroxy-β	1.00 1.13		1.72	1.97	3.33	(Hax)	3.83	(Hax)			—	6.06	6.12	6.17
(3*R*)-3,19-Dihydroxy-β	1.08		1.75	4.55 (CH_2OH)	1.48 1.78	(β, ax) (α, eq)	4.03 1.30	(Hax) (OH)	2.08 2.39	(β, ax) (α, eq)	—	6.35	6.05	6.24
4-Oxo-β	1.20		1.88	2.00	1.86		2.51				—	6.24	6.37	6.28
(3*S*)-3-Hydroxy-4-oxo-β	1.21 1.32	(β, eq) (α, ax)	1.95	2.00	1.81 2.16	(β, ax) (α, eq)	4.32 3.68	(Hax) (OH)			—	6.22	6.43	6.30
3-Hydroxy-4-oxo-2,3-didehydro-β	1.30		2.10	2.03	6.04		6.40	(OH)			—	6.31	6.52	6.33

(5*R*,6*S*)-5,6-Epoxy-5,6-dihydro-β	0.94 1.10	1.15	1.93				—	5.87	6.29	6.19
(5*R*,8*S*)-5,8-Epoxy-5,8-dihydro-β	1.11 1.18	1.46	1.80				—	5.23	5.07	
(3*S*,5*R*,6*S*)-3-Hydroxy-5,6-epoxy-5,6-dihydro-β	1.15 (β) 0.98 (α)	1.19	1.93	1.25 (β) 1.63 (α)	3.91 (α) 1.25 (OH)	1.63 (β) 2.39 (α)	—	5.88	6.29	6.20
(3*S*,5*R*,8*S*)-3-Hydroxy-5,8 epoxy-5,8-dihydro-β	1.34 (β) 1.20 (α)	1.68	1.81	1.48 (α) 1.80 (β)	4.24 (α)	1.90 (α) 2.11 (β)	—	5.31	5.07	6.19
(3*S*,5*R*,8*R*)-3-Hydroxy-5,8-epoxy-5,8-dihydro-β	1.33 (β) 1.17 (α)	1.62	1.72	1.51 (α) 1.76 (β)	4.24 (α)	1.99 (α) 2.12 (β)	—	5.25	5.17	6.19
(3*S*,5*R*,6*R*)-3,5-Dihydroxy-6,7-didehydro-5,6-dihydro-β	1.07 1.34	1.34	1.81							
(3*S*,5*R*,6*R*)-3-Acetoxy-5-hydroxy-6,7-didehydro-5,6-dihydro-β	1.07 1.36	1.39	1.82		5.38 (ax)[l]				6.06	6.13
5,6-Dioxo-5,6-seco-β	1.17	2.10	1.98			2.40	—	6.64	7.38	
3-Oxo-*retro*-β (6-*trans*)	1.39	2.16	2.03	2.40	—	5.93	—	6.91	6.80	6.47
3-Oxo-*retro*-β (6-*cis*)	1.25	2.30	2.02	2.34	—	5.94	—	6.67	6.55	6.41
(6*R*)-ϵ	0.82 0.92	1.59	1.91	1.18 1.43	2.00	5.41	2.18	5.53	6.11	6.12
(3*R*,6*R*)-3-Hydroxy-ϵ	0.85 (β, ax) 1.00 (α, eq)	1.63	1.91	1.37 1.85	4.25 (Hax) 1.48 (OH)	5.55	2.40	5.43	6.14	6.14
(6*R*)-3-Oxo-ϵ	0.97 1.05	1.91	1.92	2.10 2.38	—	5.91	2.61	5.54	6.23	6.17

Table 2.3 continued

Carotenoid end-group	16	17	18	19	2	3	4	6	7	8	10
ψ	1.61 1.69	(*cis*) (*trans*)	1.82	1.97	5.10	2.12	2.12	5.95	6.49	6.25	
3,4-Didehydro-ψ	1.82		1.94	1.97	5.94	6.47	6.21	6.19			
1,2-Epoxy-1,2-dihydro ψ	1.27 1.31		1.83	1.97	2.73	1.69	2.25	5.99	6.48		
φ	2.23 2.26[2]	(C-1Me) (C-2Me)	2.27[2]	2.08	—	6.96	6.96	—	6.60	6.28	6.22

[1]Acetate CH_3: 2.05.
[2]Assignments may be interchanged.

group does influence to some extent the positions of the signals from the olefinic protons of the polyene chain, with decreasing magnitude along the chain, so the shifts of protons and ^{13}C nuclei at positions 7,8,10 and 19 are included in these tables.

5 GEOMETRICAL ISOMERS

At low magnetic fields (e.g. 60–90 MHz) the olefinic part of a carotenoid ^{1}H NMR spectrum consists of a broad, poorly defined region from which little structural information can be obtained apart from the approximate number of olefinic protons present. In a high-field spectrum, e.g. at 400 MHz, the signals in the olefinic region are well resolved and can be assigned and their coupling relationships identified. As a general rule, for an all-*trans* (all-*E*)-carotenoid, the signals of H(11) and H(15) are usually found at lowest field near 6.7 ppm because of the additional deshielding caused by strong interactions with the hydrogen atoms of the "in-chain" methyl groups. The signals of H(12), H(14) and H(10), which occur at slightly higher fields, and especially those of H(7) and H(8), are strongly dependent on the type of end-group present.

When a natural carotenoid has been isolated and its end-groups and polyene chromophore have been identified, there remains the need to elucidate the geometrical configuration of the polyene chain, i.e. to confirm the all-*E* structure or deduce a specific *cis* (*Z*) structure. This can best be accomplished unambiguously from a ^{13}C NMR spectrum, but the limited amount of sample available frequently precludes this possibility. However, in many cases the identification of a *Z*-isomer has been achieved from the ^{1}H NMR spectrum of a relatively small sample (<100 μg). In such work the shifts of the "in-chain' methyl groups are, in general, of little use for indicating the position of a *Z* double bond; a significant change of about 0.1 ppm has been observed only with the hindered 11*Z* and 9*Z*, 11*Z* isomers. The most useful diagnostic feature is the strong downfield shift (approx. 0.5 ppm) of the signals of H(8), H(10) or H(12) for the 9*Z*, 11*Z* or 13*Z* compounds respectively. These shifts appear to be caused, at least partly, by the newly introduced strong steric interaction with the protons H(11), H(20) and H(15) respectively. The increased shielding (small upfield shifts) of other protons in the vicinity presumably arises from a release of some interactions.

From the data obtained for a range of different carotenoids and their *Z* isomers, some empirical correlations have been drawn up (Englert, 1982) which should prove very useful for elucidating the structures of *Z*-isomers of other carotenoids. These correlations are given in Table 2.5 as a list of values for an "isomerization shift", Δ, i.e. the difference

Table 2.4 ^{13}C NMR chemical shift assignments for some common carotenoid end-groups (solvent $CDCl_3$). Data from Englert (1982, 1985)

	Methyl substituents				Skeletal carbons										
Carotenoid end-group	16	17	18	19	1	2	3	4	5	6	7	8	9	10	11
β (all-*E*)	29.0		21.7	12.8	34.2	39.8	19.4	33.2	129.4	137.8	126.7	137.6	136.0	130.8	125.3
β (9*Z*)	29.0		21.9	20.8	34.3	39.8	19.4	33.2	130.0	138.3	128.5	130.1	134.7	129.4	124.0
β (7*Z*)	28.8		21.9	14.5	34.4	39.3	19.3	32.2	129.0	136.6	128.2	137.4	137.4	131.6	125.8
(3*R*)-3-Hydroxy-β (all-*E*)	30.3 28.8	(β, eq) (α, ax)	21.6	12.8	37.2	48.5	65.0	42.5	126.2	137.9	125.7	138.4	135.8	131.3	125.0
(3*R*)-3-Hydroxy-β (9*Z*)	30.4 28.8	(β, eq) (α, ax)	21.8	20.8	37.1	48.3	65.1	42.5	126.5	138.1	127.4	130.8	134.2	129.8	123.6
(3*R*)-3-Hydroxy-7,8-didehydro-β (all-*E*)	30.5 28.8	(β, eq) (α, ax)	22.5	18.1	36.6	46.7	64.9	41.5	137.3	124.3	89.1	98.6	119.1	135.1	124.3
(3*R*)-3-Hydroxy-7,8-didehydro-β (9*Z*)	30.6 28.9	(β, eq) (α, ax)	22.6	23.5	36.6	46.8	64.9	41.6	137.6	124.4	94.4	94.8	120.0	135.4	127.3
4-Hydroxy-β	27.8 29.1		18.7	12.8	34.8	34.8	28.6	70.3	130.0	141.8	125.8	138.8	135.7	131.7	125.4
4-Oxo-β	27.7		13.8	12.6	35.9	37.6	34.3	198.9	130.0	161.0	124.3	141.2	134.8	134.4	124.8
(3*S*)-3-Hydroxy-4-oxo-β	30.7 26.2	(β, eq) (α, ax)	14.0	12.6	36.8	45.5	69.3	200.4	127.0	162.1	123.3	142.4	134.3	135.2	124.4
2,3-Didehydro-3-hydroxy-4-oxo-β	28.1		13.7	12.6	39.2	125.5	144.6	182.4	128.3	161.4	123.0	142.4	134.6	135.4	124.7
5,6-Epoxy-5,6-dihydro-β	25.9		21.1	12.9	33.8	35.9	17.2	30.2	65.3	71.2	124.0	137.2	134.5	132.1	124.8
3-Hydroxy-5,6-epoxy-5,6-dihydro-β	25.1 29.6		20.3	12.9	35.4	47.9	63.0	41.8	67.0	70.3	125.2	137.2	134.5	132.5	125.2
(8*S*)-3-Hydroxy-5,8-epoxy-5,8-dihydro-β	30.6[2] 31.3[2]		28.2[2]	13.4	34.2	47.5	67.9	47.5	87.2	153.3	118.8	88.4	138.7	126.2	124.4

(8*R*)-3-Hydroxy-5,8-epoxy-5,8-dihydro-β	29.0[2] 29.1[2]		31.4[2]	12.6	33.7	47.5[3]	67.7	46.8[3]	86.9	154.1	120.0	87.8		127.2	124.5
3-Hydroxy-5,6-epoxy-5,6-dihydro-8-oxo-β	25.0 28.1		21.4	11.8	35.2	47.2[2]	64.1	41.7[2]	66.3[3]	67.2[3]	40.9	197.8	134.5	130.9	123.3
(3*S*,5*R*,6*R*)-3-Acetoxy-5-hydroxy-6,7-didehydro-5,6-dihydro-β	31.2[2] 32.1[2]		29.2[2]	14.0	35.7	45.5	68.2[1]	45.5	72.2	117.5	202.3	103.3	132.6	128.5	125.7
3-Oxo-*retro*-β(6-*trans*)	30.0		22.2	12.4	38.7	54.4	199.0	126.0	154.6	142.7	128.2	128.3	141.3	138.0	126.9
3-Oxo-*retro*-β(6-*cis*)	28.3		25.4	12.4	41.3	52.5	199.0	128.8	155.5	143.7	125.5	128.8	139.6	137.5	126.4
5,6-Dioxo-5,6-seco-β	24.5		29.7	12.9	46.5	39.4	19.3	44.0	208.4	203.7	119.9	147.8	134.0		
(6*R*)-ϵ	27.1 27.6		23.0	13.1	32.6	31.8	23.1	120.7	134.5	55.0	130.3	136.5	135.5	131.0	125.0
(3*R*,6*R*)-3-Hydroxy-ϵ	24.4 29.5	(β, ax) (α, eq)	22.8	13.1	34.1	44.7	65.9	124.6	137.9	55.2	128.8	137.7	135.1	130.8	124.9
(6*R*)-3-Oxo-ϵ	27.3 27.9		23.6	13.0	36.5	47.6	199.0	125.6	162.3	56.3	125.6	138.5	134.4	132.1	124.6
ψ	17.7 25.6	(*cis*) (*trans*)	17.0	12.8	131.6	124.0	26.8	40.3	139.1	125.8	124.7	135.4	135.8	131.6	
3,4-Didehydro-ψ	18.6 26.2	(*cis*) (*trans*)	13.0	12.9	136.0	131.2	125.5	134.9	137.1	131.2	126.6	137.5			
1,2-Epoxy-1,2-dihydro-ψ	18.8 24.9	(*cis*) (*trans*)	17.2	12.9	58.3	69.3	27.5	36.8	137.9	126.4	124.5	136.0	136.6		
3-Hydroxy-6-oxo-κ	25.2 25.9		21.4	12.8	44.0	51.0	70.4	45.4	59.0	202.9	121.1	146.8	134.1	140.5	124.7
ϕ	17.0 20.4[2]	(C-1Me) (C-2Me)	20.9[2]	12.7			127.5	126.5	126.5	139.0	135.5	132.2	124.8		

[1]Acetate CH_3:21.1, CO:170.5.
[2,3]Assignments may be interchanged.

Table 2.5 Average values of the ^{1}H NMR isomerization shift Δ for a range of carotenoid Z isomers. Δ(ppm) $= \delta Z - \delta E$ where δZ and δE are the chemical shifts recorded for the specified Z isomer and the all-E compound, respectively (solvent $CDCl_3$)

Isomer	Value for protons at positions											
	7	8	19	10	11	12	20	14	15	15′	14′	12′
9Z	0.02	0.53	—	−0.07	0.07	−0.07	—	—	—	—	—	—
11Z	—	—	−0.02	0.50	−0.31	−0.36	0.12	0.03	−0.03	—	—	—
13Z	—	—	—	0.05	—	0.53	—	−0.13	0.16	−0.07	—	—
15Z	—	—	—	—	0.03	0.07	—	0.40	−0.24	−0.24	0.40	0.07
9Z,11Z	0.05	0.55	0.04	0.46	−0.26	−0.42	0.14	0.03	—	—	—	—
9Z,13Z	—	0.52	—	−0.07	0.07	0.46	—	−0.12	0.16	−0.09	—	—
9Z,15Z	—	0.53	—	−0.06	0.08	—	—	0.42	−0.25	−0.25	0.42	0.04
13Z,15Z	—	—	—	0.04	—	0.54	—	0.31	−0.25	−0.25	0.41	0.10
9Z,13′Z	—	0.54	—	−0.08	0.05	−0.07	—	−0.04	−0.08	0.15	−0.14	0.53
13Z,13′Z	—	—	—	0.04	—	0.53	—	−0.15	0.08	0.08	−0.15	0.53

between the chemical shifts (ppm) of the *Z* and *E* isomers. The collection of △-values is a very characteristic indicator of the position of a *Z* double bond in the polyene chain. It can also be concluded that if a molecule contains two *Z* double bonds, the effects are additive if these two bonds are sufficiently distant from each other, e.g. in the 9*Z*, 9′*Z* and 9*Z*, *13*′*Z* isomers, but not if the two bonds are situated close together, e.g. 9*Z*,13*Z*. In a detailed study with synthetic model compounds, Carey *et al.* (1983) have compiled ^{13}C NMR shift data for vinyl methyl and vinyl methylene carbon atoms of polyene *Z*/*E* isomers. The ^{13}C data were considered to be much more useful than the corresponding ^{1}H NMR data in providing a general method for the unambiguous assignment of geometry in polyene isoprenoids, and have been used to elucidate the structures of the *Z* isomers of the proposed intermediates in the biosynthesis of prolycopene (XXXVI), isolated from tangerine tomatoes (Clough and Pattenden, 1983).

6 NMR IN BIOSYNTHESIS STUDIES: ISOTOPIC LABELLING

Although NMR spectroscopy is most widely used for structure determination and characterization of carotenoids, it has also proved useful for establishing the positions of ^{13}C and ^{2}H enrichment in carotenoid molecules biosynthesized from precursors labelled specifically with these stable isotopes (Britton, 1985c). In a straightforward application of such a procedure, lycopene and zeaxanthin have been biosynthesized from [2-^{13}C]-mevalonic acid by a *Flavobacterium* species (Britton *et al.*, 1979). The ^{13}C NMR signals of those carbon atoms that were enriched with ^{13}C were of substantially greater intensity than those which had no enrichment. The positions of ^{13}C labelling in the carotenoid molecule could thus be determined directly. Labelling patterns can usually be determined unambiguously in this way, provided that ^{13}C enrichments of at least 0.5% can be achieved. Much greater levels of enrichment (preferably at least 25%) are needed if deuterium (^{2}H) labelling is used. In this case the ^{1}H NMR spectrum of the biosynthetic product is determined, and the presence of ^{2}H incorporation at any position can be established from the decrease in intensity of that signal. Some examples of experiments which have used stable isotopic labelling and NMR analysis to define stereochemical features of reactions in carotenoid biosynthesis are described in Chapter 3.

C Circular dichroism (CD) and optical rotatory dispersion (ORD)

Moss and Weedon (1976) discussed briefly the ORD properties of carotenoids, particularly in terms of the empirical additivity rule, according to which the ORD curve for any carotenoid may be predicted by adding the ORD curves of its two constituent carotenoid half-molecules (Bartlett *et al.*, 1969). Some examples of the use of ORD and CD correlations to deduce the absolute configurations of some natural carotenoids were also given. Since then, almost all work on the chiroptical properties of carotenoids has used CD rather than ORD, and the CD properties of many natural carotenoids have been published and used in determinations of chirality by correlation with those of carotenoids of known configuration. [Many of the papers referred to in the first section of this chapter (new structures, stereochemistry and distribution) give CD data for the compounds described.]

Some reviews and specialized papers have been published which discuss general features such as the origin of CD in carotenoids, and factors which determine or affect the CD (Noack and Thomson, 1979, 1981; Sturzenegger *et al.*, 1980; Buchecker *et al.*, 1982; Noack, 1982). Some of the main conclusions and empirical rules derived from this work are of fundamental importance and are summarized below.

In optically active carotenoids, the asymmetric centres are located in the end-groups but the CD is observed in the electronic transitions of the main polyene chain. The chiral end-groups, especially rings, impose chirality upon the polyene chromophore. The preferred conformation of the end-group is determined by the absolute configuration of substituent groups, especially OH, together with steric hindrance around the C-6—C-7 single bond. This conformation determines the preferred angle of twist about the C-6—C-7 bond, which in turn determines the chirality of the twist imposed on the polyene. Thus, in the case of the symmetrical β-ring carotenoid zeaxanthin, steric hindrance between the H-atoms on C-7 and C-8 and the methyl groups on C-5 and C-1 prevents the ring double bond from being coplanar with the polyene chain. The presence of a chiral centre in the ring end-group, in this case the OH group at C-3 which preferentially adopts an equatorial position, determines the preferred conformation of the ring and the "handedness" of the strongly predominating twist form around C-6—C-7. As a result of this the whole conjugated system is twisted and becomes an intrinsically chiral chromophore, and a CD spectrum can be observed. In the case of carotenoids without asymmetric centres, such as β-carotene, the two distorted 6-*s-cis* conformations with the ring up or down are of identical energy and are freely interconverted, so no CD spectrum is given.

The additional chiral centres introduced by the presence of 5,6-epoxy-groups, e.g. in antheraxanthin and violaxanthin, do not greatly influence the CD. The CD spectra of these compounds are qualitatively very similar to that of zeaxanthin. The epoxy group is, however, responsible for the chirality and hence for the CD of β,β-carotene-5,6- and -5,8-epoxides.

The major leaf xanthophyll, lutein (II), has three chiral centres, at C-3 of the β-ring and C-3′ and C-6′ of the ϵ-ring. The chirality at C-3 and therefore the conformation of the 3-hydroxy-β-ring is the strongest influence which mainly determines the CD of lutein. For carotenoids which only have ϵ-rings, e.g. ϵ,ϵ-carotene-3,3′-diol (LXXXVIII), the CD properties are determined by the chiral centre at C-6; by comparison the chirality of the C-3 hydroxy groups has little influence.

Besides the obvious importance of the presence of chiral centres in the molecule, other factors profoundly affect carotenoid CD. A particularly important structural influence is the geometrical configuration of the carotenoid. Thus for a homodichiral carotenoid such as zeaxanthin the main symmetry axes of the all-*E* and mono-*Z* isomers are perpendicular and most or all of the CD bands for the all-*E* and mono-*Z* carotenoid are opposite in sign. In contrast, CD spectra of di-*Z* isomers are usually similar to those of the all-*E* compound. In any work which uses CD correlations to determine the stereochemistry of natural carotenoids the geometrical configuration of the polyene chain must therefore be known.

Carotenoid CD spectra can also be markedly temperature-dependent. For (3*S*,3′*S*)-astaxanthin bis-acetate it has been found that the CD spectra determined at room temperature and at −180°C are opposite in sign. The difference is explained on the basis that a different conformational isomer may predominate in the equilibrium mixture at the different temperatures.

Current ideas and recent experimentation on the CD of carotenoids have been discussed extensively by Noack and Thomson (1979; 1981), by Noack (1982) and by Sturzenegger *et al.* (1980), who classified carotenoid CD spectra into three categories—conservative, non-conservative and intermediate—and tabulated published data for about 50 carotenoids according to this scheme.

Finally, mention must be made of the effects of concentration and aggregation and the influence of interactions with other molecules, e.g. lipid and protein, on carotenoid CD. Noack (1982) described how, at low temperature, the CD of (9*Z*)- and (9*Z*, 9′*Z*)-astaxanthin bis-acetates depended strongly on concentration. At higher concentrations (70-90 μ*M*) surprisingly strong CD was observed under the longest wavelength absorption band, with splitting and change of sign across this band. This behaviour arises from exciton interactions between monomers and is indicative of weak molecular aggregation at the low temperature employed (−180°C). A similar effect was seen (Lemâtre *et al.*, 1980) with aqueous ethanolic solutions of lutein which exhibited strong CD in the main absorption band region, depending on the water content and affected by detergent and temperature. Many interesting effects have been noted when the CD of carotenoids in micelles or in the presence of protein etc. was measured (Takagi *et al.*, 1982a,b,c). Intrinsically achiral carotenoids may exhibit CD when examined *in vivo*, e.g. spheroidene in isolated pigment–protein complexes from chromatophores of photosynthetic bacteria shows strong CD-induced by the specific interactions with the protein etc. (J. Manwaring, E.H. Evans and G. Britton, unpublished results).

D Mass spectrometry (MS)

Moss and Weedon (1976), in their discussion of carotenoid mass spectrometry, described the most useful diagnostic fragmentations and also tabulated MS data for a considerable range of carotenoid structures. Since then, much further work has used isotopic labelling to define the mechanisms of some of these fragmentations, and the results have been published in a series of articles and reviews (Budzikiewicz, 1974, 1982; Johannes *et al.*, 1974, 1979).

Further extensive tables of data on diagnostic fragmentations of carotenoid end-groups have been presented (Enzell and Wahlberg, 1980; Enzell *et al.*, 1984). Mass spectrometry is now used routinely for the identification and characterization of carotenoids and details of the MS fragmentations of many individual carotenoids have been given. Exhaustive tabulation of these data will not be attempted in this article but some new developments and procedures will be discussed.

Almost all work on carotenoids has used direct probe electron impact (EI) MS. An examination of 18 carotenoids by field desorption MS has been reported, however (Watts *et al.*, 1975). The molecular ions were obtained as base peaks, with very few fragment ions being detected. A smaller survey of eight carotenoids by chemical ionization MS also revealed simple fragmentation patterns (Carnevale *et al.*, 1978).

An EI method which involves linked scanning of the magnetic field and the electric sector voltage of a conventional double-focusing mass spectrometer provides a promising powerful technique for distinguishing between isomeric carotenoids, even in the presence of gross contamination, by observing metastable ions without interference from normal ions

(Rose, 1982). This technique has been applied to distinguish readily between the isomeric α-carotene (V), β-carotene (I), γ-carotene (LXXXV), δ-carotene (LXXXVI), ϵ-carotene (XCII) and lycopene (LXXXI), and between lutein and zeaxanthin, and has been used to demonstrate how useful MS data can be obtained from a carotenoid (lycopene) sample which had been grossly contaminated with steroid impurity.

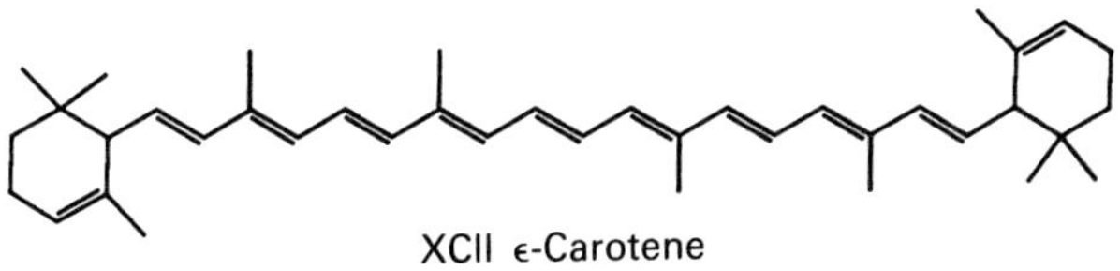

XCII ϵ-Carotene

Another area where MS has proved very valuable is in the investigation of the isotopic labelling of carotenoids biosynthesized in the presence of heavy water, ^2H_2O. The extent of isotopic enrichment in the carotenoid can readily be determined, and information about the pattern of labelling in the carotenoid molecule can sometimes be deduced from the MS fragmentation pattern. When carotenoids are biosynthesized in tissues or microorganisms maintained in water highly enriched in ^2H_2O (e.g. 50–100% enrichment) MS patterns such as those illustrated in Fig. 2.4 can be observed (Britton *et al.*, 1977a,b). From these it may be possible to deduce a great deal of information about pathways, mechanisms, time course and regulation of biosynthesis, and about the general metabolic activity of the tissue or organism under study. It is especially useful if the MS analysis is combined with a study of ^1H NMR spectra.

E Infrared (IR) and Raman spectroscopy

1 INTRODUCTION

Even in 1976 Moss and Weedon had concluded that "with the advances in other spectroscopic techniques infrared spectroscopy has become less widely used for structural studies in the carotenoid field". Although the amount of sample required has been greatly reduced by the advent of Fourier-transform (FT-IR) instrumentation, this conclusion remains valid. Major advances have been made, however, in the related technique of Raman spectroscopy, especially resonance Raman (rR) spectroscopy. The application of this method in the carotenoid field has increased enormously and holds the promise of further exciting developments in the future. In

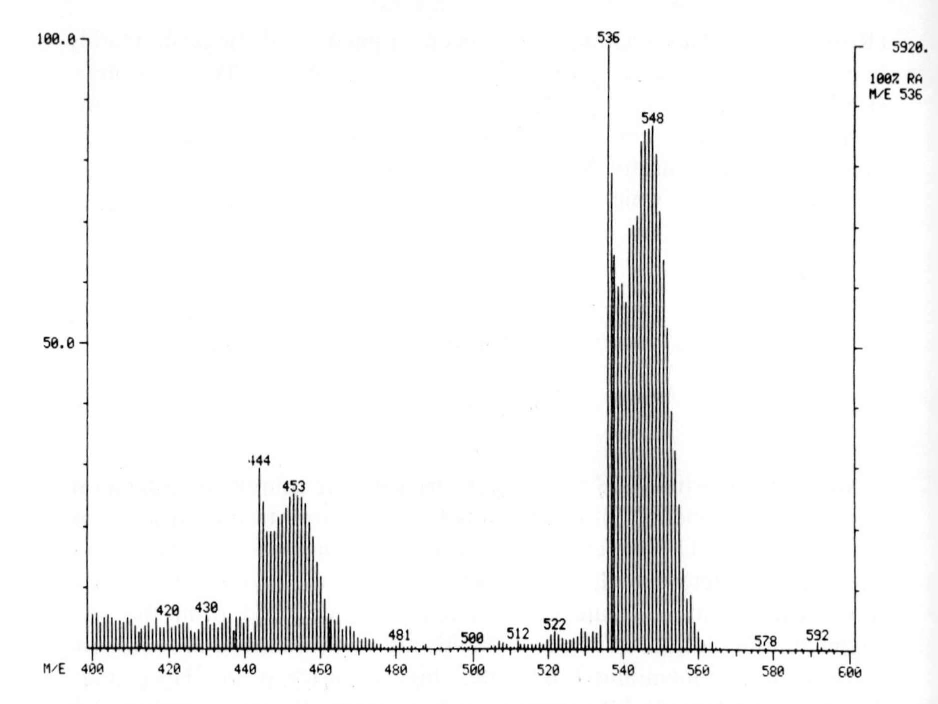

Fig. 2.4 Electron impact (EI) mass spectrum of deuterium-labelled β-carotene isolated from barley cotyledons greened for 48 hours in the presence of 50% D_2O. A VG Micromass 7070E Mass Spectrometer coupled with a Finnigan Incos Data System was employed (positive ion mode; ion source temperature of 200–220°C; emission current 200 μA; ionization potential 70eV; accelerating potential current 4 kV).

recent years several reviews and books have dealt exclusively or in part with rR spectroscopy of carotenoids, e.g. Carey (1982), Merlin (1985), Nelson (1981a,b), Warshel (1977).

2 THE RAMAN METHOD

When photons of light interact with matter, a small number of the photons will experience inelastic collisions with molecules and their energies will be symmetrically shifted to higher and lower frequences. The shifts are independent of the excitation radiation and correspond to particular molecular vibrational frequencies. Raman spectroscopy makes use of this effect. A high proton density is produced by focusing monochromatic light from a laser on to or into a sample. Light scattered by molecules in the sample is collected and focused in order to analyse the vibrational frequencies and their intensities. The information obtained is essentially the same as that provided by IR spectroscopy, though the physical processes involved are quite different. The technique can be used to examine the nature of chemical bonds, molecular structures and the interactions between molecules and their environment.

In the carotenoid field, the greatest use is made of the rR technique in which the excitation radiation used lies in an electronic absorption band of the chromophore. The strong enhancement (10^3–10^6-fold) obtained in this way allows specific vibrational modes of the chromophore to be analysed, even if the molecule is present at low concentration in a complex biological system. It is therefore possible to detect and analyse very small concentrations of carotenoid in the presence of large amounts of other molecular species which do not absorb the incident light wavelength. In particular, the analysis of carotenoids *in situ* in biological tissues can be achieved.

Resonance Raman spectroscopy offers several particular advantages over the IR technique for biochemical studies. Thus rR spectra can be obtained for molecules in an aqueous medium, since water gives very weak Raman lines. Because the incident laser beam can be focused on a very small area (e.g. 1-2 μm^2), samples of very small size can be analysed and spatial resolution at the microscopic level can be achieved. Raman spectra can be obtained on a nanosecond or picosecond time scale, allowing transient species, excited electronic states and photochemical and photobiological processes to be investigated.

3 RESONANCE RAMAN SPECTRA OF CAROTENOIDS

When determining rR spectra of carotenoids it is normal to use incident laser light of a wavelength corresponding to the main $\pi \rightarrow \pi^*$ transition of the polyene chain. This gives, in the 900–1600 cm^{-1} region, intense spectral features which arise from the vibrations of the conjugated polyene chain, but little or no information is obtained about the carotenoid end-groups. Three regions of the spectra are particularly characteristic. One very strong rR band which appears near 1520 cm^{-1} (ν_1) is assigned to the C=C stretching vibration and can be used to monitor the extent of the conjugated π-electron system and the delocalization of the π-electrons in the ground state. A second intense band appears near 1157 cm^{-1} (ν_2) and is usually attributed to the C—C stretching mode, but with some perturbation by C—H in-plane bending modes. For a wide range of all-*E* carotenoids there is an approximately linear relationship between ν_1 and ν_2, though this does not hold for several *Z* isomers.

A third, ν_3, line, assigned to the CH_3 in-plane rocking mode, is also enhanced by resonance effects. This, and other weak lines in the 1100–1400 cm^{-1} "fingerprint region" that are sensitive to the nature of the end-groups as well as to the chain conformation, contain much useful information about molecular structure, but they have not been subjected to extensive systematic study, so assignment and interpretation are difficult.

4 GEOMETRICAL ISOMERS

The rR spectra of different geometrical isomers of a carotenoid vary appreciably. The ν_1 (C=C) band correlates well with the shift in λ_{max}, and isomers with the *Z* double bond towards the centre of the chromophore generally show the largest Raman shifts. The ν_2 region is particularly sensitive to the configuration of the chain. A unique spectral pattern has been observed in the 1100–1300 cm^{-1} region for each of four mono-*Z* and four di-*Z* isomers of β-carotene (Koyama *et al.*, 1983).

5 TIME-RESOLVED rR SPECTRA: IDENTIFICATION OF TRANSIENT SPECIES

Because the time scale of the Raman effect is essentially instantaneous, spectra can be obtained of transient species with lifetimes in the nanosecond and even picosecond range. The transient species is generated by a laser pulse and a second laser is used to obtain the Raman spectrum. By this means the rR spectra of excited triplet states of β-carotene and other carotenoids have been studied in detail. These differ significantly from those of the ground state molecules, and the triplet state rR spectra of (all-*E*)- and (15*Z*)-β-carotene are similar. A decrease of the double bond order and twisting around the inner double bonds are considered to be important (Wilbrandt and Jensen, 1981).

6 SPECTRA OF MOLECULES *IN SITU*

The strong resonance enhancement of the Raman effect when the exciting light used falls within the main absorption band of the chromophore makes it possible to observe selectively vibrational modes of the chromophore essentially without interference from the non-resonant scattering from a background of complex biological material which does not absorb that wavelength. This very important feature means that it is frequently possible to obtain rR spectra from a chromophore *in situ*. This advantage has been used extensively to study the protein-complexed carotenoids in animal tissues, especially lobster shell, but has also been applied to plant systems. Thus rR spectra of carotenoids have been obtained from live carrot root and tomato fruit (Gill *et al.*, 1970) bacterial and algal cultures (Welb, 1980), and the technique has been used for quantitative analysis of carotenoids in, for example, marine phytoplankton (Hoskins and Alexander, 1977) and tobacco leaves (Forrest and Vilcins, 1979). An interesting development comes from the laser microanalysis technique of Delhaye and Dhamelincourt (1975), by which good spectra can be obtained from a very small area of the sample (2 μm^2) through a microscope objective. Thus the rR spectrum of peridinin (LI) has been recorded by illuminating the cytoplasm of a single cell of *Pyrocystis lunula*, a unicellular alga which is known to contain a peridinin–chlorophyll–protein complex (Dupaix *et al.*, 1982).

One particular problem which has been the subject of many rR investigations is the geometrical configuration of the carotenoid molecules in the photosynthetic reaction centre, especially that of photosynthetic bacteria. The rR spectra obtained have characteristic features in common with that of (15*Z*)-β-carotene, and indicate that a bent molecular shape is likely for the carotenoid in the reaction centre. Different workers have concluded that the reaction centre carotenoid must have a 15*Z* (Koyama *et al.*, 1982) or a di-*Z*-configuration (Lutz *et al.*, 1978; Agalidis *et al.*, 1980) and that this changes during the functioning of the reaction centre in photosynthesis. However, a bent conformation due to an *s-cis* conformation about one or more single bonds would also account for the observed rR properties, and seems a more likely situation (Iwata *et al.*, 1985).

REFERENCES

Agalidis, I., Lutz, M. and Reiss-Husson, F. (1980). *Biochim. Biophys. Acta* **589**, 264.

Aitzetmüller, K., Strain, H.H., Svec, W.A., Grandolfo, M. and Katz, J.J. (1969). *Phytochemistry* **8**, 1761.

Andrewes, A.G. and Starr, M.P. (1976). *Phytochemistry* **15**, 1009.

Arpin, N., Svec, W.A. and Liaaen-Jensen, S. (1976). *Phytochemistry* **15**, 529.

Baranyai, M., Molnar, P., Szabolcs, J., Radics, L. and Kajtar-Peredy, M. (1981). *Tetrahedron* **37**, 203.

Bartlett, L., Klyne, W., Mose, W.P. *et al.* (1969). *J. Chem. Soc. C*, 2527.

Ben-Amotz, A. and Avron, M. (1983). *Ann. Rev. Microbiol.* **37**, 95.

Ben-Amotz, A., Katz, A. and Avron, M. (1982). *J. Phycol.* **18**, 529.

Ben-Aziz, A., Britton, G. and Goodwin, T.W. (1973). *Phytochemistry* **12**, 2759.

Berset, D. and Pfander, H. (1984). *Helv. Chim. Acta* **67**, 964.

Berset, D. and Pfander, H. (1985). *Helv. Chim. Acta* **68**, 1149.

Bjørnland, T., Borch, G. and Liaaen-Jensen, S. (1984a). *Phytochemistry* **23**, 1711.
Bjørnland, T., Pennington, F., Haxo, F.T. and Liaaen-Jensen, S. (1984b). *Abstr. 7th IUPAC Carotenoid Symp.*, Munich.
Bodea, C., Andrewes, A.G., Borch, G. and Liaaen-Jensen, S. (1978). *Phytochemistry* **17**, 2037.
Bowden, R.D., Cooper, R.D.G., Harris, C.J., Moss, G.P., Weedon, B.C.L. and Jackman, L.M. (1983). *J. Chem. Soc. Perkin Trans. I*, 1465.
Braumann, T.H. and Grimme, L.H. (1981). *Biochim. Biophys. Acta* **637**, 8.
Britton, G. (1976). *Specialist Periodical Reports: Terpenoids and Steroids* **6**, 144. The Chemical Society, London.
Britton, G. (1977). *Specialist Periodical Reports: Terpenoids and Steroids* **7**, 155. The Chemical Society, London.
Britton, G. (1978). *Specialist Periodical Reports: Terpenoids and Steroids* **8**, 181. The Chemical Society, London.
Britton, G. (1979). *Specialist Periodical Reports: Terpenoids and Steroids* **9**, 218. The Chemical Society, London.
Britton, G. (1981). *Specialist Periodical Reports: Terpenoids and Steroids* **10**, 164. Royal Society of Chemistry, London.
Britton, G. (1982). *Specialist Periodical Reports: Terpenoids and Steroids* **11**, 133. Royal Society of Chemistry, London.
Britton, G. (1983). *Specialist Periodical Reports: Terpenoids and Steroids* **12**, 235. Royal Society of Chemistry, London.
Britton, G. (1984). *Nat. Prod. Reports* **1**, 67.
Britton, G. (1985a). *Methods Enzymol.* 111, 113.
Britton, G. (1985b). *Nat. Prod. Reports* **2**, 349.
Britton, G. (1985c). *Pure Appl. Chem.* **57**, 701.
Britton, G.(1986). *Nat. Prod. Reports* **3**, 591.
Britton, G. and Goodwin, T.W. (1969). *Phytochemistry* **8**, 2257.
Britton, G. and Goodwin, T.W. (1975). *Phytochemistry* **14**, 2530.
Britton, G., Lockley, W.J.S., Patel, N.J. and Goodwin, T.W. (1977a). *FEBS Letts* **79**, 281.
Britton, G., Lockley, W.J.S., Powls, R., Goodwin, T.W. and Heyes, L.M. (1977b). *Nature* **268**, 81.
Britton, G., Goodwin, T.W., Lockley, W.J.S., Mundy, A.P., Patel, N.J. and Englert, G. (1979). *J. Chem. Soc. Chem. Commun.* 27.
Buchecker, R. and Eugster, C.H. (1979). *Helv. Chim. Acta* **62**, 2817.
Buchecker, R. and Eugster, C.H. (1980) *Helv. Chim. Acta* **63**, 2531.
Buchecker, R., Eugster, C.H., Kjøsen, H. and Liaaen-Jensen, S. (1974). *Acta Chem. Scand.* **B28**, 449.
Buchecker, R., Liaaen-Jensen, S., Borch, G. and Siegelman, H.W. (1976). *Phytochemistry* **15**, 1015.
Buchecker, R., Marti, U. and Eugster, C.H. (1982). *Helv. Chim. Acta* **65**, 896.
Budzikiewicz, H. (1974). *Adv. Mass Spectrom.* **6**, 163.
Budzikiewicz, H. (1982). In *Carotenoid Chemistry and Biochemistry* (Eds. Britton, G. and Goodwin, T.W.) p. 155, Pergamon, Oxford.
Burczyk, J. (1987). *Phytochemistry* **26**, 113, 121.
Bushway, R.J. (1985). *J. Liquid Chromatogr.* **8**, 1527.
Cadosch, H. and Eugster, C.H. (1974). *Helv. Chim. Acta* **57**, 1466.
Cardini, F., Ginanneschi, M., Selva, A. and Chelli, M. (1987). *Phytochemistry* **26**, 2029.

Carey, P.R. (1982). *Biochemical Applications of Raman and Resonance Raman Spectroscopies*. Academic Press, New York.
Carey, L., Clough, J.M. and Pattenden, G. (1983). *J. Chem. Soc. Perkin Trans.* 1, 3005.
Carnevale, J., Cole, E.R., Nelson, D. and Shannon, J.S. (1978). *Biomed. Mass Spectrom.* **5**, 641.
Clough, J.M. and Pattenden, G. (1979). *Chem. Commun.* 616.
Clough, J.M. and Pattenden, G. (1983). *J. Chem. Soc. Perkin Trans.* 1, 3011.
Cooper, R.D.G., Jackman, L.M. and Weedon, B.C.L. (1962). *Proc. Chem. Soc.* 215.
Curl, A.C. and Bailey, G.F. (1954). *J. Agric. Food Chem.* **2**, 685.
Czeczuga, B. (1986) *Biochem. Syst. Ecol.* **14**, 13.
Dabbagh, A.-G. and Egger, K. (1974). *Z. Pflanzenphysiol.* **72**, 177.
Davies, B.H. (1976). In *Chemistry and Biochemistry of Plant Pigments*, 2nd Edn. (Ed. Goodwin, T.W.) Vol. 2, p. 38, Academic Press, London.
De Almeida, L.B., Penteado, M.De V.C., Simpson, K.L., Britton, G., Acemoglu, M. and Eugster, C.H. (1986). *Helv. Chim. Acta* **69**, 1554.
Delhaye, M. and Dhamelincourt, P. (1975). *J. Raman Spectroscopy* **3**, 33.
De Ritter, E. and Purcell, A.E. (1981). In *Carotenoids as Colorants and Vitamin A Precursors* (Ed. Bauernfeind, J.C.) p. 815, Academic Press, New York.
Diallo, B. and VanHaelen, M. (1987) *Phytochemistry* **26**, 1491.
Dupaix, A., Arrio, B., Lecuyer, B. *et al.* (1982). *Biol. Cell.* **43**, 157.
Englert, G. (1982). In *Carotenoid Chemistry and Biochemistry* (Eds. Britton, G. and Goodwin, T.W.) p. 107, Pergamon, Oxford.
Englert, G. (1985). *Pure Appl. Chem.* **57**, 701.
Englert, G., Brown, B.O., Moss, G.P. *et al.* (1979). *Chem. Commun.* 545.
Enzell, C.R. and Wahlberg, I. (1980). *Biochem. Appl. Mass Spectrom.* (suppl. 1) **407**.
Enzell, C.R., Wahlberg, I. and Ryhage, R. (1984) *Mass Spectrom. Rev.* **3**, 395.
Eugster, C.H. (1982) In *Carotenoid Chemistry and Biochemistry* (Eds. Britton, G. and Goodwin, T.W.), p. 1, Pergamon, Oxford.
Eugster, C.H. (1985). *Pure Appl. Chem.* **57**, 639.
Fiksdahl, A., Mortensen, J.T. and Liaaen-Jensen, S. (1978). *J. Chromatogr.* **157**, 111.
Fiksdahl, A., Bjørnland, T. and Liaaen-Jensen, S. (1984a). *Phytochemistry* **23**, 649.
Fiksdahl, A., Withers, N., Guillard, R.R.L. and Liaaen-Jensen, S. (1984b). *Comp. Biochem. Physiol* **78B**, 265.
Forrest, G. and Vilcins, G. (1979). *J. Agric. Food Chem.* **27**, 609.
Francis, G.W., Knutsen, E. and Lien, T. (1973). *Acta Chem. Scand.* **22**, 1054.
Gill, D., Kilponen, R.G. and Rimai, L. (1970). *Nature* **227**, 743.
Goodwin, T.W. (1976). *Chemistry and Biochemistry of Plant Pigments*, 2nd edn. 2 vols. Academic Press, London.
Goodwin, T.W. (1980). *Biochemistry of the Carotenoids*, 2nd edn. Vol. 1, Chapman and Hall, London.
Goodwin, T.W. (1984). *Biochemistry of the Carotenoids*, 2nd edn. Vol. 2, Chapman and Hall, London.
Gross, J. and Eckhardt, G. (1981). *Phytochemistry* **20**, 2267.
Hallenstvet, M., Buchecker, R., Borch, G. and Liaaen-Jensen, S. (1977). *Phytochemistry* **16**, 583.
Hansmann, P. and Kleinig, H. (1982). *Phytochemistry* **21**, 238.

Hoskins, L.C. and Alexander, V. (1977). **49**, 695.
Ida, K. (1981). *Bot. Mag. Tokyo* **94**, 181.
Ikuno, Y., Maoka, T., Shimizu, M., Komori, T. and Matsuno, T. (1985). *J. Chromatogr.* **328**, 387.
Iwata, K., Hayashi, H. and Tasumi, M. (1985). *Biochim. Biophys. Acta* **810**, 269.
Johannes, B., Brzezinka, H. and Budzikiewicz, H. (1974). *Org. Mass Spectrom.* **9**, 1095.
Johannes, B., Brzezinka, H. and Budzikiewicz, H. (1979) *Z. Naturforsch.* **34B**, 300.
Johansen, J.E., Svec, W.A., Liaaen-Jensen, S. and Haxo, F.T. (1974). *Phytochemistry* **13**, 2261.
Kamber, M., Pfander, H. and Noack, K. (1984). *Helv. Chim. Acta* **67**, 968.
Kjøsen, H., Norgard, S., Liaaen-Jensen, S. *et al.* (1976). *Acta Chem. Scand.* **30B**, 107.
Kleinig, H. (1969). *J. Phycol.* **5**, 281.
Koyama, Y., Kito, M., Takii, T., Saiki, K., Tsukida, K. and Yamashida, J. (1982). *Biochim. Biophys. Acta* **680**, 109.
Koyama, Y., Takii, T., Saiki, K. and Tsukida, S. (1983). *Photobiochem. Photobiophys.* **5**, 209.
Krinsky, N.I. and Welankiwar, S. (1984). *Methods Enzymol.* **105**, 155.
Lambert, W.E., Nelis, H.J., De Ruyter, M.G.M. and De Leenheer, A.P. (1985). *Chromatogr. Sci.* **30**, 1.
Lemâtre, J., Maudinas, B. and Ernst, C. (1980). *Photochem. Photobiol.* **31**, 201.
Liaaen-Jensen, S. (1977). In *Marine Natural Products Chemistry* (Eds. D.J. Faulkner and W.H. Fenical), p. 239, Plenum, New York.
Liaaen-Jensen, S. (1978). In *Marine Natural Products: New Perspectives*, (Ed. P. Scheuer) vol 2, p. 1 Academic Press, New York.
Liaaen-Jensen, S. (1985). *Pure Appl. Chem.* **57**, 649.
Liaaen-Jensen, S. and Andrewes, A.G. (1985). *Methods Microbiol.* **18**, 235.
Lichtenthaler, H.K., Prenzel, U. and Kuhn, G. (1982). *Z. Naturforsch.* **37C**, 10.
Lugtenburg, J. (1985). *Pure Appl. Chem.* **57**, 753.
Lutz, M., Agalidis, I., Hervo, G., Cogdell, R.J. and Reiss-Husson, F. (1978). *Biochim. Biophys. Acta* **503**, 287.
Märki-Fischer, E. and Eugster, C.H. (1985a). *Helv. Chim. Acta* **68**, 1704.
Märki-Fischer, E. and Eugster, C.H. (1985b). *Helv. Chim. Acta* **68**, 1708.
Märki-Fischer, E., Buchecker, R., Eugster, C.H., Englert, G., Noack, K. and Vecchi, M. (1982). *Helv. Chim. Acta* **65**, 2198.
Märki-Fischer, E., Marti, U., Buchecker, R. and Eugster, C.H. (1983). *Helv. Chim. Acta* **66**, 494.
Märki-Fischer, E., Buchecker, R. and Eugster, C.H. (1984a). *Helv. Chim. Acta* **67**, 461.
Märki-Fischer, E., Buchecker, R. and Eugster, C.H. (1984b). *Helv. Chem. Acta* **67**, 2143.
Matsuno, T., Tani, Y., Maoka, T., Matsuo, K. and Komori, T. (1986). *Phytochemistry* **25**, 2837.
Merlin, J.C. (1985). *Pure Appl. Chem.* **57**, 785.
Molnar, P. and Szabolcs, J. (1980). *Phytochemistry* **19**, 623.
Molnar, P., Szabolcs, J. and Radics, L. (1986). *Phytochemistry* **25**, 195.
Molnar, P., Szabolcs, J. and Radics, L. (1987). *Phytochemistry* **26**, 1493.

Moss, G.P. and Weedon, B.C.L. (1976). In *Chemistry and Biochemistry of Plant Pigments*, 2nd edn. (Ed. Goodwin T.W.) vol. 1, p. 149, Academic Press, London.
Nelis, H.J.C.F. and de Leenheer, A.P. (1983). *Anal. Chem.* **55**, 270.
Nelson, W.H. (1981a). *Am. Lab.* **13**, 94.
Nelson, W.H. (1981b). *Int. Lab.* **11**, 12.
Nitsche, H. (1973). *Z. Naturforsch.* **28C**, 481.
Nitsche, H. (1974a), *Z. Naturforsch.* **29C**, 659.
Nitsche, H. (1974b). *Arch. Microbiol.* **95**, 79.
Noack, K. (1982). In *Carotenoid Chemistry and Biochemistry* (Eds. Britton, G. and Goodwin, T.W.), p. 135, Pergamon. Oxford.
Noack, K. and Thomson, A.J. (1979). *Helv. Chim. Acta.* **62**, 1902.
Norgard, S., Liaaen-Jensen, S., Haxo, F.T., Wegfahrt, P. and Rapoport, H. (1971). *J. Amer. Chem. Soc.* **93**, 1823.
Parris, N.A. (1978). *J. Chromatogr.* **157**, 161.
Pennington, F.C., Haxo, F.T., Borch, G. and Liaaen-Jensen, S. (1985). *Biochem. Syst. Ecol.* **13**, 215.
Ragan, M.A. and Chapman, D.J. (1978). *A Biochemical Phylogeny of the Protists.* Academic Press, New York.
Ricketts, T.R. (1971). *Phytochemistry* **10**, 155, 161.
Rønneberg H., Foss, P., Ramdahl, T., Borch, G., Skulberg, O. and Liaaen-Jensen, S. (1980). *Phytochemistry* **19**, 2167.
Rønneberg, H., Borch, G., Buchecker, R., Arpin, N. and Liaaen-Jensen, S. (1982). *Phytochemistry* **21**, 2087.
Rønneberg, H., Andrewes, A., Borch, G., Berger, R. and Liaaen-Jensen, S. (1985). *Phytochemistry* **24**, 309.
Rose, M.E. (1982). In *Carotenoid Chemistry and Biochemistry* (Eds. Britton, G. and Goodwin, T.W.) p. 167, Pergamon, Oxford.
Ruddat, M. and Will, O.H. (1985). *Methods Enzymol.* **111**, 189.
Rüedi, P. (1985). *Pure Appl. Chem.* **57**, 793.
Rüttimann, A., Schiedt, K. and Vecchi, M. (1983). *J. High Resol. Chromatogr., Chromatogr. Commun.* **6**, 612.
Rychener, M., Bigler, P. and Pfander, H. (1984). *Helv. Chim. Acta* **67**, 386.
Siefermann-Harms, D. (1985). *Biochim. Biophys. Acta* **811**, 325.
Siefermann-Harms, D., Hertzberg, S., Borch, G. and Liaaen-Jensen, S. (1981). *Phytochemistry* **20**, 85.
Stewart, I. and Wheaton, T.A. (1971). *J. Chromatogr.* **55**, 325.
Strain, H.H. (1951). In *Manual of Phycology* (Ed. Smith, G.M.), Chronica Botanica, Waltham, Mass.
Strain, H.H., Svec, W.A., Aitzetmüller, K. *et al.* (1971). *J. Amer. Chem. Soc.* **93**, 1823.
Stransky, H. and Hager, A. (1970). *Arch. Mikrobiol.* **72**, 84.
Sturzenegger, V., Buchecker, R. and Wagniere, G. (1980). *Helv. Chim. Acta* **63**, 1074.
Swift, I.E. and Milborrow, B.V. (1981). *Biochem. J.* **199**, 69.
Takagi, S., Takeda, K., Kameyama, K. and Takagi, T. (1982a) *Agric. Biol. Chem.* **46**, 2035.
Takagi, S., Takeda, K. and Shiroishi, M. (1982b). *Agric. Biol. Chem.* **46**, 2217.
Takagi, S., Takeda, K. and Takagi, T. (1982c). *Agric. Biol. Chem.* **46**, 399.
Tang, X.-S. and Satoh, K. (1985). *FEBS Letts* **179**, 64.

Taylor, R.F. (1983). *Adv. Chromatogr.* **22**, 157.
Taylor, R.F. and Ikawa, M. (1980). *Methods Enzymol.* **67**, 233.
Thornber, J.P. (1986). *Enc. Pl. Physiol.* (new series) **19**, 98.
Thornber, J.P., Markwell, J.P. and Reinman, S. (1979). *Photochem. Photobiol.* **29**, 1205.
Toth, G. and Szabolcs, J. (1981). *Phytochemistry* **20**, 2411.
Toth, G. Kajtar, J. Molnar, P. and Szabolcs, J. (1978). *Acta Chim. Acad. Sci. Hung.* **97**, 359.
Tsukida, K., Saiki, K., Takii, T., and Koyama, Y. (1982). *J. Chromatogr.* **245**, 359.
Vecchi, M. (1978). *Kontron Symposium on HPLC, Zürich*, cited by Rüedi (1985).
Vecchi, M. and Müller, R.K. (1979). *J. High Resol. Chromatogr. Chromatogr. Commun.* **2**, 195.
Vecchi, M., Englert, G., Maurer, R. and Meduna, V. (1981). *Helv. Chim. Acta* **64**, 2747.
Vecchi, M., Englert, G. and Mayer, H. (1982). *Helv. Chim. Acta* **65**, 1050.
Walton, T.J., Britton, G. and Goodwin, T.W. (1970). *Phytochemistry* **9**, 2545.
Warshel, A. (1977). *Ann. Rev. Biophys. Bioeng.* **6**, 273.
Watts, C.D., Maxwell, J.R., Games, D.E. and Rossiter, M. (1975). *Org. Mass Spectrom.* **10**, 1102.
Welb, S.J. (1980). *Phys. Rep.* **60**, 201.
Wilbrandt, R. and Jensen, N.-H. (1981). *Ber. Bunsenges. Phys.Chem.* **85**, 508.
Withers, N.W., Fiksdahl, A., Tuttle, R.C. and Liaaen-Jensen, S. (1981). *Comp. Biochem. Physiol.* **68B**, 345.
Wright, S.W. and Shearer, J.D. (1984). *J. Chromatogr.* **294**, 281.
Zechmeister, L. (1962). *Cis-trans Isomeric Carotenoids, Vitamins A and Arylpolyenes.* Springer-Verlag, Vienna.
Zechmeister, L. and Pinckard, J.H. (1960). *Fort. Chem. Org. Naturstoffe* **18**, 284.
Zonta, F., Stancher, B. and Marletta, G.P. (1987). *J. Chromatogr.* **403**, 207.

3
Biosynthesis of Carotenoids

GEORGE BRITTON

Department of Biochemistry, University of Liverpool, P.O. Box 147, Liverpool L69 3BX, U.K.

1 INTRODUCTION

The carotenoids of higher plants, algae and fungi, and those of most bacteria, are C_{40} tetraterpenes biosynthesized by the well known isoprenoid pathway, as summarized in Fig. 3.1. The early stages in the pathway, i.e. those by which the C_5 isoprenoid units are constructed and then used to build the required chain length of prenyl diphosphate intermediates, are

PLANT PIGMENTS
ISBN 0-12-289847-8

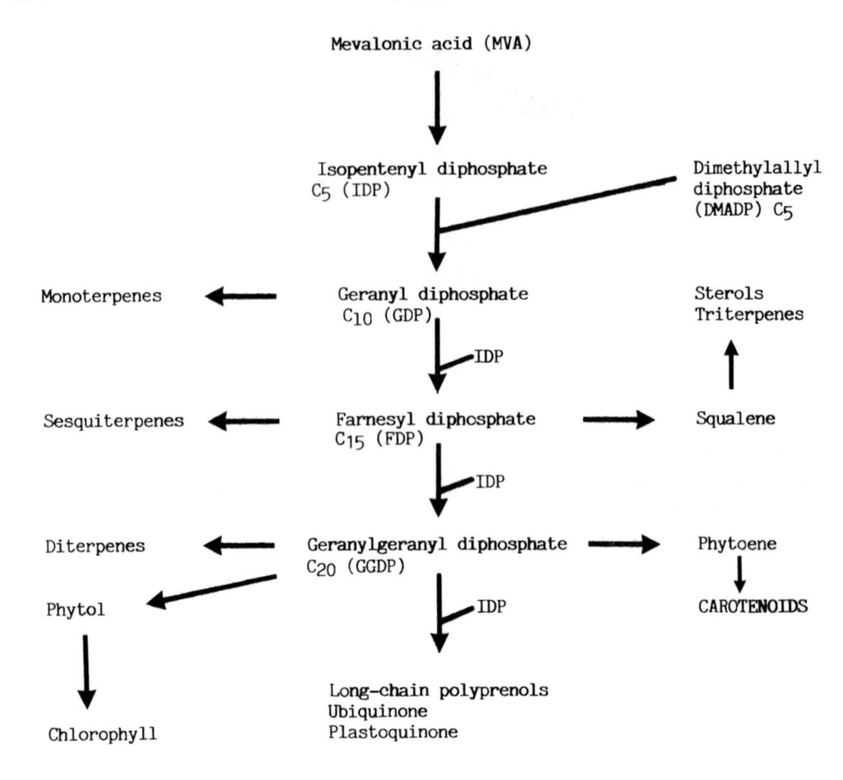

Fig. 3.1 Summary of the biosynthesis of isoprenoid compounds.

common to the biosynthesis of all classes of terpenoids. Only the later stages, after geranylgeranyl diphosphate, are unique to the formation of carotenoids. It is these processes, namely formation of phytoene (7,8,11,12,7′,8′,11′,12′-octahydro-ψ,ψ-carotene), desaturation, cyclization and later modifications, as outlined in Fig. 3.2, that will be the subject of this article. The pathways and reaction sequences are generally not in doubt, and information about them, the mechanisms of the reactions and the enzymes responsible was presented extensively in the previous edition of this work. This article will now survey progress on these topics, and on the regulation of carotenoid biosynthesis, since the contribution to the previous edition was written in 1974. The survey cannot be fully comprehensive, but must concentrate on those areas where major advances have been made or which are considered potentially most important for future developments. Further details are available in the numerous other review articles on carotenoid biosynthesis published in recent years (Britton, 1976, 1979a, 1982, 1983, 1985, 1986; Davies and Taylor, 1976;

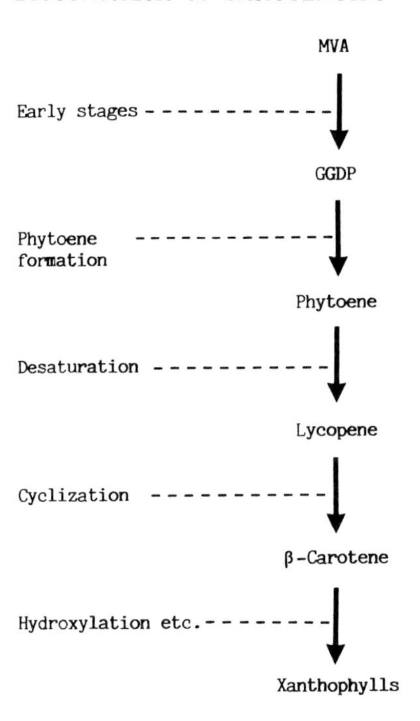

Fig. 3.2 Summary of the stages of carotenoid biosynthesis.

Porter and Spurgeon, 1979, 1983; Davies, 1980; Goodwin, 1980; Spurgeon and Porter 1983; Bramley, 1985a).

2 REACTIONS AND PATHWAYS

A Formation of Phytoene

In the early stages of carotenoid biosynthesis, the characteristic isoprenoid precursor mevalonic acid (MVA, I) is converted into the C_5 compound isopentenyl diphosphate (IDP), some of which undergoes isomerization

MVA (I)

to dimethylallyl diphosphate (DMADP). The isoprenoid chain is then built from these precursors by means of prenyl transferase enzymes to give, successively, the C_{10} geranyl diphosphate (GDP), the C_{15} farnesyl diphosphate (FDP) and the C_{20} geranylgeranyl diphosphate (GGDP, II). These reactions will not be discussed further. The key early reaction in carotenoid biosynthesis is that in which the first C_{40} hydrocarbon phytoene (IV) is formed from two molecules of GGDP. This reaction proceeds via a C_{40} intermediate, prephytoene diphosphate (PPDP, III). Phytoene is formed direct from this intermediate with the stereochemistry of hydrogen loss illustrated in Fig. 3.3, and not via the squalene analogue lycopersene (Gregonis and Rilling, 1973, 1974). The chirality (1*R*, 2*R*, 3*R*) previously assumed for PPDP by analogy with presqualene diphosphate has now been proved (Altman *et al.*, 1977, 1978). The conversion of PPDP [either racemic or the natural (1*R*, 2*R*, 3*R*)-isomer] into phytoene has been demonstrated with enzyme preparations from *Mycobacterium* (Altman *et al.*, 1972), *Phycomyces blakesleeanus* (Altman *et al.*, 1977, 1978), a *Flavobacterium* species (N.J. Patel and G. Britton, unpublished results), chloroplasts of *Triticum sativum* leaves (Camara, 1984) and chromoplasts of *Capsicum annuum* (Camara, 1984; Camara *et al.*, 1985b). The phytoene that can be isolated from most natural sources, especially higher plants and fungi, is the (15*Z*)- or (15,15′-*cis*)-isomer. This is also the isomer which is formed in most of the cell-free systems that have been shown to have phytoene-synthesizing ability. (All-*E*-) or (all-*trans*)-phytoene has only been found as the predominant isomer in some bacteria, and only with some of these bacteria, e.g. *Mycobacterium* species (Gregonis and Rilling, 1973) and *Halobacterium cutirubrum* (Kushwaha *et al.*, 1976), has the formation of (all-*E*)-phytoene by cell-free systems been demonstrated.

B Desaturation

Phytoene undergoes a series of sequential desaturation reactions (Fig. 3.4) to give phytofluene (7,8,11,12,7′,8′-hexahydro-ψ,ψ-carotene, V), ζ-carotene (7,8,7′,8′-tetrahydro-ψ,ψ-carotene, VI) or its ‘unsymmetrical’ isomer 7,8,11,12-tetrahydro-ψ,ψ-carotene VII), neurosporene (7,8-dihydro-ψ,ψ-carotene, VIII) and finally lycopene (ψ,ψ-carotene, IX). At each stage, two hydrogen atoms are removed by trans-elimination from adjacent positions (McDermott *et al.*, 1973b) to introduce a new double bond and extend the conjugated polyene chromophore by two double bonds (Fig. 3.5).

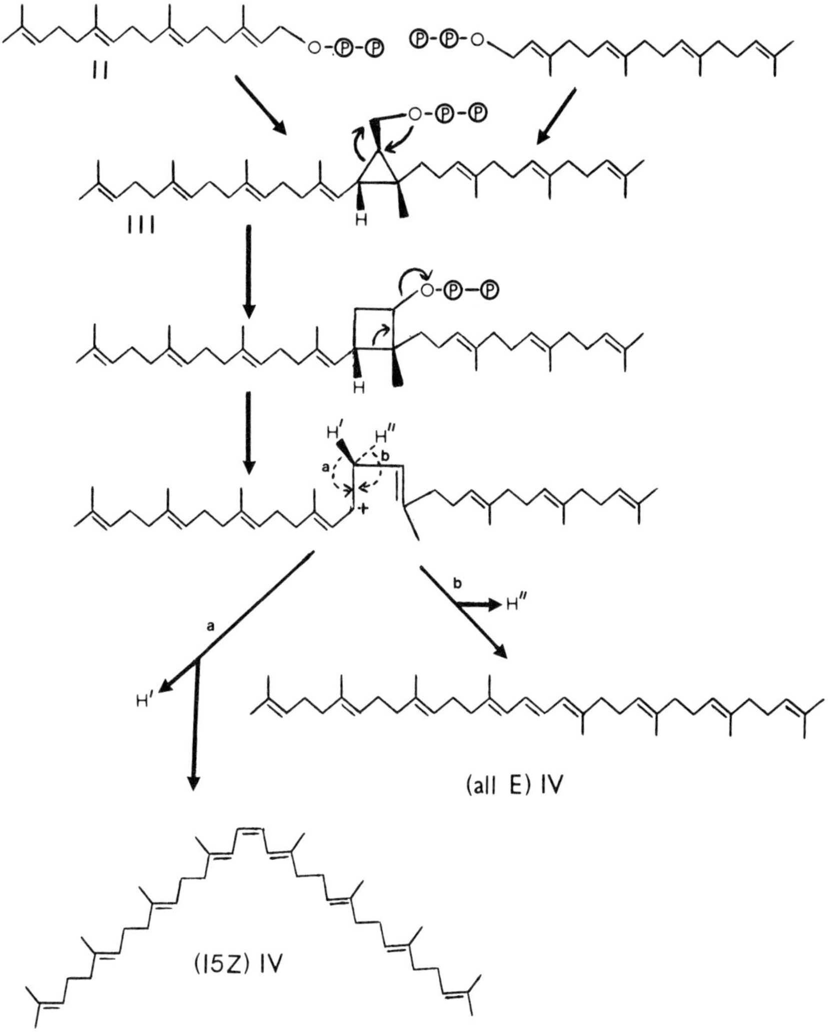

Fig. 3.3 Mechanism and stereochemistry of the formation of (all-*E*)- and (15*Z*)-phytoene.

Since the phytoene that has been isolated from most plant tissues and from incubations with plant cell-free systems is the (15*Z*)isomer, whereas the final fully desaturated carotenoids which accumulate are usually all-*trans* (all-*E*), it is generally assumed that isomerization of the 15,15′

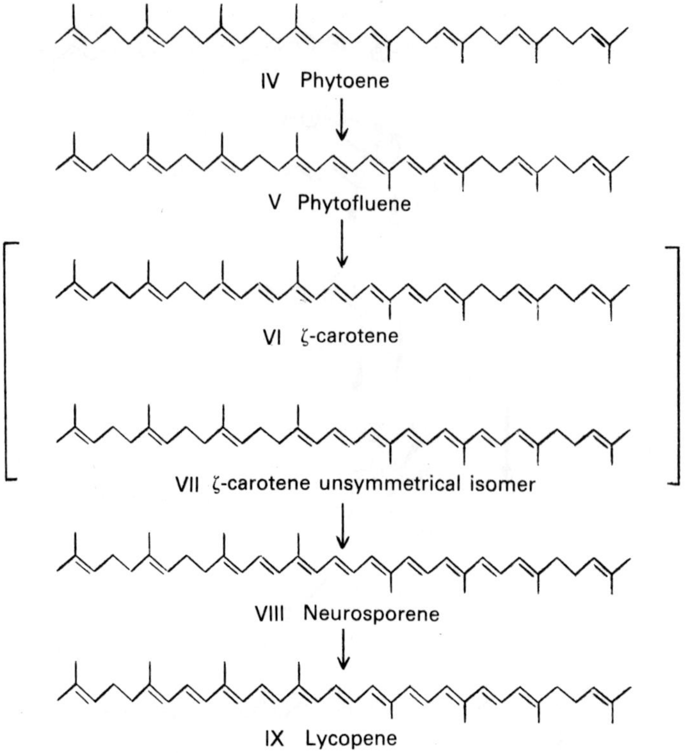

IV Phytoene

V Phytofluene

VI ζ-carotene

VII ζ-carotene unsymmetrical isomer

VIII Neurosporene

IX Lycopene

Fig. 3.4 Sequence of the desaturation reactions.

H_{5R} H_{5S} H_{2R} H_{2S} → H_{5S} H_{2R}

Fig. 3.5 Stereochemistry of desaturation. (H_{5R}, H_{5S}, H_{2R}, H_{2S} are the hydrogen atoms derived from the 5-*proR*, 5-*pro-S*, 2-*pro-R* and 2-*pro-S* hydrogens of MVA, respectively.)

double bond must occur at some stage of the desaturation sequence. The problem of isomerization and the stage at which it occurs has not, however, been satisfactorily resolved. Evidence has been obtained to indicate that, in a *Flavobacterium*, isomerization occurs at the level of phytoene (Brown *et al.*, 1975), whereas in tomato and *Capsicum* chromoplasts isomerization of phytofluene has been proposed (Kushwaha *et al.*, 1970; Camara *et al.*, 1980), and a study of carotenoid transformations in a mutant strain of *Scenedesmus obliquus* suggests that isomerization at the ζ-carotene level may be important (Britton and Powls, 1977; Britton *et al.*, 1977d).

The direct enzymic conversion of phytoene into coloured carotenoids has proved to be very difficult to achieve. Porter's group reported the incorporation of *cis*-phytoene, *cis*-phytofluene and *trans*-ζ-carotene into lycopene and other unsaturated carotenoids by an enzyme preparation from tomato plastids (Kushwaha *et al.*, 1970) and similar results have been obtained with a preparation from chromoplasts of *Capsicum annuum* (Camara and Monéger, 1982). A preparation of membrane-bound enzymes from chromoplasts of daffodil (*Narcissus pseudonarcissus*), solubilized by the detergent CHAPS, converted *cis*-phytoene into β-carotene in high yield (Beyer *et al.*, 1985).

With all these systems, the presence of the oxidized coenzymes FAD or NADP (or both) was essential, though their direct involvement in the desaturation reactions has not been proved. The involvement of some kind of simple electron transport system related to cytochrome P_{450} in the desaturations has been suggested (Britton, 1979b); the oxidized coenzymes could be required to maintain the electron transport components in the required oxidation state. Desaturation is easily inhibited by many compounds, including diphenylamine and bleaching herbicides, such as norflurazon.

1 POLY-*CIS*-CAROTENOIDS

The tangerine mutant of tomato (*Lycopersicon esculentum*) accumulates prolycopene, a poly-*cis* isomer of lycopene, as its main carotenoid in place of the normal all-*trans* compound. The presence of a similar pigment, assumed but not proved to be prolycopene, has also been reported in some other fruits and in some mutant strains of the green algae *Chlorella* and *Scenedesmus*. Prolycopene from the tangerine tomato has been characterized as the (7*Z*, 9*Z*, 7′*Z*, 9′*Z*) isomer (Clough and Pattenden, 1979; Englert *et al.*, 1979), and a series of *Z*-isomers of the

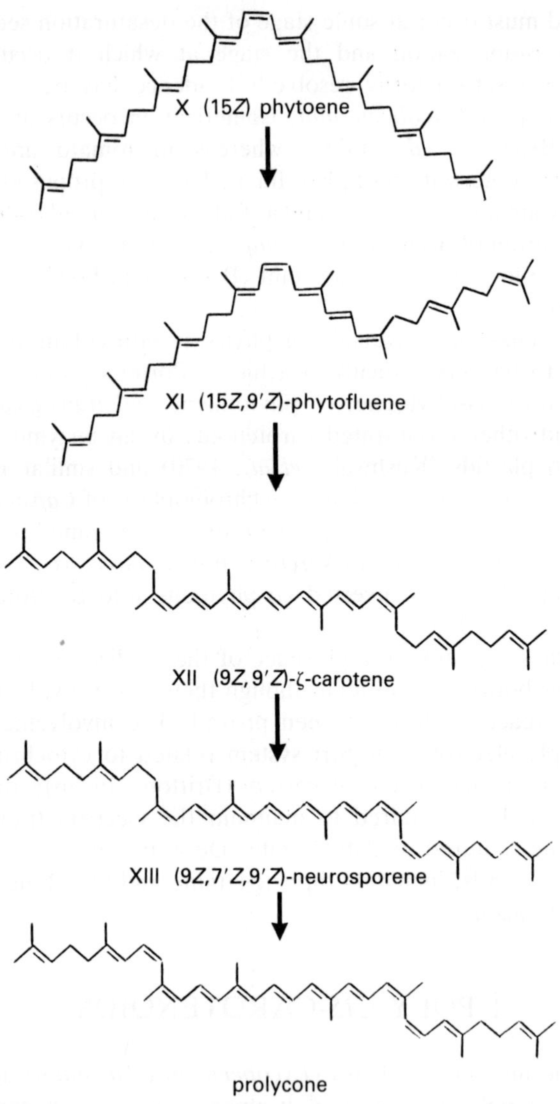

X (15*Z*) phytoene

XI (15*Z*,9′*Z*)-phytofluene

XII (9*Z*,9′*Z*)-ζ-carotene

XIII (9*Z*,7′*Z*,9′*Z*)-neurosporene

prolycone

Fig. 3.6 Scheme summarizing the biosynthesis of the poly-*cis* carotenoid, prolycopene.

desaturation intermediates, (15*Z*)-phytoene, (15*Z*, 9′*Z*)-phytofluene, (9*Z*, 9′*Z*)-ζ-carotene and (9*Z*, 7′*Z*, 9′*Z*)-neurosporene has also been identified (Clough and Pattenden, 1979, 1983; Frecknall and Pattenden, 1984). Based upon these structures, a scheme for the biosynthesis of prolycopene from (15*Z*)-phytoene was suggested (Fig. 3.6). The reported conversions of precursor carotenoids into prolycopene and other products by cell-free extracts of tangerine tomatoes (Qureshi *et al.*, 1974c) need to be reassessed in the light of these obvious structural relationships.

Experiments involving the incorporation of the stereospecifically labelled substrates [(4*R*)-4-$^{3}H_{1}$]-MVA, [(4*S*)-4-$^{3}H_{1}$]-MVA, [(5*R*)-5-$^{3}H_{1}$]-MVA and [5,5-$^{3}H_{2}$]-MVA have shown that the stereochemistry of hydrogen loss from C-7 (derived from C-5 of MVA) and C-10 (derived from C-4 of MVA) is the same as in the biosynthesis of all-*trans*-lycopene (R.J.H. Williams, G. Britton and T.W. Goodwin, unpublished results). Unfortunately no information is available about the loss of hydrogen from C-8 (derived from C-2 of MVA), so absolute comparison of the stereochemistry of biosynthesis of *trans*-lycopene and prolycopene is not yet possible.

C Cyclization

Arguments still persist about whether the acyclic intermediate that is cyclized is lycopene or neurosporene. Perhaps this question has been given too much importance; the cyclization reaction would be the same in both cases, and it is more satisfactory to consider that the requirement for cyclization is simply that one half of the carotenoid molecule, i.e. one end-group, should have reached the lycopene level of desaturation.

The chirality at C-6′ in β,γ-carotene (XIV) from the fungus *Caloscypha fulgens* has been shown to be (*S*), i.e. opposite to that in α-carotene [(6′*R*)-β,ε-carotene, XV] from most plant sources (Hallenstvet *et al.*, 1977). It is generally believed that the β-, γ-, and ε-end-groups are formed by proton loss from alternative positions in the same transient carbonium ion intermediate (Fig. 3.7) but no biosynthetic work on the γ-ring has been reported.

XIV β,γ-Carotene

XV α-Carotene

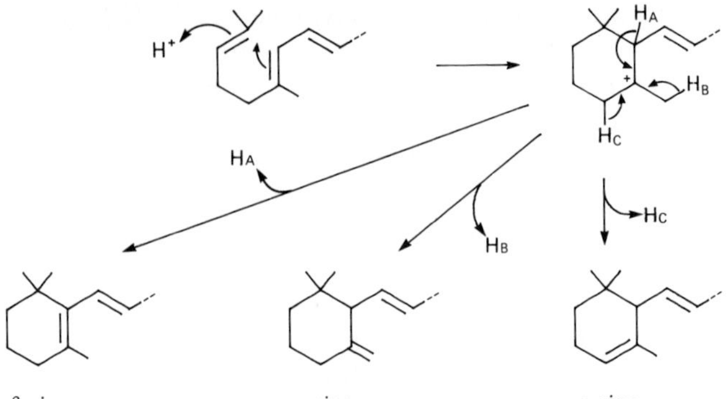

Fig. 3.7 Mechanism of the formation of β-, γ- and ε-rings from a common precursor.

1 INHIBITORS OF CYCLIZATION

Several compounds have been identified as efficient inhibitors of the cyclization reaction in carotenoid biosynthesis. The most useful of these has been nicotine; in many organisms its presence causes lycopene to accumulate in place of the normal cyclic carotenoids (Howes and Batra, 1970). The inhibition is usually reversible to a considerable extent; on removal of the nicotine the normal cyclic carotenoids are formed, initially at the expense of the accumulated lycopene, direct utilization of which has now been demonstrated (McDermott *et al.*, 1973a,c, 1974). At lower concentrations of nicotine, when the inhibition of cyclization is not complete, it is common for a monocyclic carotenoid to accumulate in place of the normal major bicyclic compound, e.g. γ-carotene [(β,ψ-carotene), XVI] in place of β-carotene [(β,β-carotene), XVII] or rubixanthin [(β,ψ-caroten-3-ol), XVIII] instead of zeaxanthin [(β,β-carotene-3,3′-diol), XIX]. The lower amounts of nicotine thus result in the inhibition of cyclization in one half of the carotenoid molecule rather than simply in a lower degree of inhibition of both cyclizations to give a mixture of acyclic and bicylic compounds. Similar inhibition of cyclization has been achieved with amines such as 2-(4-chlorophenylthio)-triethylammonium chloride (CPTA) (Coggins *et al.*, 1970). These compounds are more effective than nicotine in higher plants, but the inhibition they cause is usually not reversible.

Use of the cyclization inhibitors, especially nicotine, has made it possible to prove directly the proposed mechanism of H^+-initiated cyclization, and to deduce the stereochemical course of the reactions.

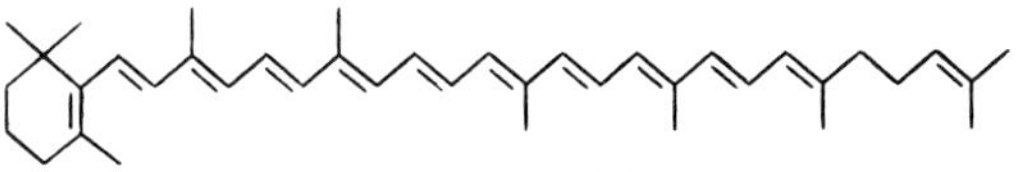

XVI γ-Carotene

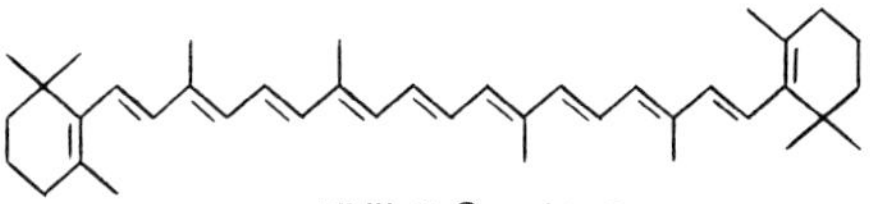

XVII β-Carotene

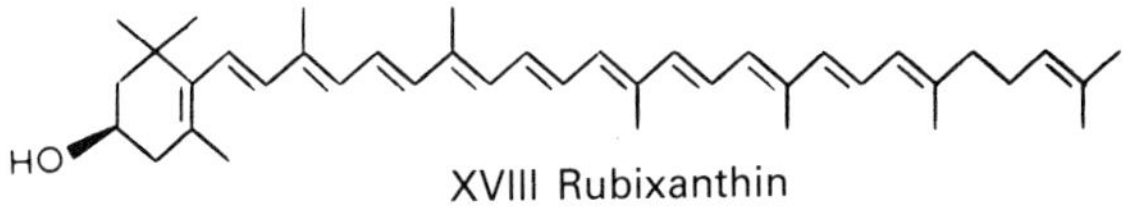

HO

XVIII Rubixanthin

OH

HO

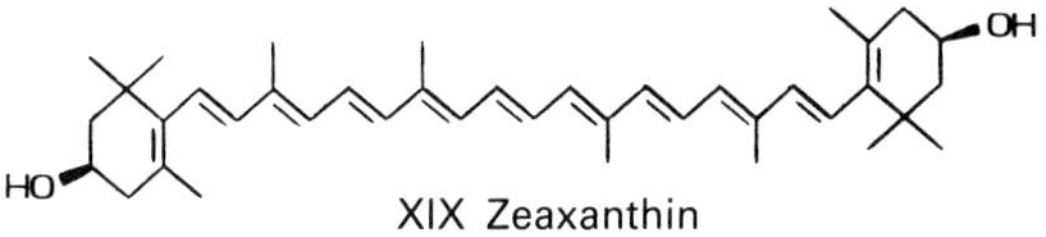

XIX Zeaxanthin

2 MECHANISM AND STEREOCHEMISTRY

The proposed mechanism for cyclization involves initial proton attack at C-2 of the acyclic precursor (Fig. 3.7). The incoming hydrogen atom would be retained at C-2 of the cyclic carotenoids formed. This has now been proved by experiments in which accumulated acyclic precursor carotenoids were allowed to cyclize *in situ* in conditions where ^2H_2O replaced the water of the medium. Thus a *Flavobacterium* species, grown in the presence of nicotine, accumulated lycopene. When the nicotine was removed and the cells were resuspended in medium prepared from ^2H_2O, the direct cyclization of lycopene and its conversion into zeaxanthin (XIX) were demonstrated. The zeaxanthin isolated was a $^2H_2$ species with a deuterium atom incorporated specifically at C-2 of each ring (Britton *et al.*, 1977a). An alternative experiment used a mutant strain, PG1, of the green alga *Scenedesmus obliquus*. This mutant, when grown in the dark, accumulates ζ-carotene. On illumination, the normal cyclic carotenes and xanthophylls are formed, initially at the expense of the accumulated ζ-carotene (Britton and Powls, 1977; Britton *et al.*, 1977d). When cells were transferred to ^2H_2O before illumination, cyclization was initiated by $^2H^+$, and $^2H_2$-species of α-carotene, β-carotene, lutein (β,ε-carotene-3,3′-diol, XX) and zeaxanthin were isolated, confirming the

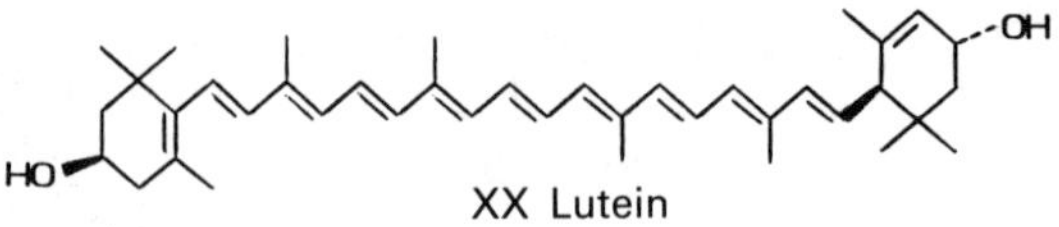

XX Lutein

proposed mechanism for formation of both the β- and ε-rings (Britton *et al.*, 1977c). These experimental systems have allowed the stereochemistry of the H^+ attack which initiates the cyclization to be determined. The ^2H introduced at C-2 of each ring during cyclization in the conditions described above is introduced stereospecifically, and analysis by ^1H nuclear magnetic resonance (NMR) spectroscopy proved that, in both *Flavobacterium* and *Scenedesmus*, and in the formation of both β- and ε-rings, the ^2H is introduced in the β-position at C-2, i.e. attack occurs from above the plane of the C-1,2 double bond of the acyclic precursor when this is folded, as illustrated in Fig. 3.8 (Britton *et al.*, 1977b; Britton and Mundy, 1980).

A second feature of the stereochemistry of cyclization that must be resolved is the behaviour of the methyl substituents at C-1. This has now been determined by a series of procedures which used labelling with the stable isotope, ^{13}C. In a simple experiment, [2-^{13}C]-MVA was incorporated into lycopene (in the presence of nicotine) and zeaxanthin by *Flavobacterium*. Analysis of the carotenoids by ^{13}C-NMR spectroscopy showed that the ^{13}C-enrichment in the lycopene was in the expected positions, i.e. C-4, C-8, C-12 and C-16. In zeaxanthin, C-4, C-8 and C-12 were again enriched, as was the axial (α) methyl substituent at C-1. This result therefore allows the stereochemistry of formation of the β-rings in zeaxanthin to be defined as illustrated in Fig. 3.8 (Britton *et al.*, 1979).

Scenedesmus obliquus will not incorporate MVA efficient into its chloroplast carotenoids, so alternative labelling procedures have had to be devised and implemented in order to elucidate the stereochemistry of formation of the ε-ring of lutein. The problem was overcome by culturing the organism in the dark with 20%-enriched [$^{13}C_6$]-glucose as carbon source substrate. Under these circumstances, the glucose is metabolized to a substantial extent to intact acetate units, equivalent to [$^{13}C_2$-acetate] which in turn is incorporated into MVA within the cell. In isoprene units derived from this MVA, four carbon atoms originate as two intact acetate units so that ^{13}C–^{13}C coupling can therefore be expected in the ^{13}C NMR spectra of any compounds made from these units. The remaining carbon atoms, (those derived from C-2 of MVA) will be single carbon atoms which will not participate in similar coupling. In any isoprenoid compound, carbon atoms which originate from C-2 of MVA produced endogenously

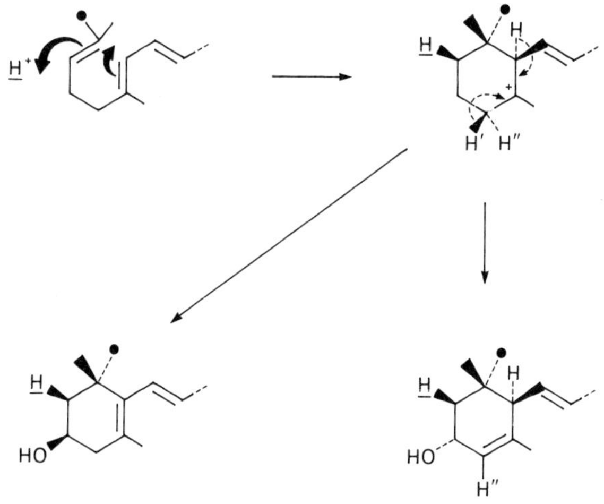

Fig. 3.8 Stereochemistry of cyclization to form the β- and ε-rings in zeaxanthin and lutein.

in this way can easily be distinguished in the ^{13}C-NMR spectrum by the lack of ^{13}C–^{13}C coupling (Fig. 3.9). In the case of lutein in *Scenedesmus*, the 1α (equatorial) methyl substituent of the ε-ring showed no coupling, whereas the 1β (axial) methyl group was coupled with C-1 of the ring. This confirmed that the 1α, equatorial methyl group was the one derived from C-2 of MVA, i.e. from C-16 of lycopene and ζ-carotene, and therefore defines the stereochemistry of ε-ring formation, as illustrated in Fig. 3.8 (Britton, 1985).

A series of alternative schemes has been proposed, describing the different possible modes of folding of the acyclic precursor and the directions of electrophilic attack at C-2 and of C-1,6-bond formation. The main features included were the shape of folding of the ring precursor (chair or boat), and whether the C-1,2 double bond folds above or below the rest of the carotenoid end-group (Britton, 1971). This scheme was extended by Eugster (1979) to include the alternatives of (5*E*) and (5*Z*) stereochemistry of the end-group precursor, a feature which may be particularly important in determining the chirality of γ- and ε-rings. The (5*Z*)-isomers of neurosporene and lycopene, amongst others, have been isolated from fruits of *Rosa pomifera* (Märki-Fischer *et al.*, 1983). The results of the labelling experiments with zeaxanthin and lutein, reported above, are in agreement with either of the folding patterns illustrated in Fig. 3.10. The formation of the (6*S*)-γ-ring obviously does not conform to the same stereochemical pattern, but no biosynthetic studies on the γ-ring have been reported.

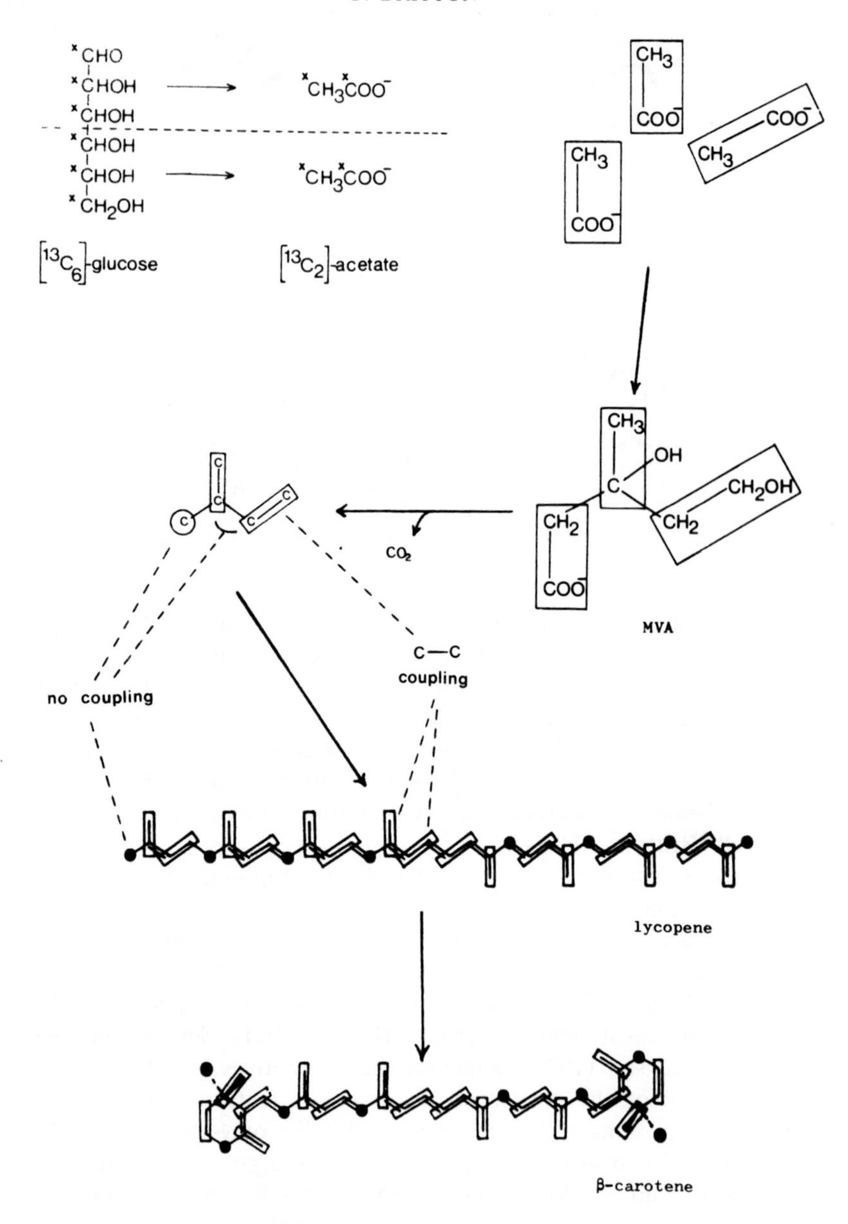

Fig. 3.9 Expected pattern of labelling and ^{13}C–^{13}C coupling in isoprene units and carotenoids biosynthesized from [$^{13}C_6$]-glucose (20% enrichment).

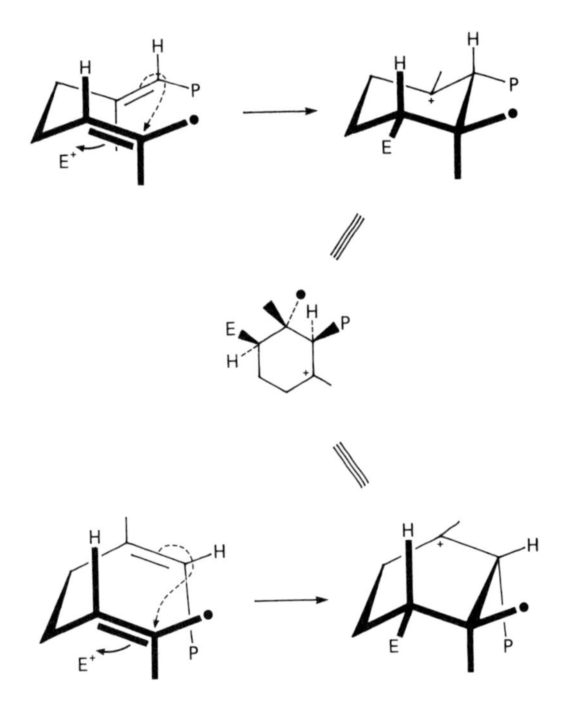

Fig. 3.10 Alternative possibilities for the stereochemistry of folding of the acyclic end-group during carotenoid cyclization.

D Reactions alternative to cyclization

The cyclization reaction may be considered as an addition reaction to the C-1,2-double bond of a suitably folded acyclic precursor end-group. It is typical of the non-sulphur photosynthetic bacteria that, with few exceptions, they are not able to make cyclic carotenoids but instead add water or hydrogen to the C-1,2 double bond to produce the 1-hydroxy-, 1-methoxy- or 1,2-dihydro-carotenoids which are characteristic especially of the Rhodospirillaceae. These reactions are considered to be analogous to the first stage of the cyclization and are blocked by the inhibitors, such as nicotine and CPTA, that inhibit cyclization (McDermott *et al.*, 1973a). The reactions, and their relationship to cyclization, are illustrated in Fig. 3.11. As with cyclization, this hydration mechanism for introduction of the C-1 hydroxy group has been proved directly by demonstrating the introduction of ^{2}H at C-2 when cultures were grown in the presence of nicotine to accumulate hydrocarbon precursors and then resuspended in

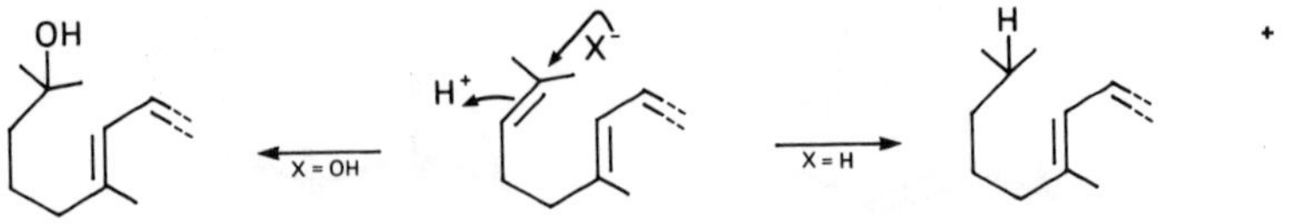

Fig. 3.11 Hydroxylation and hydrogenation at C-1,2 as alternatives to cyclization.

^2H_2O after removal of the inhibitor (Patel *et al.*, 1983). The direct conversion of lycopene into rhodopin (1-hydroxy-1,2-dihydro-ψ,ψ-carotene, XXI) by a cell-free preparation of chromatophores of *Rhodomicrobium vannielii* has been demonstrated (R.K. Singh and G. Britton, unpublished results). The formation of the 1,2-dihydrocarotenoids of *Rhodopseudomonas viridis* seems to proceed in an analogous manner to the hydration, but with the addition of H_2 to the C-1,2-double bond. Nicotine and CPTA are again inhibitory (Britton *et al.*, 1977e).

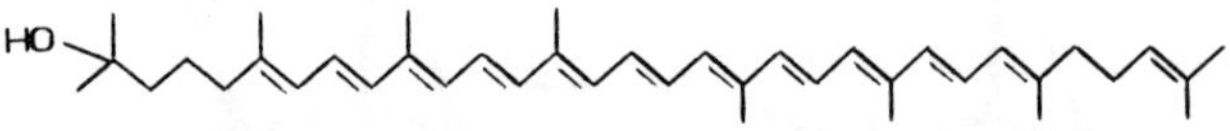

XXI Rhodopin

E Biosynthesis of C_{45} and C_{50} carotenoids

The carotenoids of several Gram-positive non-photosynthetic bacteria have been shown to have C_{45} and C_{50} structures. Biosynthetically these are substituted C_{40} skeletons, with the extra one or two C_5 units being added on at a late stage. In the acyclic series, *Halobacterium* species accumulate bacterioruberin [2,2′-bis (3-hydroxy-3-methylbutyl)-3,4,3′,4′-tetradehydro-1,2,1′,2′-tetrahydro- ψ,ψ-carotene-1,1′-diol, XXII], formation of which is considered to occur by a reaction which is analogous to C-1,2-hydration, but in which a C_5 electrophile replaces H^+ as the initiating species in the electrophilic attack at C-2 (Fig. 3.12). The

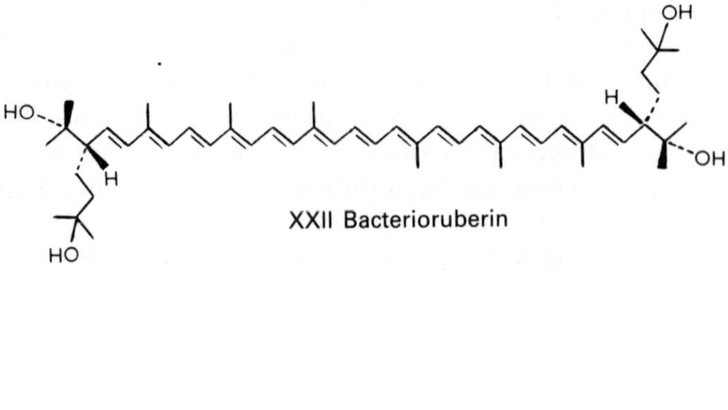

XXII Bacterioruberin

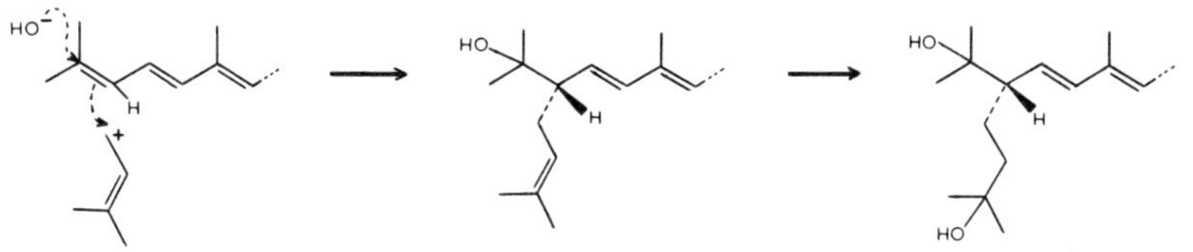

Fig. 3.12 Stereochemistry of formation of the end-group of the acyclic C_{50} carotenoid bacterioruberin.

introduction of the C_5 group is stereospecific, as illustrated, and gives the (2*S*)-isomer (Johansen and Liaaen-Jensen, 1976).

Three different ring types with C_5 substituents at C-2 have been identified in the C_{45} and C_{50} carotenoids, i.e. substituted β-, γ- and ε-rings in C.p. 450[2,2′-bis(4-hydroxy-3-methylbut-2-enyl)-β,β-carotene, XXIII], sarcinaxanthin [2,2′-bis (4-hydroxy-3-methylbut-2-enyl)-γ,γ-carotene, XXIV] and decaprenoxanthin [2,2′-bis(4-hydroxy-3-methylbut-2-enyl)-ε,ε-carotene, XXV], respectively. As in the C_{40} series, labelling studies with [2,2-$^3H_2$]-MVA and (4*R*)-[4-$^3H_1$]-MVA have shown that the different ring types are not interconverted (Swift and Milborrow, 1981b). The chirality at C-2 of all these C_{50} carotenoids is opposite to that which would be expected if the C_5 group were introduced with the same stereochemistry as the initiating H^+ in the C_{40} series (Andrewes *et al.*, 1974, 1975; Hertzberg and Liaaen-Jensen, 1977). Decaprenoxanthin also differs from its C_{40} ε-ring counterparts in its chirality (*S*) at C-6 and C-6′. Labelling studies with ^{13}C (Britton, 1985) have indicated that the

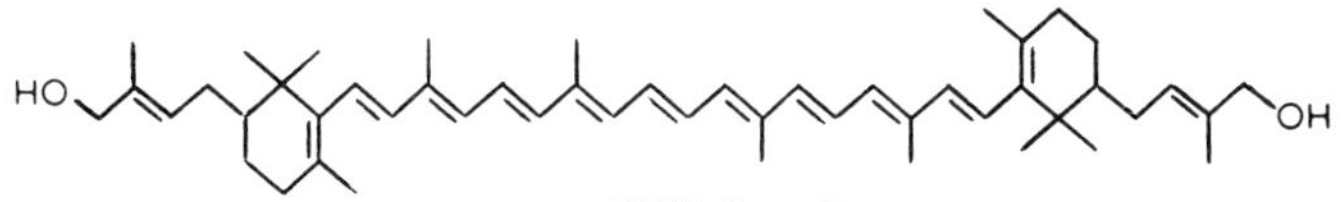

XXIII C.p. 450

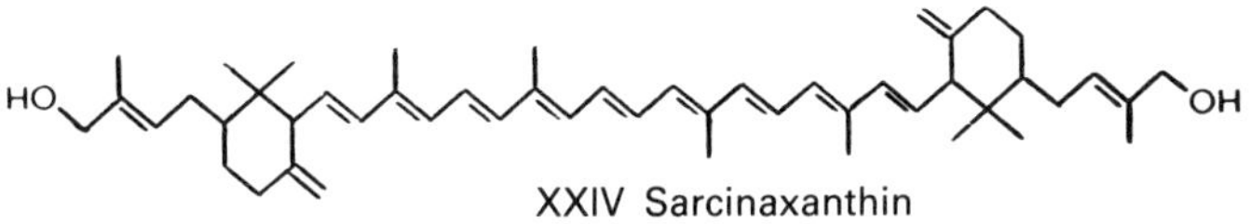

XXIV Sarcinaxanthin

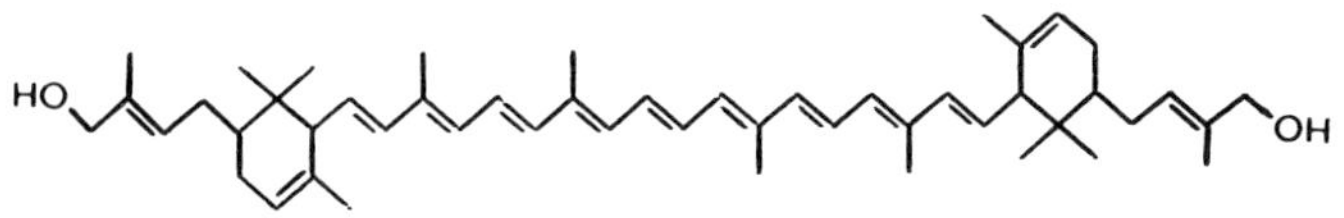

XXV Decaprenoxanthin

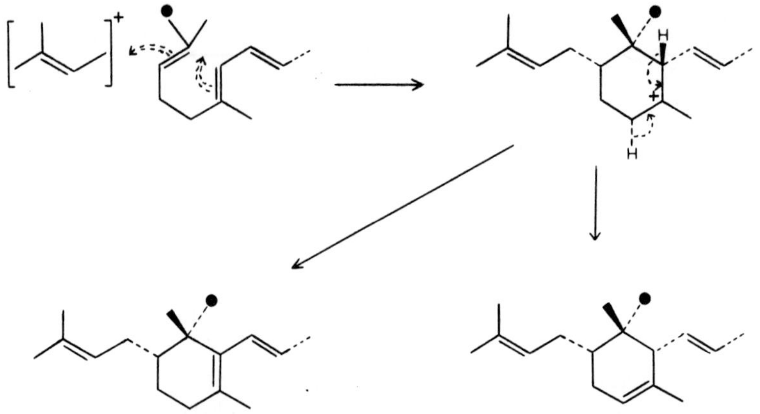

Fig. 3.13 Mechanism and stereochemistry of formation of the end-groups of cyclic C_{50} carotenoids.

stereochemical behaviour of the C-1 methyl substituents during cyclization to form β- and ϵ-rings in the C_{50} series appears to be the same as for the C_{40} carotenoids (Fig. 3.13). In the formation of the ϵ-ring of decaprenoxanthin, the hydrogen atom lost from C-4 has been reported to be the one which was originally the (2-*pro-R*)-hydrogen of MVA (Fahey and Milborrow, 1978). This contrasts with the result previously obtained for α-carotene (Britton, 1976) which showed the loss of the (2-*pro-S*)-hydrogen of MVA. These findings, when taken with the established chirality at C-2 and C-6, do not correspond to any of the extended series of possible patterns of folding illustrated by Eugster (1979).

F Later reactions

1 HYDROXYLATION

Apart from the hydroxylation at C-1 described above, which can take place as early as the phytoene stage, oxygen functions and other structural modifications in carotenoids are believed to be introduced at the end of the biosynthetic sequences. Some studies relevant to the introduction of the hydroxy group at C-3 of a carotenoid ring have been described. Hydroxylation at C-3 of the β-ring, e.g. in zeaxanthin or lutein, normally proceeds by direct replacement of the (3-*pro-R*)-hydrogen atom of the hydrocarbon precursor (i.e. the one which was originally the (5-*pro-R*)-hydrogen of MVA) by OH (Britton, 1976) (Fig. 3.14). The reaction is

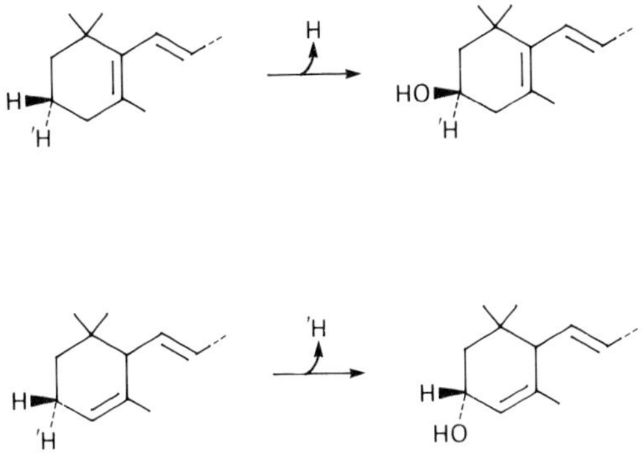

Fig. 3.14 Stereochemistry of hydroxylation at C-3 in the β- and ε-rings.

assumed to be catalysed by a mixed-function oxidase enzyme, involving cytochrome-P450.

The chirality at C-3′ of lutein has been established as (*R*), i.e. opposite to that at C-3 (also designated *R*) of lutein and at C-3 and C-3′ of zeaxanthin. In *Calendula officinalis* the introduction of the hydroxy group into the ε-ring of lutein also occurs by direct replacement of the appropriate hydrogen atom at C-3 of the hydrocarbon precursor, in this case the one derived from the (5-*pro*-*S*)-hydrogen atom of MVA (Milborrow *et al.*, 1982) (Fig. 3.14).

The ketocarotenoid astaxanthin (3,3′-dihydroxy-β,β-carotene-4,4′-dione) occurs naturally in different enantiomeric forms; that obtained from algae is the expected (3*S*, 3′*S*)-isomer (XXVI), but the yeast *Phaffia rhodozyma* accumulates the (3*R*, 3′*R*) form (XXVII). The assumption is made that, in the algae, the C-3 hydroxy groups are introduced first, with normal stereochemistry, to give (3*R*, 3′*R*)-zeaxanthin, into which the C-

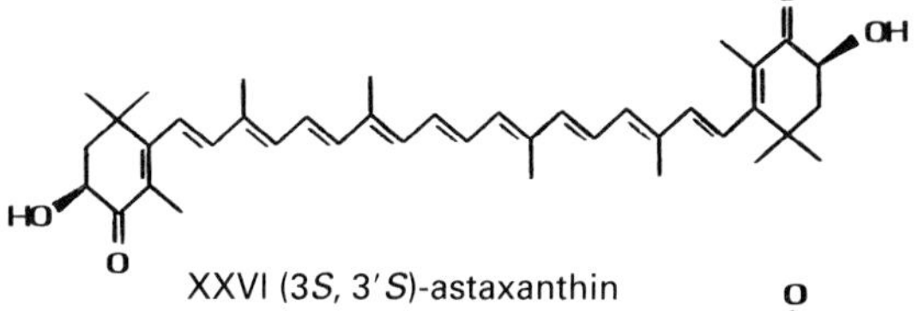

XXVI (3*S*, 3′*S*)-astaxanthin

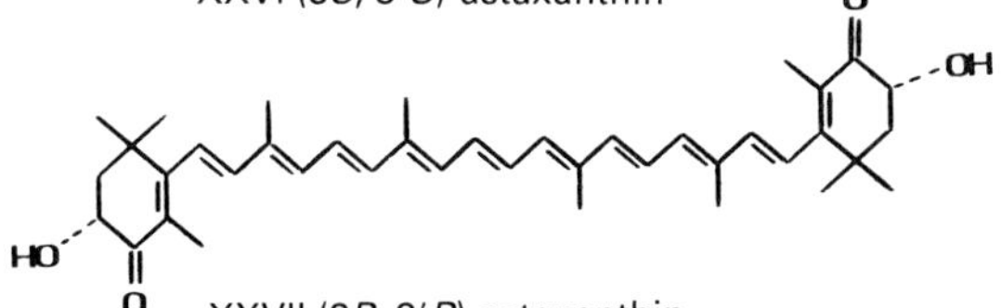

XXVII (3*R*, 3′*R*)-astaxanthin

4-keto groups are then introduced. It has been proposed (Andrewes and Starr, 1976; Andrewes *et al.*, 1976) that, in *Phaffia*, the keto groups are introduced first, and the conformation imposed on the ring by these C-4 keto groups determines that the subsequent introduction of the C-3 hydroxy groups then occurs by direct hydroxylation but with the opposite stereochemistry.

Nothing is known of the mechanisms or stereochemistry of hydroxylation at other positions in the carotenoid molecule.

2 CYCLOPENTANE CAROTENOIDS

The proposal that the cyclopentane (κ) ring end-group of capsanthin [(3,3′-dihydroxy-β,κ-caroten-6′-one), XXVIII] and capsorubin [(3,3′-dihydroxy-κ,κ-carotene-6,6′-dione) XXIX] is formed by rearrangement of a 5,6-epoxy-β-ring end-group (Fig. 3.15) has received support from experiments which have demonstrated the incorporation of radioactive antheraxanthin [(5,6-epoxy-5,6-dihydro-β,β-carotene-3,3′-diol), XXX] and violaxanthin [(5,6,5′,6′-diepoxy-5,6,5′,6′-tetrahydro-β,β-carotene-3,3′-diol), XXXI] into capsanthin and capsorubin, respectively, by chromoplasts of red pepper, *Capsicum annuum* (Camara, 1980a,b; Camara and Monéger, 1981). It is interesting that the epoxycarotenoids used as substrates were made chemically from zeaxanthin and would therefore have been largely the (5*S*, 6*R*)-isomers, not the normal natural (5*R*, 6*S*)-forms that would be the expected intermediates in the reaction.

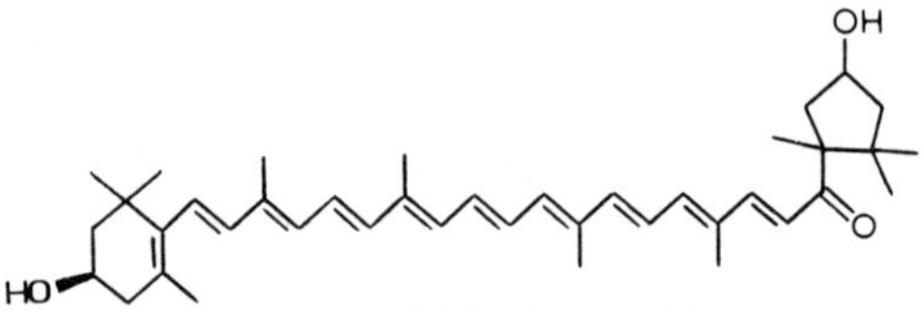

XXVIII Capsanthin

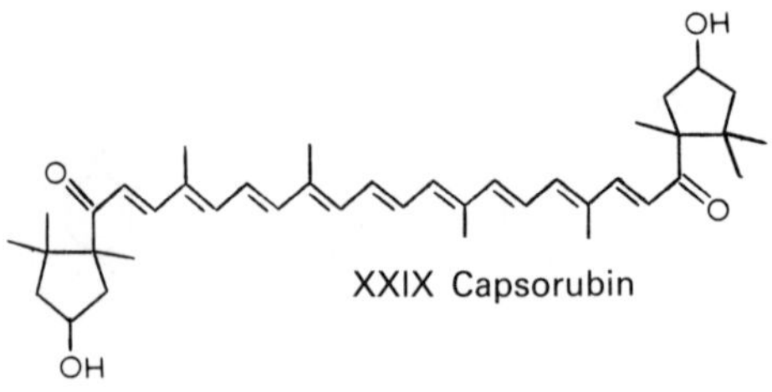

XXIX Capsorubin

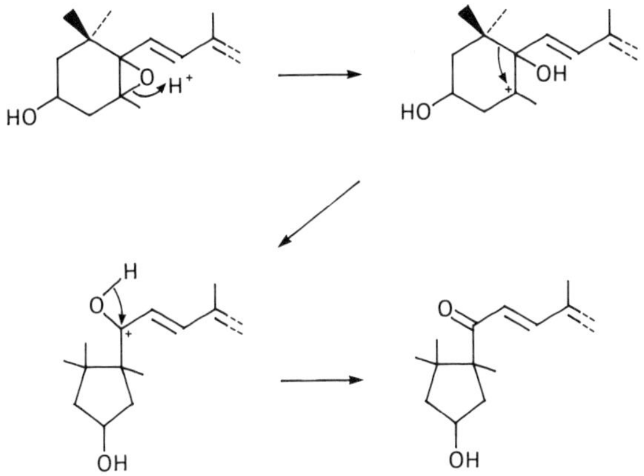

Fig. 3.15 Proposed mechanism for formation of the κ-ring in capsanthin and capsorubin.

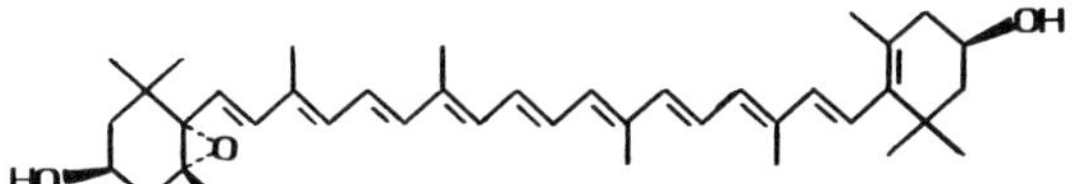

XXX Antheraxanthin

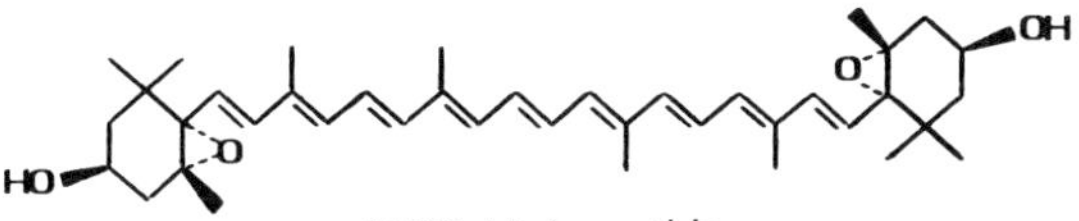

XXXI Violaxanthin

3 ACETYLENIC AND ALLENIC CAROTENOIDS

The dinoflagellate *Amphidinium carterae* has been used to study the biosynthesis of the acetylenic carotenoid diadinoxanthin [(5,6-epoxy-7′,8′-didehydro-5,6-dihydro-β,β-carotene-3,3′-diol), XXXII], the allenic carotenoid neoxanthin [(5′,6′-epoxy-6,7-didehydro-5,6,5′,6′-tetrahydro-β,β-carotene- 3,5,3′-triol), XXXIII] and the allenic norcarotenoid perid-

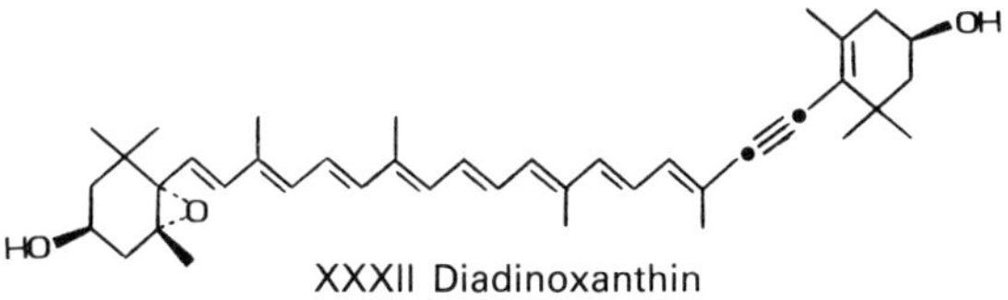

XXXII Diadinoxanthin

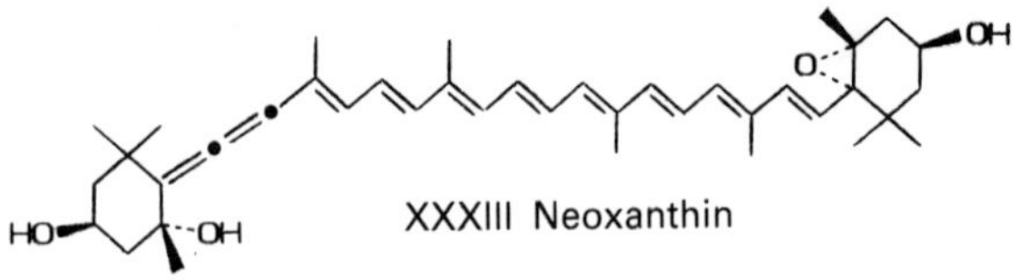

XXXIII Neoxanthin

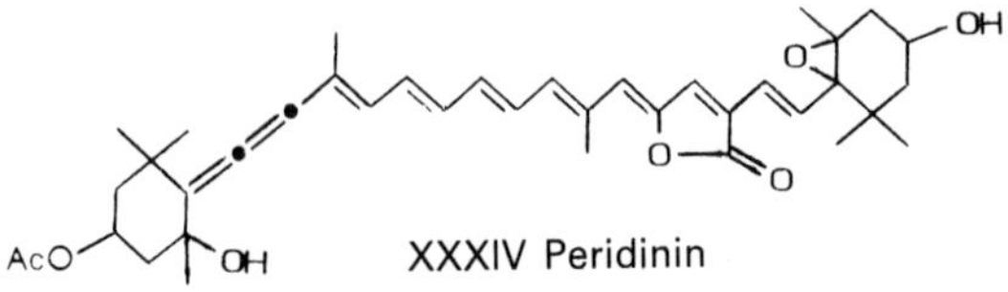

XXXIV Peridinin

inin [(5′,6′-epoxy-3,5,3′-trihydroxy-6,7-didehydro-5,6,5′,6′- tetrahydro-10,11,20-trinor-β,β-caroten-19′,11′-olide 3-acetate), XXXIV] (Swift and Milborrow, 1981a; Milborrow, 1982; Swift *et al.*, 1982). The conversion of zeaxanthin into all of these carotenoids by a cell-free system from *A. carterae* was reported. This shows that the acetylenic and allenic groups are introduced at the end of the biosynthetic sequence.

Comparison of ^{14}C:^{3}H ratios in zeaxanthin, neoxanthin and diadinoxanthin biosynthesized from [2-^{14}C, 2,2-$^{3}H_2$]-MVA indicated that tritium was retained at C-8 of neoxanthin. This would preclude the possibility that the allenic neoxanthin is formed from the acetylenic diadinoxanthin. In confirmation of this, the conversion of labelled neoxanthin into diadinoxanthin was demonstrated, in agreement with the previously proposed sequence of reactions illustrated in Fig. 3.16.

The major carotenoid present in dinoflagellates and diatoms is the C_{37} compound, peridinin, which is assumed to be formed from a C_{40} precursor by a route which includes the loss of a C_3 fragment. This was confirmed by the demonstrated incorporations of zeaxanthin and neoxanthin into peridinin. A series of incorporation experiments with specifically labelled species of MVA (or with lycopene biosynthesized from these substrates) suggested that the carbon atoms lost were C-13, C-14 and C-20 of the C_{40} carotenoid skeleton, and that they were lost as an intact C_3 unit that

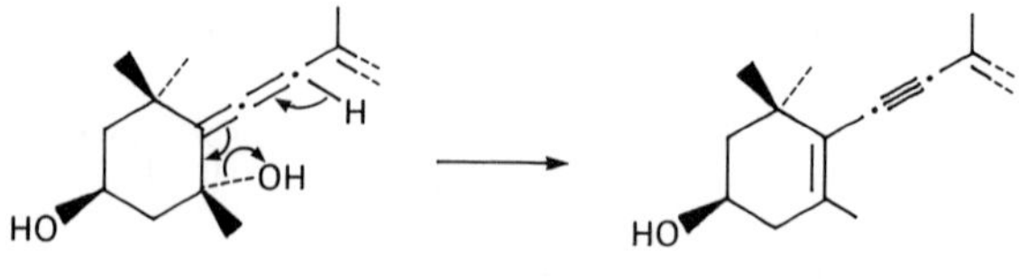

Fig. 3.16 Proposed mechanism for the conversion of the allenic end-group of neoxanthin into the acetylenic end-group of diadinoxanthin.

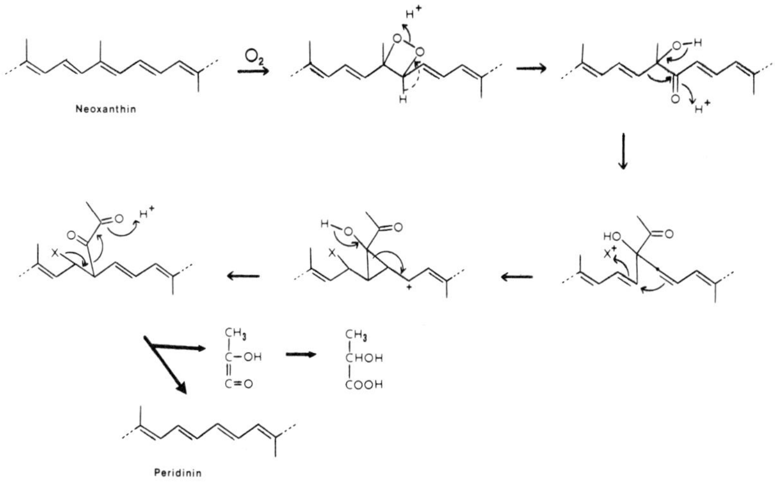

Fig. 3.17 Mechanism proposed by Milborrow (1982) for the extrusion of a C_3 unit in the biosynthesis of the C_{37} norcarotenoid peridinin.

could be trapped as lactate. A tentative mechanism has been proposed for the excision reaction (Fig. 3.17).

4 ARYL CAROTENOIDS

Sulphur photosynthetic bacteria, both green (Chlorobiaceae) and purple (Chromatiaceae) characteristically contain carotenoids having trimethylphenyl rings in place of the conventional trimethylcyclohexenyl rings. Two different methylation patterns occur, the 1,2,5-trimethylphenyl or ϕ-end-group, as in chlorobactene [(ϕ,ψ-carotene), XXXV] and the 1,2,3-trimethylphenyl or χ-end-group found in okenone [(1′-methoxy-1′,2′-dihydro-χ,ψ-caroten-4′-one), XXXVI]. It is generally accepted that these aromatic ring systems are derived biosynthetically by further dehydrogenation of a 3,4-didehydro-β-ring. Thus formation of the 1,2,5-

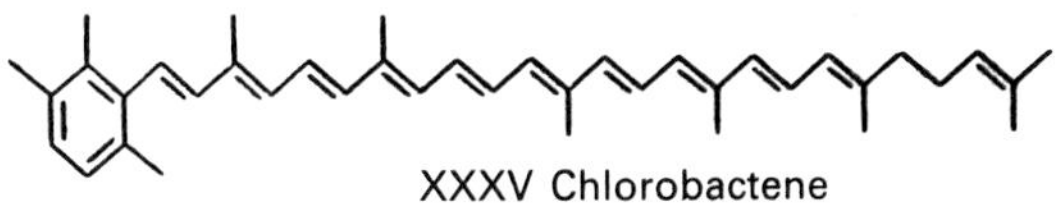

XXXV Chlorobactene

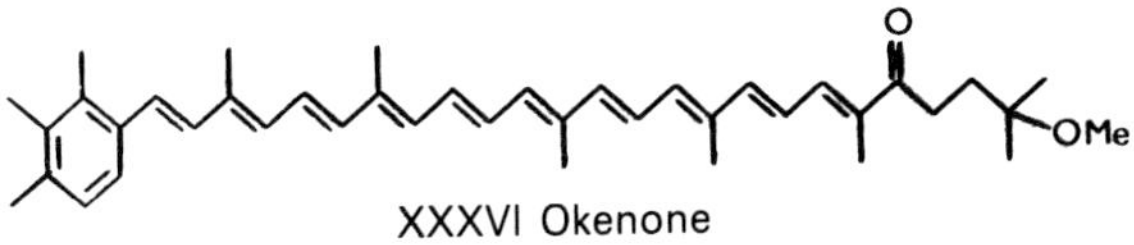

XXXVI Okenone

trimethylphenyl end-group must involve migration of one of the C-1 methyl substituents of the β-ring precursor (Fig. 3.18). In *Chloropseudomonas ethylica*, the methyl group migrating from C-1 to C-2 has been shown to be that which arises from the C-3′ methyl group of MVA; the one originating from C-2 of MVA remains at C-1 of the ring (Moshier and Chapman, 1973). If the stereochemical disposition of the C-1 methyl substituents in the β-ring precursor is the same as that which has been established for zeaxanthin (see Fig. 3.8), the stereochemical course of the aromatization reaction is defined as illustrated in Fig. 3.18.

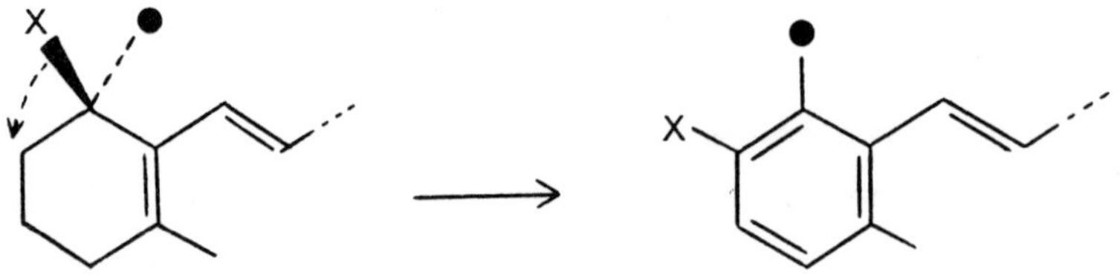

X X

Fig. 3.18 Mechanism and stereochemistry of rearrangement to form the 1,2,5-trimethylphenyl (ϕ) end-group of chlorobactene.

No information is available about the considerably more complex series of rearrangements that must be required to produce the 1,2,3-trimethylphenyl end-group. A 'Ladenburg prism' mechanism has been proposed (Liaaen-Jensen, 1969) (Fig. 3.19), but there is no experimental evidence to support this.

5 MODIFICATIONS OF THE ACYCLIC CAROTENOIDS OF PHOTOSYNTHETIC BACTERIA

The basic reaction in the biosynthesis of the end-groups of the acyclic xanthophylls of photosynthetic bacteria is the C-1,2 hydration described above. After this, however, other reactions are used to make the wide range of carotenoids present in the different species. The most characteristic reaction is methylation of the C-1 tertiary hydroxy groups to give the normal major pigments, such as spheroidene [(1-methoxy-3,4-didehydro-1,2,7′,8′-tetrahydro-ψ,ψ-carotene), XXXVII] and spirilloxanthin [(1,1′-dimethoxy-3,4,3′,4′-tetradehydro-1,2,1′,2′-tetrahydro-ψ,ψ-carotene), XXXVIII]. Schemes have been proposed for the biosynthesis of these compounds from neurosporene and lycopene, respectively (Fig. 3.20). Apart from normal desaturation, these pathways involve only

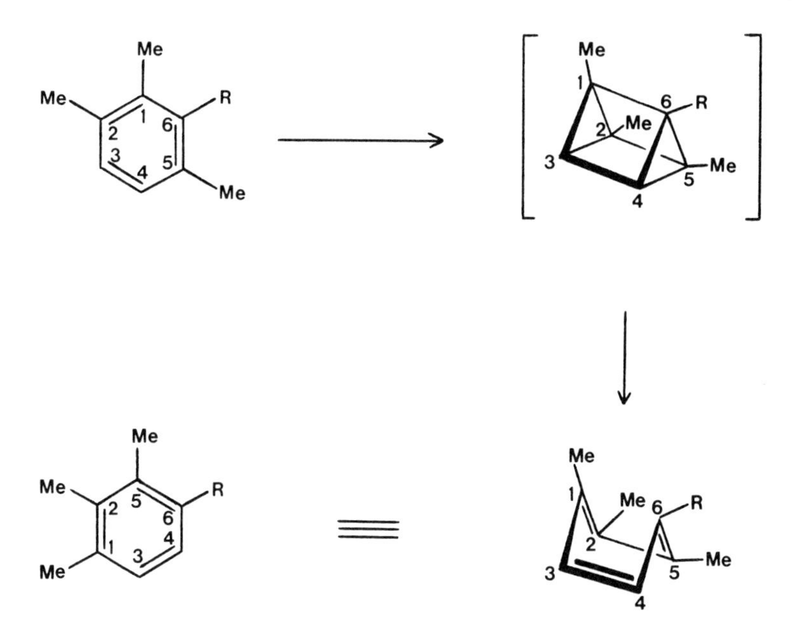

Fig. 3.19 Proposed Ladenburg prism mechanism for the rearrangement to form the 1,2,3-trimethylphenyl (χ) end-group.

three reactions, namely C-1,2-hydration, C-1 *O*-methylation, and C-3,4 desaturation. It is the relative sequences in which these reactions occur in the two halves of the carotenoid molecule that give rise to a wide range of possible intermediates and a variety of "alternative pathways". The *O*-methylation reaction has been studied in Rhodobacter (formerly *Rhodopseudomonas) sphaeroides* (Singh *et al.*, 1973). The methyl group introduced arises conventionally from *S*-adenosylmethionine, as demonstrated by labelling experiments and specific degradation of the methoxy carotenoid, spheroidenone [(1-methoxy-3,4-didehydro-1,2,7′,8′-tetrahydro-ψ,ψ-caroten-2-one), XXXIX].

The introduction of carbonyl groups into these acyclic compounds also occurs commonly. The rapid and efficient conversion of spheroidene into spheroidenone when anaerobic cultures of *R. sphaeroides* and other species are exposed to air is well known. The oxygen introduced is derived from molecular oxygen (Schneour, 1962), but the possible 2-hydroxy-intermediate in the reaction has not been detected. How such a conversion, presumably enzyme-catalysed, can occur when the carotenoids are in their functional locations in the photosynthetic complexes is not known.

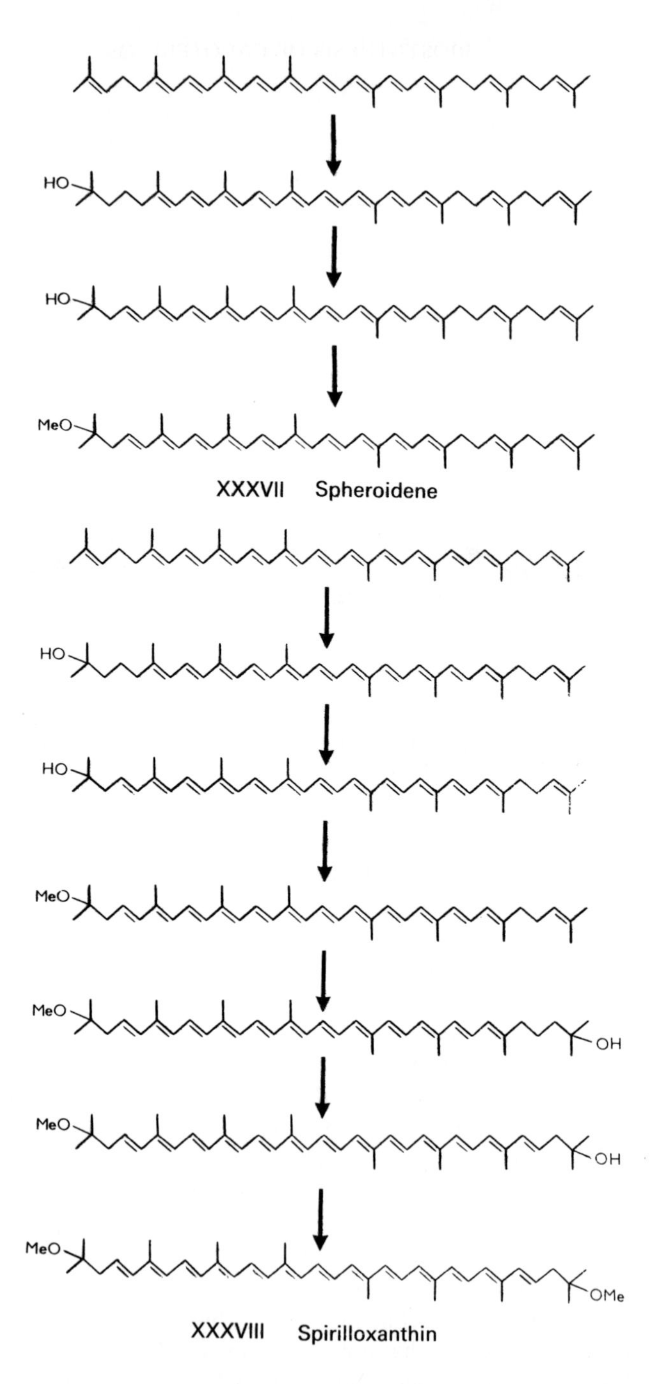

Fig. 3.20 Sequence of reactions in the biosynthesis of spheroidene (XXXVII) and spirilloxanthin (XXXVIII) from neurosporene and lycopene respectively, in photosynthetic bacteria.

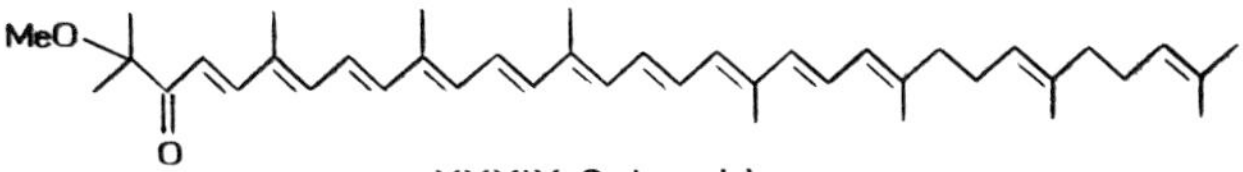

MeO

O

XXXIX Spheroidenone

In some photosynthetic bacteria, cross-conjugated C-13 aldehyde derivatives are readily formed, under anaerobic conditions (Francis and Liaaen-Jensen, 1970). This oxidation reaction is usually accompanied by isomerization to the 13-*cis*-form, but the mechanism of the transformation is not known.

G C_{30} Diapocarotenoids

The carotenoids of several bacteria, notably *Streptococcus faecium*, *Staphylococcus aureus* and *Pseudomonas rhodos*, have been shown to be C_{30} compounds, designated 4,4′-diapocarotenoids. In addition to a series of hydrocarbons [(4,4′-diapophytoene, 4,4′-diapophytofluene, 4,4′-diapo-ζ-carotene, and 4,4′-diaponeurosporene), XL-XLIII] analogous to the C_{40} desaturation series, related xanthophylls such as 4,4′-diaponeurosporen-4-ol (XLIV) and the complex carbohydrate derivative, staphyloxanthin, have been identified (Taylor and Davies, 1973, 1974a,b, 1976; Marshall and Wilmoth, 1981a; Kleinig and Schmitt, 1982).

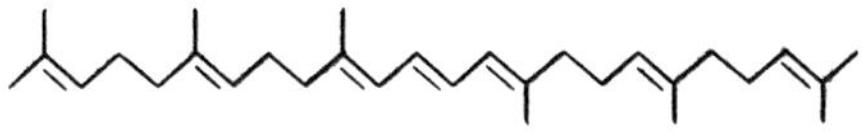

XL 4,4′-Diapophytoene

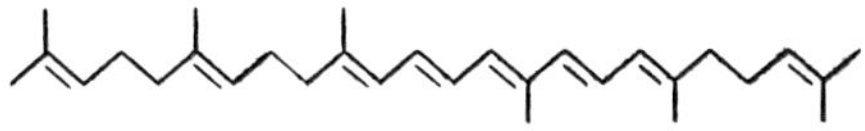

XLI 4,4′-Diapophytofluene

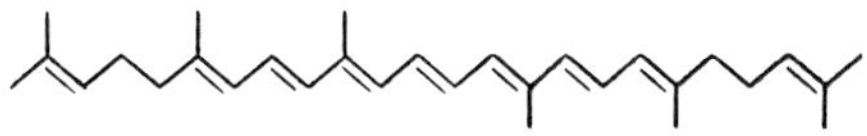

XLII 4,4′-Diapo-ζ-carotene

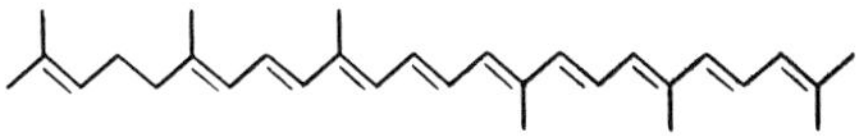

XLIII 4,4′-Diaponeurosporene

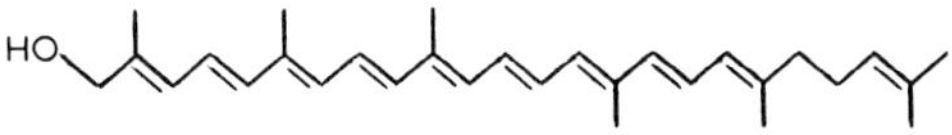

HO

XLIV 4,4′-Diaponeurosporen-4-ol

Cell-free systems from these bacteria have been used to incorporate labelled MVA, isopentenyl diphosphate and farnesyl diphosphate into the diapocarotenoids, but geranylgeranyl diphosphate was not incorporated. The results obtained proved that the bacterial 4,4′-diapocarotenoids are triterpenes, biosynthesized from two C_{15} farnesyl diphosphate molecules, via the C_{30} 4,4′-diapophytoene (XL) (dehydrosqualene), and are not derived by degradative loss of two C_5 units from a standard C_{40} carotenoid skeleton (Davies and Taylor, 1982; Taylor and Davies, 1982). The mechanisms of the reactions, namely diapophytoene formation and desaturation, are assumed to be the same as in the C_{40} series (Marshall and Wilmoth, 1981b; Kleinig and Schmitt, 1982), but no experimental evidence for this has been obtained.

The several apocarotenoids [e.g. β-citraurin (3-hydroxy-8′-apo-β-caroten-8′-al)] and C_{20} diapocarotenoids (e.g. crocetin-derivatives) of higher plants are degradation products of conventional C_{40} carotenoids (Pfander and Schurtenberger, 1982).

3 CAROTENOGENIC ENZYME SYSTEMS

In 1974, almost all the work on carotenoid biosynthesis in cell-free systems had been performed with extracts of tomato chromoplasts or a preparation of the mould *Phycomyces blakesleeanus*. These were crude preparations and no real progress had been made towards purifying the carotenogenic enzymes. This remains a very difficult area of study and, although several crude cell-free preparations from other natural sources have been described, pure enzymes have not been obtained and the characteristics of the enzyme-catalysed reactions have not been established.

It is relatively easy to isolate cell-free preparations which are capable of converting GGDP, IDP or MVA into phytoene. The desaturation of phytoene *in vitro* is much more difficult to achieve. The desaturase enzymes are assumed to be membrane-bound, along with the enzymes for cyclization, hydroxylation and other later modifications. It is generally believed that carotenoid biosynthesis takes place on a multienzyme complex which is bound to, or may even be an integral part of, a membrane. Phytoene synthetase is considered to be peripheral to the membrane and is easily dissociated from it, but the enzymes responsible for the later reactions are tightly associated and very difficult to dislodge or solubilize without completely destroying their activity. There is likely to be variation in the organization, tightness of association etc. among different organisms.

In studies of carotenoid biosynthesis in cell-free systems, factors such as the correct presentation of water-insoluble substrates, in a solubilized form, to the enzyme are crucial; dispersion by means of detergent (usually Tween 80) has been widely used and the administration of an ethanolic solution has also proved successful. The rigorous purification of labelled substrates and products is obviously essential but has probably not always been achieved. The fact that small amounts of one labelled carotenoid, of high specific activity, can be "carried" by another unlabelled carotenoid through several purification stages means that misleading results are easily obtained.

Progress with those cell-free systems that have been studied most intensively is outlined below. The subject of the biosynthesis of carotenoids *in vitro* has been reviewed by Bramley (1985a).

A Higher plants

In plants, the biosynthesis of carotenoids is restricted to plastids, e.g. chromoplasts of fruit and flowers, chloroplasts of photosynthetic tissues. The work with chromoplasts is outlined below; carotenoid biosynthesis in chloroplasts will be discussed later.

1 CHROMOPLASTS OF TOMATO (*LYCOPERSICON ESCULENTUM*)

The pioneering work on the biosynthesis of carotenoids in cell-free systems came from Porter's laboratory (Porter and Spurgeon, 1979). An acetone powder was made from chromoplasts of tomato fruit, and this preparation retained enzyme activity for the synthesis of carotenes. Over a period of several years these acetone powders, made from several different strains of tomato, were used to demonstrate first the incorporation of IDP into phytoene, lycopene and cyclic carotenes and then the direct conversion of precursor carotenoids into products. Extraction of the acetone powder with buffer gave a soluble enzyme system which could synthesize phytoene from IDP. This enzyme was purified substantially (350-fold) and obtained as a 2×10^5 dalton complex which contained all the enzyme activities necessary for converting IDP into phytoene, i.e. isopentenyl diphosphate: dimethylallyl diphosphate isomerase, prenyl transferase, prephytoene diphosphate synthetase and phytoene synthetase (Maudinas *et al.*, 1977). The instability of the complex precluded its purification to homogeneity. When the complex was subjected to ion-exchange chromatography, the ability to synthesize phytoene was lost, but the preparation still contained a 40 000-dalton protein which actively converted IDP into PPDP (Islam *et al.*, 1977). This enzyme is remarkably small, considering that it contains three different enzyme activities, but it has been purified to apparent homogeneity. The metal ion, Mn^{2+}, which could partly be replaced by Mg^{2+}, was required by this enzyme, and for phytoene synthesis, for which ATP was markedly stimulatory though it is not directly involved in the reaction.

Enzyme systems capable of desaturation have been obtained from acetone powders of plastids from normal red tomatoes and mutant strains by precipitation with ammonium sulphate. The preparation from red tomatoes converted *cis*-phytoene into *cis*-phytofluene, *trans*-phytofluene, ζ-carotene, neurosporene, lycopene and γ-carotene (Kushwaha *et al.*, 1970). Later studies also demonstrated the incorporation of *cis*-phytofluene, *trans*-phytofluene and *trans*-ζ-carotene into neurosporene and lycopene (Qureshi *et al.*, 1974a), whereas preparations from tangerine tomatoes converted *cis*-phytofluene, *trans*-phytofluene, *cis*-ζ-carotene and *trans*-ζ-carotene into prolycopene and the related poly*cis*-neurosporene (Qureshi *et al.*, 1974b,c). For desaturation, anaerobic conditions were used and a requirement for $NADP^+$, FAD and Mn^{2+} was reported.

Similar tomato systems have also been used to demonstrate the incorporation of radioactive lycopene into the cyclic α-carotene, β-carotene, γ-carotene and δ-carotene [(ε,ψ-carotene), XLV], especially with the high-β and high-δ strains (Kushwaha *et al.*, 1969; Papastephanou, 1973). The reported requirement for FAD is surprising since the cyclization reaction does not involve any oxidation step. The incorporations of substrate carotenes were achieved with crude preparations; no purification of the desaturase or cyclase enzymes was achieved.

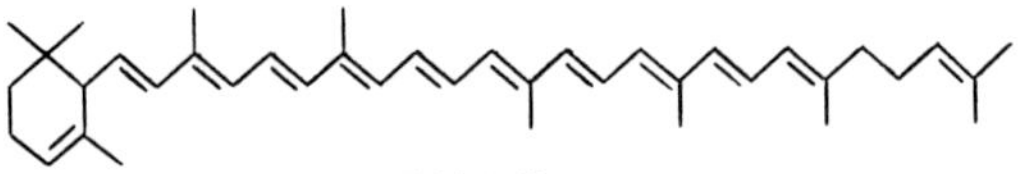

XLV δ-Carotene

2 CHROMOPLASTS FROM RED PEPPER (*CAPSICUM ANNUUM*)

Fruit of the red pepper, *Capsicum annuum*, accumulate large amounts of the cyclopentane carotenoids capsanthin (XXVIII) and capsorubin (XXIX). These carotenoids, and the enzymes responsible for their biosynthesis, are located entirely in the chromoplasts. A crude enzyme system from *Capsicum* chromoplasts has been used extensively by Camara and coworkers to study various stages in carotenoid biosynthesis (Camara and Monéger, 1982; Camara *et al.*, 1982a,b,c,d), and the partial purification of some of the enzymes has been achieved (Camara, 1985). The soluble fraction from the chromoplasts was able only to incorporate IDP into phytoene. Later steps occurred exclusively in the membrane fraction. A 1.9×10^5 dalton phytoene synthetase similar to that from tomatoes has been isolated (Camara, 1985). The synthesis of phytoene from IDP by this complex was stimulated by polyethyleneglycol and inhibited by allylic diphosphates. Solubilization of chromoplast membranes with Tween 80 or via an acetone powder gave preparations with lycopene cyclase activity. These preparations converted lycopene into β-carotene, though no formation of γ-carotene was detected: they did not require $NADP^+$ or FAD (Camara *et al.*, 1984, 1985c; Camara and Dogbo, 1986). Desaturation of phytoene to phytofluene and ζ-carotene has been reported (Camara and Monéger, 1982) but the desaturase enzyme system has not been studied in detail.

Preparations of *Capsicum* chromoplast membranes have also been used to demonstrate the incorporation of antheraxanthin and violaxanthin into capsanthin and capsorubin respectively, and look promising for studies of the enzymes of xanthophyll biosynthesis (Camara and Monéger, 1980, 1981, 1982).

3 CHROMOPLASTS OF DAFFODIL (*NARCISSUS PSEUDONARCISSUS*)

Chromoplasts from flowers (daffodil, *Narcissus pseudonarcissus*) have been used as the basis of another very promising cell-free carotenogenic system. The chromoplasts will incorporate IDP into β-carotene (Beyer *et al.*, 1980; Kreuz *et al.*, 1982). The membrane-bound enzymes have been solubilized with the detergent CHAPS; the soluble system thus produced converted *cis*-phytoene into β-carotene in high yield. The enzymes are considered to behave as a tight assembly line (Beyer *et al.*, 1985).

B Fungi

The most extensive work on the biosynthesis of carotenoids by cell-free extracts of fungi has been performed with the moulds *Phycomyces blakesleeanus* and *Neurospora crassa*.

1 *PHYCOMYCES BLAKESLEEANUS*

The wild-type of *Phycomyces* synthesizes β-carotene, and many mutants have been isolated which accumulate, instead, the intermediates phytoene or lycopene, or in which β-carotene production is very greatly increased. Carotenogenic cell-free extracts have been prepared from the wild-type and from several of these mutant strains (Bramley *et al.*, 1975, 1977; De la Concha *et al.*, 1983; Bramley, 1985b). Carotene biosynthesis in *Phycomyces* is associated with the membranes, though the enzymes for phytoene synthesis are peripheral to the membrane and are easily dislodged (Bramley, 1985b; Bramley and Taylor, 1985). Active cell-free preparations are obtained by extraction of lyophilized mycelia with buffer or detergent solution. The crude systems are capable of incorporating MVA and prenyl diphosphates into carotene, and have been used to demonstrate the conversion of precursor carotenoids into products. Thus, the isomerization of 15-*cis*-phytoene to the all *trans*-isomer (Bramley and Davies, 1976; De la Concha *et al.*, 1983) and the incorporations of radioactive phytoene and neurosporene into lycopene and β-carotene, and of lycopene into γ-carotene and β-carotene have been described. The incorporation of [^{14}C]-neurosporene into β-carotene was reduced to an approximately equal extent by the addition of unlabelled samples of either of the proposed alternative intermediates, lycopene or β-zeacarotene [(7′,8′-dihydro-β-ψ-carotene), XLVI]. From this, it was concluded that the two alternative pathways from neurosporene to β-carotene via lycopene or β-zeacarotene are of equal importance (Bramley *et al.*, 1977).

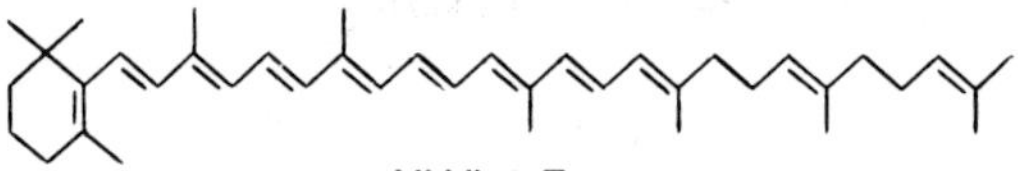

XLVI β-Zeacarotene

Both routes, however, would require the same reactions, occurring in a different order.

The enzymes which catalyse the incorporation of MVA into phytoene could be dissolved from the membrane by buffers of high ionic strength, whereas detergent treatment was required to solubilize the later enzymes which convert phytoene into β-carotene (Bramley, 1985b; Bramley and Taylor, 1985). Only two of the many detergents used gave a preparation that retained its enzyme activity. With the solubilized and partly purified *Phycomyces* enzyme system, as with that from tomato chromoplasts, ATP had a stimulatory effect on phytoene synthesis, and desaturation required NADP (Bramley, 1985a).

2 *NEUROSPORA CRASSA*

Neurospora crassa accumulates β-carotene and neurosporaxanthin [(4′-apo-β-caroten-4′-oic acid), XLVII]. The conversion of MVA into GGDP

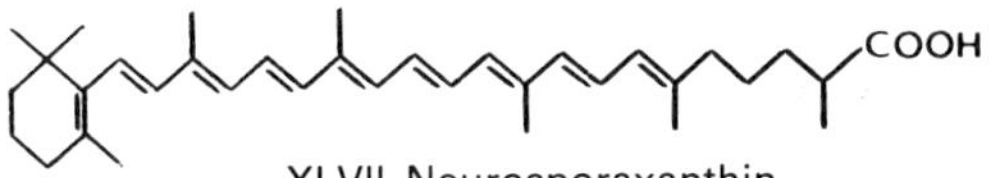

COOH

XLVII Neurosporaxanthin

is catalysed by soluble enzymes, but the later carotenogenic enzymes, including phytoene synthetase, are located in the plasma membranes and the endoplasmic reticulum (Spurgeon *et al.*, 1979; Mitzka-Schnabel and Rau, 1981; Mitzka-Schnabel, 1985). They can be solubilized by the detergents CHAPS or sodium cholate, though the addition of lipid is necessary to restore full enzyme activity. With GGDP as substrate, phytoene synthetase has no specific cofactor requirements, but Mg^{2+} ions are stimulatory (Mitzka-Schnabel, 1985).

C Bacteria

Most bacteria apparently will not take up or metabolize exogenous MVA. There have, however, been several reports of the incorporation of IDP into carotenoids by crude cell-free preparations or homogenates of a number of bacteria, e.g. into β-carotene by *Halobacterium cutirubrum* (Kushwaha *et al.*, 1976), into C_{45} and C_{50} carotenoids by *Flavobacterium dehydrogenans, Corynebacterium poinsettiae* and *Micrococcus luteus* (Fahey and Milborrow, 1978; Evans and Prebble, 1980, 1982; Swift and Milborrow, 1981b), and into β-carotene and zeaxanthin by a *Flavobacterium* strain R1560 (Brown *et al.*, 1975; Britton *et al.*, 1980). The conversion of phytoene into β-carotene by *H. cutirubrum* has also been described (Kushwaha *et al.*, 1976).

A cell-free system from *Flavobacterium* R1560, an organism which accumulates large amounts of zeaxanthin, has been used extensively (Brown *et al.*, 1975; Britton *et al.*, 1980). Active preparations can be obtained when the cells are broken by several methods, particularly sonication and disruption by the French press. These preparations will incorporate MVA extremely efficiently into phytoene and, to a lesser extent, into phytofluene, lycopene, β-carotene and zeaxanthin. Similar incorporations into phytoene have also been achieved with IDP and PPDP.

The isomerization of 15-*cis*-phytoene to *trans*-phytoene has been demonstrated with this system, but the direct desaturation of phytoene was achieved only to a very small extent. A solubilized membrane-containing fraction gave very efficient cyclization of radioactive lycopene to β-carotene (up to 20% incorporation); neurosporene was also incorporated into β-carotene (M. Browne and G. Britton, unpublished results). The direct introduction of oxygen functions in the biosynthesis of xanthophylls has proved notoriously difficult to demonstrate, but the hydroxylation of β-carotene to zeaxanthin has been achieved at a low level with the *Flavobacterium* system.

The partial purification of the enzyme system responsible for the synthesis of phytoene from MVA has been achieved by ammonium sulphate precipitation from a soluble fraction; the efficiency of synthesis of phytoene was greatly improved by the inclusion of detergent in the incubations.

D Algae

There have been very few reports of carotenoid biosynthesis by cell-free systems from algae, and no systematic studies. A homogenate from the dinoflagellate *Amphidinium carterae* was used to demonstrate the incorporation of zeaxanthin into neoxanthin, diadinoxanthin and peridinin (Swift *et al.*, 1982). Recently Sandmann and Bramley (1985a) have used the novel approach of incubating with a coupled enzyme system, containing a cell extract of a suitable strain of *Phycomyces* which, it is assumed, provided the early enzymes for synthesis of phytoene or β-carotene, and a membrane fraction from the blue-green alga (cyanobacterium) *Aphanocapsa* 6714 which then converted the phytoene or β-carotene into cyclic carotenoids, especially β-cryptoxanthin [(β,β-caroten-3-ol), XLVIII]. The biosynthesis of xanthophylls from GGDP and other substrates, including β-carotene, has been achieved with cell-free extracts and membrane preparations of *Aphanocapsa* alone (Clarke *et al.*, 1982; Bramley and Sandmann, 1985; Sandmann and Bramley, 1985b).

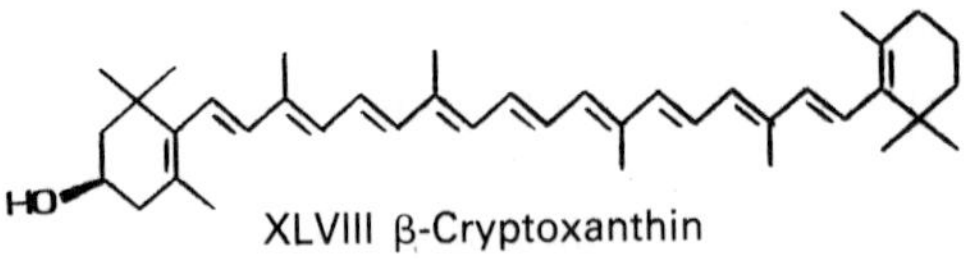

HO

XLVIII β-Cryptoxanthin

4 REGULATION OF CAROTENOID BIOSYNTHESIS

A Fungi

1 GENETICS

Many colour mutants have been produced from carotenogenic fungi, notably *Phycomyces blakesleeanus*, *Neurospora crassa*, and *Ustilago violacea* (Cerda-Olmedo and Torres-Martinez, 1979; Roncero and Cerda-Olmedo, 1982). Extensive genetic complementation studies have been performed in *Phycomyces* and *Neurospora*. In the case of *Phycomyces*, two structural genes, *car B* and *car R*, have been assigned to the desaturation and cyclization reactions respectively (de la Guardia *et al.*, 1971; Aragon *et al.*, 1976). Four copies of the product of *car B* and two copies of the product of *car R* are suggested to form a multi-enzyme complex which converts phytoene into β-carotene (Cerda-Olmedo and Torres-Martinez, 1979). A third structural gene, *car A*, has been identified; this is thought to play a role in the transfer of the intermediate substrate, lycopene, to the cyclase enzyme (Murillo *et al.*, 1981). Genetic mapping experiments place *car A* very close to *car R*; the two may represent segments of the same gene (Torres-Martinez *et al.*, 1980). Other genes, such as *car S*, are involved in regulating the amount of carotenoid produced (Murillo and Cerda-Olmedo, 1976; Revuelta and Eslava, 1983).

The earliest genetic studies on carotenogenic fungi were performed with *Neurospora* (Huang, 1964; Subden and Threlkeld, 1970; Subden and Goldie, 1973; Kushwaha *et al.*, 1978). The results obtained are broadly similar to those with *Phycomyces*, but a complementation group defining a gene *ylo* is involved in the formation of the xanthophyll neurosporaxanthin. Recent work on *Neurospora* has been much less extensive than that with *Phycomyces*, although electrophoresis has revealed that the wild-type and an *al-2* white mutant differ in one protein band which is assumed to be a protein complex involved in carotenogenesis (Subden and Turian, 1970a,b).

2 CHEMICAL BIOINDUCTION

Many further studies on carotenoid biosynthesis in the heterothallic fungus, *Blakeslea trispora*, have been reported and the subject has been reviewed (Lampila *et al.*, 1985). When the (+) and (−) strains of this fungus are cultured together, the amount of β-carotene produced is enormously increased; a "hormone", trisporic acid, has been identified as the stimulatory factor (Bu'Lock *et al.*, 1976). Trisporic acid is derived from β-carotene and not incorporated into it (Austin *et al.*, 1970).

A stimulatory effect can be produced by various other compounds structurally related to trisporic acid, including abscisic acid, β-ionone and retinol (Dandekar *et al.*, 1980; Yakovleva *et al.*, 1980; Mehta *et al.*, 1981). An inhibitor protein and a membrane-bound protease appear to be important mediators of the regulatory effects. The involvement of a phosphatase enzyme, regulated by phospate levels (Govind *et al.*, 1981; Dholakia and Modi, 1983, 1984) and of cyclic AMP (Dandekar and Modi,

1980; Govind and Modi, 1981), as well as a stimulatory effect of Cu^{2+} ions (Govind *et al.*, 1982) have also been reported.

The accumulation of carotenoids by *Neurospora crassa* is also influenced by cyclic AMP (Kritsky *et al.*, 1982; Sokolovsky *et al.*, 1983).

3 PHOTOINDUCTION

In fungi which form some carotenoid in the dark, e.g. *Phycomyces*, it is common for carotenogenesis to be stimulated substantially by light. With other fungi, and some bacteria, in which little or no carotenoid is present in the dark, carotenoid synthesis is photoinduced by exposure to light and oxygen (Rau, 1976, 1980, 1983, 1985; Harding and Shropshire, 1980). Only a brief period of illumination is needed, after which carotenoid synthesis will take place in the dark. Rau (1985) suggests that such photoregulation provides a "saving' mechanism by which the photoprotective carotenoids are synthesized only when they are needed, thus avoiding wastage of energy and materials.

The mechanism of photoinduction has been studied intensively in the yeast *Rhodotorula minuta* (Tada and Shiroishi, 1982a,b,c,d; Tada *et al.*, 1982) and in several moulds, notably *Neurospora crassa* and *Fusarium aquaeductuum*—an outline of work with these moulds is given below. A brief period of illumination is followed by a lag phase, after which carotenoids begin to accumulate, even in the dark. The synthesis of carotenoids will stop after a time, in the absence of further illumination. The effective light is blue light (around 450 nm), and the photoreceptor is generally believed to be a flavoprotein (Rau, 1980), though β-carotene itself has been suggested for this role (Whitaker and Shropshire, 1981). Oxygen is necessary for the photoinduction process, but illumination in the absence of oxygen, followed by exposure to air in the dark, can also be effective. The photoinductive effect is biphasic, but both phases have a lag period and require enzyme synthesis *de novo* (Rau and Rau-Hund, 1977; Schrott, 1980a,b, 1981). It appears that all the enzymes of the carotenogenic pathway, especially phytoene synthetase, are photoinduced. Work with inhibitors suggests that the production of carotenogenic enzymes is regulated at the level of transcription, and synthesis of mRNAs precedes enzyme synthesis (Schrott and Rau, 1977). The appearance of at least four new peptides, after photoinduction, has been seen (Mitzka-Schnabel, 1985) but any connection between these and carotenogenesis remains to be proved. The involvement of cyclic AMP in the synthesis of the carotenogenic enzymes in *Neurospora* has been reported (Kritsky *et al.*, 1981; Kritsky, 1985).

Similar photoinduction of carotenoid biosynthesis by processes involving enzyme synthesis *de novo* has been observed in bacteria, such as *Mycobacterium* (Johnson *et al.*, 1974; Kato *et al.*, 1981) and *Brevibacterium* (Koyama *et al.*, 1981).

B Algae

Under unfavourable nutritional conditions, especially nitrogen deficiency, some green algae may become red because of the formation of large amounts of β-carotene or combinations of its keto-derivatives, echinenone [(β,β-caroten-4-one), XLIX], hydroxyechinenone [(3-hydroxy-β,β-carot-en-4-one), L], canthaxanthin [(β,β-carotene-4,4′-dione), LI] and astaxanthin (XXVI). These 'secondary' carotenoids are located outside the chloroplast in globules or in the cell wall in association with sporopollenin, the structural polymer, of which they may be precursors (Burczyk, 1987a,b; Burczyk *et al.*, 1981). *Dunaliella bardawil* has been investigated extensively as a potential commercial source of β-carotene. When subjected to high salt concentration, light intensity and temperature, and depleted in nitrate, this species accumulates extraordinarily large amounts of β-carotene in globules. Carotene concentrations can reach as high as 8% of cell dry weight (Ben-Amotz *et al.*, 1982).

The biosynthesis of the ketocarotenoids has not been studied in detail but is assumed to follow the normal pathways (Donkin, 1976). Studies

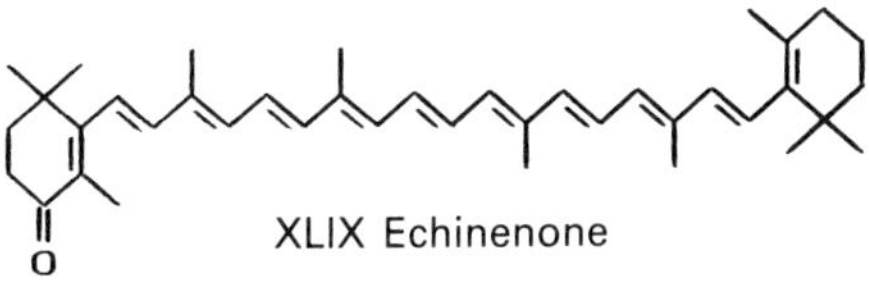

O

XLIX Echinenone

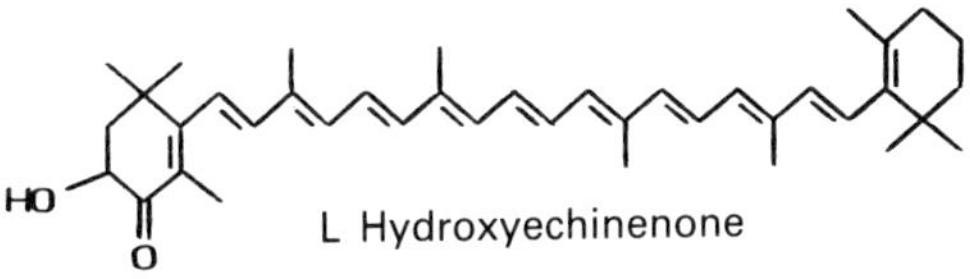

HO

O

L Hydroxyechinenone

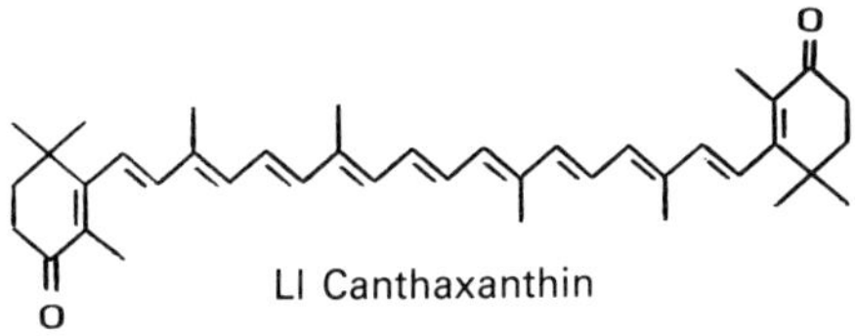

O

O

LI Canthaxanthin

of carotenoid changes that occur when ζ-carotene-accumulating mutants of *Scenedesmus obliquus* are transferred to deuterium oxide confirmed that the secondary carotenoids, like the normal chloroplast cyclic carotenoids, were $^{2}H_2$-species. This proves that the accumulated ζ-carotene can be used to make both the chloroplast carotenoids and the extra-chloroplastic secondary carotenoids (Britton *et al.*, 1977c).

C Higher plants

1 CAROTENOID BIOSYNTHESIS AND CHLOROPLAST DEVELOPMENT

(a) Carotenoids in chloroplasts

In all higher plants the carotenoids are synthesized and accumulate in plastids, e.g. chloroplasts in photosynthetic tissues or chromoplasts in flowers, fruit and roots. The carotenoids in the chloroplast are located, together with the chlorophylls, in functional pigment–protein complexes (PPC) in the thylakoid membranes. (For a review see Siefermann-Harms, 1985.) The different PPC have different and characteristic carotenoid compositions. Thus, the reaction centre core complexes of photosystem I and photosystem II are rich in, and may contain only, β-carotene (Omata *et al.*, 1984), whereas the more peripheral light-harvesting chlorophyll-proteins (LHCP) associated with the two photosystems contain the xanthophylls lutein, violaxanthin and neoxanthin. There have been reports that some xanthophyll, especially violaxanthin, is present in the chloroplast envelope (Douce and Joyard, 1979). Other workers, however, have found no evidence of carotenoids in the chloroplast envelope (Grumbach, 1983).

Carotenoid biosynthesis is an integral part of the construction of the PPC in the thylakoids, and is closely regulated to the formation of other components, e.g. chlorophyll, proteins, lipids. The whole process of chloroplast development is thought to be under direct nuclear control by a block of genes which is switched on as a unit. The supply of the various components, including carotenoids, would then be regulated by control mechanisms which operate on the individual biosynthetic pathways.

(b) Control by light

Plants that are grown or maintained in the dark become etiolated. They are yellow in colour and have etioplasts instead of the chloroplasts normally present in green leaves. The etioplasts are usually devoid of chlorophyll but do contain substantial amounts of carotenoids, almost entirely xanthophylls (Goodwin, 1958). On illumination, etiolated plants develop normal chloroplasts and become green. The normal chloroplast carotenoids are synthesized along with chlorophyll. It is not known whether the carotenoids present in the etioplasts become incorporated into the thylakoids of the chloroplasts. The greening of etiolated seedlings provides a convenient system for studying chloroplast development. The normal course of plant growth, however, does not involve etiolation and etioplasts. Under a normal light regime chloroplasts are formed *de novo*, from proplastids.

Light is obviously the main regulatory requirement for chloroplast development and hence for carotenoid biosynthesis. The major stimulus for carotenoid biosynthesis is red light, acting via both phytochrome and chlorophyll (Oelmüller and Mohr, 1985). It is thought that the reaction centre complexes, including their carotenoids, are made first, and then the LHCP containing the xanthophylls (Argyroudi-Akoyunoglou and Akoyunoglou, 1979; Tanaka and Tsuji, 1985; Burkey, 1986). Chloroplast carotenoid compositions are influenced by light, as the relative amounts of the various PPC may change in response to changes in environmental conditions. Thus 'sun' and 'shade' chloroplasts, with different PPC distributions and carotenoid contents and compositions, are present in many plants grown in strong and dim light, respectively (Wild, 1979; Grumbach and Lichtenthaler, 1982; Lichtenthaler *et al.*, 1982). With beech trees (*Fagus sylvatica*), normally shaded leaves have a significantly higher xanthophyll:carotene ratio than do sunlit leaves, a consequence of their increased relative amounts of LHCP (Lichtenthaler *et al.*, 1981). Similar results have been obtained with pea plants (*Pisum sativum*) grown under low or high light (Chu and Anderson, 1984).

(c) Regulation of biosynthesis within the chloroplast

There is no doubt that the carotenoids that accumulate in the chloroplast are biosynthesized within that organelle. There are conflicting views, however, about whether the chloroplast is completely autonomous, or whether early biosynthetic intermediates need to be imported. Thus there are reports that isolated chloroplasts can synthesize carotenoids from MVA (Buggy *et al.*, 1969), and that the formation of MVA itself, from acetyl-coenzyme A, takes place within the chloroplast; the key enzyme 3-hydroxy-3-methylglutaryl-CoA reductase has been identified in chloroplasts (Grumbach and Bach, 1979). Other workers, however, have concluded that the chloroplast cannot carry out these early reactions, but must import the later intermediate, IDP (Kreuz and Kleinig, 1981).

The location of the later stages of carotenoid biosynthesis within the chloroplast is also disputed. In spinach, for example, phytoene synthetase and desaturase activities were found in the envelope fraction, and the thylakoid fraction was devoid of activity (Lütke-Brinkhaus *et al.*, 1982). This contrasts with the situation in radish (*Raphanus sativus*), where carotenoid biosynthesis, including phytoene formation, was restricted to the thylakoid fraction (Grumbach and Britton, 1984).

(d) Organization of carotenoid biosynthesis in the chloroplast

Because the carotenoids are localized and function in different PPC within the chloroplast, it is not sufficient to think only of the biosynthesis of total leaf carotenoid or total chloroplast carotenoid as a single entity. The factors which determine the different carotenoid compositions of the individual PPC must also be understood (Britton, 1986). There is evidence, for example, that the time course of xanthophyll synthesis (especially lutein) is not the same as that of bulk carotene synthesis in the same developing system. Since β- and ε-rings are not interconverted, the final stages of lutein synthesis must be independent of those of β-carotene and the other xanthophylls. It is not known, however, at which stage the two pathways diverge, i.e. whether lutein and β-carotene are made *de novo*, by two distinct enzyme assemblies, or whether the divergence involves only the final cyclization and hydroxylation reactions, after phytoene synthesis and desaturation have taken place on a common enzyme assembly.

When either light-grown or dark-grown seedlings are incubated, in the light, with [^{3}H]-MVA and $^{14}CO_2$ simultaneously, different ^{3}H:^{14}C ratios are observed for the individual carotenoids, particularly β-carotene and lutein, indicating variations in the time courses of their synthesis (Sergeant and Britton, 1984). However, no great differences were observed in either total incorporations or ^{3}H:^{14}C ratios when the carotenoids of a heavy membrane fraction, containing stacked thylakoids enriched in photosystem II and LHCP, were compared with those of a light fraction enriched in photosystem I (J.M. Sergeant and G. Britton, unpublished results). These results confirm the different behaviour of lutein and β-carotene, but suggest that the carotenoid molecules destined for the PPC of stacked and unstacked thylakoids, respectively, are not made at separate sites, either during the initial construction of the photosynthetic apparatus or when a steady state has been reached. Nothing is known of the delicate regulatory mechanisms that must exist to determine the specific destination within the complexes of the thylakoid for each carotenoid molecule that is made. In relation to this question it may be significant that carotenoid molecules which are labelled in a short incubation with various labelled substrates lose their labelling much more quickly than those which are labelled in a longer-term incubation (K.H. Grumbach and G. Britton, unpublished results). This result suggests that newly formed carotenoid molecules may enter a labile pool from which either they are taken for incorporation into the PPC or rapidly destroyed. The overall story of carotenoid biosynthesis and its regulation in chloroplasts is very complicated (Britton, 1986). At least three phases of carotenoid biosynthesis must be considered:

(i) the bulk synthesis that occurs during the initial construction of the photosynthetic apparatus;
(ii) the synthesis that continues in mature chloroplasts as part of turnover;
(iii) the synthesis that occurs as a response to or consequence of changes in environmental conditions, especially light intensity.

The factors which regulate these three phases may be different. In addition, the enzymes responsible for making the carotenoids in etiolated plants may continue to function when dark-grown plants are illuminated, especially in the early stages before enzyme synthesis and chloroplast development *de novo* become established.

It must also be appreciated that chloroplast development is not likely to follow the same course universally. Substantial differences can be expected in the characteristics of chloroplast development and carotenoid biosynthesis in different plant species and even varieties of the same species. Factors such as the age of the plant or tissue under investigation and the growth conditions employed are very important, and cotyledons and true leaves, even of the same plant, may behave quite differently and are not strictly comparable. Finally the course of chloroplast development, and its associated carotenoid biosynthesis, in plants growing normally in the light is not the same as in a greening etiolated system, so comparisons and extrapolations should be made with extreme caution.

(e) Chemical inhibitors and regulation of carotenoid biosynthesis

Carotenoid biosynthesis in plants, fungi and bacteria is affected by many chemicals. Some of these have been used extensively to gain valuable information about the biosynthetic pathways. The classical work, since about 1960, used diphenylamine (DPA) which inhibits desaturation. In the presence of DPA, phytoene and other intermediates accumulate. Kinetic studies of the transformations that occur after removal of the DPA have allowed the subsequent pathways to be elucidated. More recently, nicotine and 2-(4-chlorophenylthio)-triethylammonium chloride (CPTA) have been widely used as inhibitors of cyclization and related reactions such as 1,2-hydration in the photosynthetic bacteria. The inhibition by nicotine is reversible, and its use has allowed the direct demonstration of these reactions and reaction sequences and the elucidation of the stereochemistry of the cyclization reactions.

The effects of these compounds in higher plants are usually much smaller and less specific than in fungi or bacteria. However, in recent years, many new chemical regulators have been evaluated for their effects on carotenoid biosynthesis in citrus fruits and other plant systems. Many amines have been studied and shown to have a range of different effects on the carotenoid biosynthetic pathway (Yokoyama *et al.*, 1982). Some greatly increased the total amount of carotenoid produced; others caused alterations in the carotenoid composition, especially by blocking desaturation or cyclization. In some cases, production of a poly*cis* isomer of lycopene was stimulated. The identification of this isomer as prolycopene [(7,9,7′,9′-tetra*cis*), X] was not confirmed by NMR spectroscopy.

Chemicals which block carotenoid biosynthesis, especially those which prevent the desaturation of phytoene, have been widely studied and are in use as bleaching herbicides.

2 CAROTENOID BIOSYNTHESIS AND HERBICIDE ACTION

The most important function of carotenoids, especially β-carotene, in plants is to protect the chloroplast against chlorophyll-sensitized photooxidative damage. If carotenoids are not available, this protection is removed and exposure to light will kill the plant. Chemicals which block carotenoid biosynthesis, especially by preventing the desaturation of phytoene, are therefore potentially lethal, and some such compounds are in use as bleaching herbicides. The topic of carotenoid biosynthesis as a target for herbicide activity has been reviewed by Britton (1979b) and by Ridley (1982).

The most effective compounds are obviously those which cause the accumulation of phytoene or ζ-carotene in place of the normal, coloured carotenoids. The chromophores of phytoene and ζ-carotene are too short to enable them to protect against chlorophyll-sensitized photooxidation. The best known of these bleaching herbicides are probably the pyridazinones, especially metflurazon (SAN 6706) and norflurazon (SAN 9789) (Eder, 1979). Plants treated with these compounds die because they accumulate phytoene instead of the normal, coloured, cyclic carotenoids of their pigment–protein complexes (Bartels and McCullough, 1972; Ben-Aziz and Koren, 1974). More recently, similar inhibitory effects have been reported for several other substances and classes of compounds, including fluridone (Bartels and Watson, 1978), difunone (Urbach *et al.*,

1976), substituted 4-hydroxypyridines (Ridley, 1982) and substituted diphenylethers (Lambert and Böger, 1983), whereas some substituted 6-methylpyrimidines cause the accumulation of ζ-carotene (Ridley, 1982).

It is usually assumed that these bleaching herbicides act by directly blocking the active site of the phytoene desaturase enzyme (Bramley *et al.*, 1984). However, there are many alternative possibilities which need to be considered. Thus the compound may prevent synthesis of the carotenogenic enzymes or the organization of these enzymes into a multi-enzyme complex in the membrane. It could also restrict the availability of substrates or cofactors, or interfere with the organization or functioning of the electron transport system that is believed to be associated with desaturation.

Compounds which affect cyclization and cause the accumulation of lycopene can also be lethal, e.g. CPTA, aminotriazole (amitrole) (Grumbach, 1981), and undecyl- and tridecyldiethylamines (Camara *et al.*, 1985a). Although, theoretically, the lycopene that accumulates in plants treated with these compounds should protect against photooxidation, it apparently cannot occupy the correct structural sites in the thylakoid PPC, so that photodynamic damage and death cannot be prevented. Any compound which inhibits any step in carotenoid biosynthesis or which prevents the proper incorporation of the carotenoid into its functional site within the chloroplast is, therefore, potentially useful as a bleaching herbicide.

D Photosynthetic bacteria

1 CAROTENOIDS IN PIGMENT–PROTEIN COMPLEXES

In the photosynthetic bacteria, as in higher plants and algae, the carotenoids are located, together with bacteriochlorophylls, in functional pigment–protein complexes (PPC) in the photosynthetic membranes (Siefermann-Harms, 1985). These bacteria have only one photosystem, but the carotenoids are located specifically in the photosynthetic reaction centre and in one or two light-harvesting antenna complexes, designated B800-850 and B875-890 on the basis of their main light-absorption maxima. The pigment compositions of reaction centres are quite uniform, with one carotenoid molecule present per four bacteriochlorophyll and two bacteriophaeophytin molecules. The actual carotenoids present obviously depend on the organism. For the antenna complexes, bacteriochlorophyll:carotenoid molar ratios of 1:1, 2;1 and 3:1 have been reported, but recent re-analysis by modern high-performance liquid chromatography methods has shown that the ratio is normally 2:1 (M.B. Evans, R. J. Cogdell and G. Britton, unpublished results).

2 NUTRITIONAL AND ENVIRONMENTAL REGULATION

Overall carotenoid compositions are influenced by many factors, including availability of nutrients and especially light intensity and oxygen concentration. Cell growth and pigment synthesis are independently regulated processes (Cohen-Bazire *et al.*, 1957). Changes in light intensity can greatly affect pigment contents. When the light intensity is reduced, the organisms usually respond by synthesizing new photosynthetic membranes or expanding the size of their light-harvesting antennae (Firsow and Drews, 1977), a process which obviously requires the synthesis of additional carotenoid. Oxygen partial pressure is also a major regulatory factor (Schmidt, 1978). If anaerobic cultures of *Rhodobacter sphaeroides* etc. are exposed to increasing oxygen partial pressure, the first response is to convert a substantial proportion of the spheroidene (XXXVII) into its 2-oxo-derivative spheroidenone (XXXIX).

In this and other species, as the oxygen concentration increases further, carotenoid synthesis is progressively inhibited. A recent report indicates that oxygen does not directly regulate carotenoid production, but its effect on carotenoid synthesis is a consequence of the regulation of bacteriochlorophyll synthesis (Biel and Marrs, 1985).

3 GENETIC CONTROL

A detailed genetic analysis of carotenogenesis in *Rhodopseudomonas capsulata* has been undertaken (Marrs, 1982). Seven genes which specifically affect carotenogenesis have been identified: *crt H* is correlated with normal desaturation, *crt C* with C-1,2 hydration, *crt D* with C-3,4 dehydrogenation, *crt F* with O-methylation and *crt A* with the introduction of the oxo-group at C-2. Mutants affected in *crt B* or *crt E* accumulate no carotenoids and appear to be blocked before the synthesis of phytoene. A further gene, *crt I*, which also mediates the conversion of phytoene into coloured carotenoids has recently been reported (Giuliano *et al.*, 1986). Mapping studies showed that these genes are in a tight cluster. Moreover, the syntheses of all components of the PPC are genetically intimately related, and all genes for photosynthesis and the synthesis of the photosynthetic pigments, both carotenoids and bacteriochlorophylls, are tightly clustered.

REFERENCES

Altman, L.J., Ash, L., Kowerski, R.C. *et al.*,(1972). *J. Am. Chem. Soc.* **94**, 3257–9.

Altman, L.J., Laungani, D.R., Rilling, H.C. and Vasak, J. (1977). *J. Chem. Soc., Chem. Commun.* 860–1.

Altman, L.J., Kowerski, R.C. and Laungani, D.R. (1978). *J. Am. Chem. Soc.* **100**, 6174–82.

Andrewes, A.G. and Starr, M.P. (1976). *Phytochemistry* **15**, 1009–11.

Andrewes, A.G., Liaaen-Jensen, S. and Borch, G. (1974). *Acta Chem. Scand.* **B28**, 737–42.

Andrewes, A.G., Liaaen-Jensen, S. and Weeks, O.B. (1975). *Acta Chem. Scand.* **B29**, 884–6.

Andrewes, A.G., Phaff, M.J. and Starr, M.P. (1976). *Phytochemistry* **15**, 1003–7.

Aragon, C.M.G., Murillo, F.J., de la Guardia, M.D. and Cerda-Olmedo, E. (1976). *Eur. J. Biochem.* **63**, 71–5.

Argyroudi-Akoyunoglou, J.H. and Akoyunoglou, G. (1979). *FEBS Letts* **104**, 78–84.

Austin, D.J., Bu'Lock, J.D. and Drake, D. (1970). *Experientia* **26**, 348–9.

Bartels, P. and McCullough, C. (1972). *Biochem. Biophys. Res. Commun.* **48**, 16–22.

Bartels, P.G. and Watson, C.W. (1978). *Weed. Sci.* **26**, 198–203.

Ben-Amotz, A., Katz, A. and Avron, M. (1982). *J. Phycol.* **18**, 529–37.

Ben-Aziz, A. and Koren, E. (1974). *Plant Physiol.* **54**, 916–20.

Beyer, P., Kreuz, K. and Kleinig, H. (1980). *Planta* **150**, 435–8.

Beyer, P., Weiss, G. and Kleinig, H. (1985). *Eur. J. Biochem.* **153**, 341–6.

Biel, A.J. and Marrs, B.L. (1985). *J. Bacteriol.* **162**, 1320–1.

Bramley, P.M. (1985a). *Adv.Lipid Res.* **21**, 243–79.

Bramley, P.M. (1985b). *Pure Appl. Chem.* **57**, 671–3.

Bramley, P.M. and Davies, B.H. (1975). *Phytochemistry* **14**, 463–9.

Bramley, P.M. and Davies, B.H. (1976). *Phytochemistry* **15**, 1913–16.

Bramley, P.M. and Sandmann, G. (1985). *Phytochemistry* **24**, 2919–22.

Bramley, P.M. and Taylor, R.F. (1985). *Biochim. Biophys. Acta* **839**, 155–60.

Bramley, P.M., Aung Than and Davies, B.H. (1977). *Phytochemistry* **16**, 235–8.

Bramley, P.M., Clarke, I.E., Sandmann, G. and Böger, P. (1984). *Z. Naturforsch.* **39C**, 460–3.

Britton, G. (1971). In *Aspects of Terpenoid Chemistry and Biochemistry* (Ed. Goodwin, T.W.), pp. 255–289. Academic Press, London.

Britton, G. (1976). *Pure Appl. Chem.* **47**, 223–36.

Britton, G. (1979a). In *Comprehensive Organic Chemistry* vol. 5 (Ed. Haslam, E.) pp. 1025–1042. Pergamon, Oxford.

Britton, G. (1979b). *Z. Naturforsch.* **34C**, 979–85.

Britton, G. (1982). *Physiol. Végétale.* **20**, 735–55.

Britton, G. (1983). *The Biochemistry of Natural Pigments.* Cambridge University Press, Cambridge.

Britton, G. (1985). *Pure Appl. Chem.* **57**, 701–8.

Britton, G. (1986). In *Regulation of Chloroplast Differentiation* (Eds. Akoyunoglou, G. and Senger, H.), pp. 125–134. A.R. Liss, New York.

Britton, G. and Mundy, A.P. (1980). *Dev. Plant Biol.* **6**, 345–50.

Britton, G. and Powls, R. (1977). *Phytochemistry* **16**, 1253–5.
Britton, G., Lockley, W.J.S., Patel, N.J. and Goodwin, T.W. (1977a) *FEBS Letts* **79**, 281–3.
Britton, G., Lockley, W.J.S., Patel, N.J., Goodwin, T.W. and Englert, G. (1977b). *J. Chem. Soc., Chem. Commun.* 655–6.
Britton, G., Lockley, W.J.S., Powls, R., Goodwin, T.W. and Heyes, L.M. (1977c). *Nature* **268**, 81–2.
Britton, G., Powls, R. and Schulze, R.M. (1977d). *Arch. Microbiol.* **113**, 281–4.
Britton, G., Singh, R.K., Malhotra, H.C., Goodwin, T.W. and Ben-Aziz, A. (1977e). *Phytochemistry* **16**, 1561–6.
Britton, G., Goodwin, T.W., Lockley, W.J.S., Mundy, A.P., Patel, N.J. and Englert, G. (1979). *J. Chem. Soc., Chem. Commun.* 27–28.
Britton, G., Goodwin, T.W., Brown, D.J. and Patel, N.J. (1980). *Methods Enzymol.* **67**, 264–70.
Brown, D.J., Britton, G. and Goodwin, T.W. (1975). *Biochem. Soc. Trans.* **3**, 741–2.
Buggy, M.J., Britton, G. and Goodwin, T.W. (1969). *Biochem. J.* **114**, 641–3.
Bu'Lock, J.D., Jones, B.E. and Winskill, N. (1976). *Pure Appl. Chem.* **47**, 191–202.
Burczyk, J. (1987a). *Phytochemistry* **26**, 113–19.
Burczyk, J. (1987b). *Phytochemistry* **26**, 121–8.
Burczyk, J., Szkawran, H., Zontek, I. and Czygan, F.-C. (1981). *Planta* **151**, 247–50.
Burkey, K.O. (1986). *Photosynthesis Res.* **10**, 37–49.
Camara, B. (1980a). *FEBS Letts* **118**, 315–18.
Camara, B. (1980b). *Biochem. Biophys. Res. Commun.* **93**, 113–17.
Camara, B. (1984). *Plant Physiol.* **74**, 112–16.
Camara, B. (1985). *Pure Appl. Chem.* **57**, 675–7.
Camara, B. and Dogbo, O. (1986) *Plant Physiol.* **80**, 172–4.
Camara, B. and Monéger, R. (1980). *Dev. Plant Biol.* **6**, 363–7.
Camara, B. and Monéger, R. (1981). *Biochem. Biophys. Res. Commun.* **99**, 1117–22.
Camara, B. and Monéger, R. (1982). *Physiol. Vég.* **20**, 757–73.
Camara, B., Payan, C., Escoffier, A. and Monéger, R. (1980). *C.R. Hebd. Séances Acad. Sci., Sér. D.* **291**, 303–6.
Camara, B., Bardat, F., and Monéger, R. (1982a). *C.R. Séances Acad. Sci., Sér. 3*, **294**, 339–42.
Camara, B., Bardat, F. and Monéger, R. (1982b). *C.R. Séances Acad. Sci., Sér. 3*, **294**, 549–51.
Camara, B., Bardat, F. and Monéger, R. (1982c). *C.R. Séances Acad. Sci., Sér. 3*, **294**, 649–52.
Camara, B., Bardat, F. and Monéger, R. (1982d). *Eur. J. Biochem.* **127**, 255–8.
Camara, B., Bardat, F., Dogbo, O., D'Harlingue, A., Villat, F. and Monéger, R. (1984). *Dev. Plant Biol.* **9**, 241–4.
Camara, B., D'Harlingue, A., Dogbo, O., Agnes, C. and Henaff, J. (1985a). *C.R. Séances Acad. Sci., Sér. 3* **300**, 297–300.
Camara, B., Dogbo, O., Bardat, F. and Monéger, R. (1985b). *C.R. Séances Acad. Sci. Sér. 3* **301**, 749–52.
Camara, B., Dogbo, O., D'Harlingue, A., Kleinig, H. and Monéger, R. (1985c). *Biochim. Biophys. Acta* **836**, 262–6.

Cerda-Olmedo, E. and Torres-Martinez, S. (1979). *Pure Appl. Chem.* **51**, 631–7.
Chu, Z.-X. and Anderson, J.M. (1984). *Photobiochem. Photobiophys.* **8**, 1–10.
Clarke, I.E., Sandmann, G., Bramley, P.M. and Böger, P. (1982). *FEBS Letts* **140**, 203–6.
Clough, J.M. and Pattenden, G. (1979). *J. Chem. Soc., Chem. Commun.* 616–619.
Clough, J.M. and Pattenden, G. (1983). *J. Chem. Soc., Perkin Trans.* **1**, 3011–18.
Coggins, C.W., Henning, G.L. and Yokoyama, H. (1970). *Science* **168**, 1589–90.
Cohen-Bazire, G., Sistrom, W.R. and Stanier, R.Y. (1957). *J. Cellular Comp. Physiol.* **49**, 25–51.
Dandekar, S. and Modi, V.V. (1980). *Biochim. Biophys. Acta* **628**, 398–406.
Dandekar, S., Modi, V.V. and Jani, U.K. (1980). *Phytochemistry* **19**, 795–8.
Davies, B.H. (1980). In: *Pigments in Plants*, 2nd edn. (Ed. Czygan, F.-C.), pp. 31–56. Gustav Fischer, Stuttgart.
Davies, B.H. and Taylor, R.F. (1976). *Pure Appl. Chem.* **47**, 211–21.
Davies, B.H. and Taylor, R.F. (1982). *Can. J. Biochem.* **60**, 684–92.
De la Concha, A., Murillo, F.J., Skone, E.J. and Bramley, P.M. (1983). *Phytochemistry* **22**, 441–5.
De la Guardia, M.D., Aragon, C.M.G., Murillo, F.J. and Cerda-Olmedo, E. (1971). *Proc. Natl. Acad. Sci., U.S.A.*. **68**, 2012–15.
Dholakia, J.N. and Modi, V.V. (1983). *Biochem. Int.* **7**, 599–603.
Dholakia, J.N. and Modi, V.V. (1984). *J. Gen. Microbiol.* **130**, 2043–9.
Donkin, P. (1976). *Phytochemistry* **15**, 711–15.
Douce, R. and Joyard, J. (1979). *Adv. Bot. Res* **7**, 1–116.
Eder, F.A. (1979). *Z. Naturforsch.* **34C**, 1052–4.
Englert, G., Brown, B.O., Moss, G.P. *et al.* (1979). *J. Chem. Soc., Chem. Commun.* 545–547.
Eugster, C.H. (1979). *Pure Appl. Chem.* **51**, 463–506.
Evans, J.A. and Prebble, J.N. (1980). *Biochem. Soc. Trans.* **8**, 125–6.
Evans, J.A. and Prebble, J.N. (1982). *Microbios Letters* **21**, 149–159.
Fahey, D. and Milborrow, B.V. (1978). *Phytochemistry* **17**, 2077–82.
Firsow, N.N. and Drews, G. (1977). *Arch. Microbiol.* **115**, 299–306.
Francis, G.W. and Liaaen-Jensen, S. (1970). *Acta Chem. Scand.* **24**, 2705–12.
Frecknall, E.A. and Pattenden, G. (1984). *Phytochemistry* **23**, 1707–10.
Giuliano, G., Pollock, D. and Scolnik, P.A. (1986). *J. Biol. Chem.* **261**, 12925–9.
Goodwin, T,.W. (1958). *Biochem. J.* **70**, 612–17.
Goodwin, T.W. (1980). *The Biochemistry of the Carotenoids.* vol. 1, Plants. Chapman and Hall, London.
Govind, N.S. and Modi, V.V. (1981). *Indian J. Exp. Biol.* **19**, 544–6.
Govind, N.S., Mehta, B., Sharma, M. and Modi, V.V. (1981). *Phytochemistry* **20**, 2483–5.
Govind, N.S., Amin, A.R. and Modi, V.V. (1982). *Phytochemistry* **21**, 1043–4.
Gregonis, D.E. and Rilling, H.C. (1973). *Biochem. Biophys. Res. Commun.* **54**, 449–54.
Gregonis, D.E. and Rilling, H.C. (1974). *Biochemistry* **13**, 1538–42.
Grumbach, K.H. (1981). In: *Proceedings of the Fifth International Photosynthesis Congress* (Ed. Akoyunoglou, G.), pp. 625–636. Balaban International Science Services, Philadelphia.
Grumbach, K.H. (1983). *Z. Naturforsch.* **38C**, 996–1002.
Grumbach, K.H. and Bach, T.J. (1979). *Z. Naturforsch.* **34C**, 941–3.
Grumbach, K.H. and Britton, G. (1984). In: *Advances in Photosynthesis Research*

(Ed. C. Sybesma) vol. IV, pp. 69–72. Martinus Nijhoff/Dr. W. Junk, Dordrecht, Netherlands.

Grumbach, K.H. and Lichtenthaler, H.K. (1982). *Photochem. Photobiol.* **35**, 209–12.

Hallenstvet, M., Buchecker, R., Borch, G. and Liaaen-Jensen, S. (1977). *Phytochemistry* **16**, 583–5.

Harding, R.W. and Shropshire, W. (1980). *Ann. Rev. Plant Physiol.* **31**, 217–38.

Hertzberg, S. and Liaaen-Jensen, S. (1977). *Acta Chem. Scand.* **B31**, 215–18.

Howes, C.D. and Batra, P.P. (1970). *Biochim. Biophys. Acta* **222**, 174–8.

Huang, P.C. (1964). *Genetics* **49**, 453–69.

Islam, M.A., Lyrene, S.A., Miller, E.M. and Porter, J.W. (1977). *J. Biol. Chem.* **252**, 1523–5.

Johansen, J.E. and Liaaen-Jensen, S. (1976). *Tetrahedron Letters*, 955–958.

Johnson, J.H., Reed, B.C. and Rilling, H.C. (1974). *J. Biol. Chem.* **249**, 402–6.

Kato, F., Koyama, Y., Muto, S. and Yamagishi, S. (1981). *Chem. Pharm. Bull.* **29**, 1674–80.

Kleinig, H. and Schmitt, R. (1982). *Z. Naturforsch.* **37C**, 758–60.

Koyama, Y., Yazawa, Y., Kato, K. and Yamagishi, S. (1981). *Chem. Pharm. Bull.* **29**, 176–80.

Kreuz, K. and Kleinig, H. (1981). *Planta* **153**, 578–81.

Kreuz, K., Beyer, P. and Kleinig, H. (1982). *Planta* **154**, 66–9.

Kritsky, M.S. (1985). *FEMS Symp.* **23**, 137–8.

Kritsky, M.S., Sokolovsky, V.Y., Belozerskaya, T.A. and Chernysheva, E.K. (1981). *Dokl. Akad. Nauk SSSR* **258**, 759–62.

Kritsky, M.S., Sokolovsky, V.Y., Belozerskaya, T.A. and Chernysheva, E.K. (1982). *Arch. Microbiol.* **133**, 206–8.

Kushwaha, S.C., Subbarayan, C., Beeler, D.A. and Porter, J.W. (1969). *J. Biol. Chem.* **244**, 3635–42.

Kushwaha, S.C., Suzue, G., Subbarayan, C. and Porter J.W. (1970). *J. Biol. Chem.* **245**, 4708–17.

Kushwaha, S.C., Kates, M. and Porter, J.W. (1976). *Can. J. Biochem.* **54**, 816–23.

Kushwaha, S.C., Kates, M., Renaud, R.L. and Subden, R.E. (1978). *Lipids* **13**, 352–5.

Lambert, R. and Böger, P. (1983). *Pesticide Biochem. Physiol.* **20**, 183–7.

Lampila, L.E., Wallen, S.E. and Bullerman, L.B. (1985). *Mycopathologia* **90**, 65–85.

Liaaen-Jensen, S. (1969). *Pure Appl. Chem.* **20**, 421–48.

Lichtenthaler, H.K., Buschmann, C., Döll, M. *et al.* (1981). *Photosynth. Res.* **2**, 115–41.

Lichtenthaler, H.K., Burgstahler, R., Buschmann, C., Meier, D., Prenzel, U. and Schönthal, A. (1982). In: *Effects of Stress on Photosynthesis* (Ed. Marcelle, R.), pp. 353–370. Dr. W. Junk, Dordrecht, Netherlands.

Lütke-Brinkhaus, F., Liedvogel, B., Kreuz, K. and Kleinig, H. (1982). *Planta* **156**, 176–80.

McDermott, J.C.B., Ben-Aziz, A., Singh, R.K., Britton, G. and Goodwin, T.W. (1973a). *Pure Appl. Chem.* **35**, 29–45.

McDermott, J.C.B., Britton, G. and Goodwin, T.W. (1973b). *Biochem. J.* **134**, 1115–7.

McDermott, J.C.B., Britton, G. and Goodwin, T.W. (1973c). *J. Gen. Microbiol.* **77**, 161–71.

McDermott, J.C.B., Brown, D.J., Britton, G. and Goodwin, T.W. (1974). *Biochem. J.* **144**, 231–43.
Märki-Fischer, E., Marti, U., Buchecker, R. and Eugster, C.H. (1983). *Helv. Chim. Acta* **66**, 494–513.
Marrs, B.L. (1982). In: *Carotenoid Chemistry and Biochemistry* (Eds. Britton, G. and Goodwin, T.W.) pp. 273–277. Pergamon, Oxford.
Marshall, J.H. and Wilmoth, G.J. (1981a). *J. Bacteriol.* **147**, 900–13.
Marshall, J.H. and Wilmoth, G.J. (1981b). *J. Bacteriol.* **147**, 914–19.
Maudinas, B., Bucholtz, M.L., Papastephanou, C., Katiyar, S.S., Briedis, A.V. and Porter, J.W. (1977). *Arch. Biochem. Biophys.* **180**, 354–62.
Mehta, B., Govind, N.S., Sharma, M. and Modi, V.V. (1981). *Ind. J. Exp. Biol.* **19**, 1142–5.
Milborrow, B.V. (1982). In: *Carotenoid Chemistry and Biochemistry* (Eds. Britton, G. and Goodwin, T.W.) pp. 279–295. Pergamon, Oxford.
Milborrow, B.V., Swift, I.E. and Netting, A.G. (1982). *Phytochemistry* **21**, 2853–7.
Mitzka-Schnabel, U. (1985). *Pure Appl. Chem.* **57**, 667–9.
Mitzka-Schnabel, U. and Rau, W. (1981). *Phytochemistry* **20**, 63–9.
Moshier, S.E. and Chapman, D.J. (1973). *Biochem. J.* **136**, 395–404.
Murillo, F.J. and Cerda-Olmedo, E. (1976). *Mol. Gen. Genetics* **148**, 19–24.
Murillo, F.J., Torres-Martinez, S., Aragon, C.M.G. and Cerda-Olmedo, E. (1981). *Eur. J. Biochem.* **119**, 511–16.
Oelmüller, R. and Mohr, H. (1985). *Planta* **164**, 390–5.
Omata, T., Murata, N. and Satoh, K. (1984). *Biochim. Biophys. Acta* **765**, 403–5.
Papastephanou, C., Barnes, F.J., Briedis, A.V. and Porter, J.W. (1983). *Arch. Biochem. Biophys.* **157**, 415–25.
Patel, N.J., Britton, G. and Goodwin, T.W. (1983). *Biochim. Biophys. Acta* **760**, 92–6.
Pfander, H. and Schurtenberger, H. (1982). *Phytochemistry* **21**, 1039–42.
Porter, J.W. and Spurgeon, S.L. (1979). *Pure Appl. Chem.* **51**, 609–22.
Qureshi, A.A., Andrewes, A.G., Qureshi, N. and Porter, J.W. (1974a). *Arch. Biochem. Biophys.* **162**, 93–107.
Qureshi, A.A., Kim, M., Qureshi, N. and Porter, J.W. (1974b). *Arch. Biochem. Biophys.* **162**, 108–16.
Qureshi, A.A., Qureshi, N., Kim, M. and Porter, J.W. (1974c). *Arch. Biochem. Biophys.* **162**, 117–25.
Rau, W. (1976). *Pure Appl. Chem.* **47**, 237–43.
Rau, W. (1980). In: *Pigments in Plants*, 2nd edn. (Ed. F.-C. Czygan), pp. 80–103. Gustav Fischer Verlag, Stuttgart.
Rau, W. (1983). In: *Biosynthesis of Isoprenoid Compounds*, vol. 2 (Eds. Porter, J.W. and Spurgeon, S.L.), pp. 123–157. Wiley, New York.
Rau, W. (1985). *Pure Appl. Chem.* **57**, 777–84.
Rau, W. and Rau-Hund, A. (1977). *Planta* **136**, 49–52.
Revuelta, J.L. and Eslava, A.P. (1983). *Mol. Gen. Genetics* **192**, 225–9.
Ridley, S.M. (1982). In: *Carotenoid Chemistry and Biochemistry* (Eds. Britton, G. and Goodwin, T.W.) pp. 353–369. Pergamon, Oxford.
Roncero, M.I.G. and Cerda-Olmedo, E. (1982). *Current Genetics* **5**, 5–8.
Sandmann, G. and Bramley, P.M. (1985a). *Planta* **164**, 259–63.
Sandmann, G. and Bramley, P.M. (1985b). *Biochim. Biophys. Acta* **843**, 73–6.
Schmidt, K. (1978). In: *The Photosynthetic Bacteria* (Eds. Clayton, R.K. and Sistrom, W.R.), pp. 729–750. Plenum, New York.

Schneour, E.A. (1962). *Biochim. Biophys. Acta* **62**, 534–40.
Schrott, E.L. (1980a). *Planta* **150**, 174–9.
Schrott, E.L. (1980b). In: *Blue Light Syndrome* (Ed. Senger, H.) pp. 309–318. Springer, Berlin.
Schrott, E.L. (1981). *Planta* **151**, 371–4.
Schrott, E.L. and Rau, W. (1977). *Planta* **136**, 45–8.
Sergeant, J.M. and Britton, G. (1984). In: *Advances in Photosynthesis Research* (Ed. Sybesma, C.), vol. IV, pp. 779–782. Martinus Nijhoff/Dr. W. Junk, Dordrecht, The Netherlands.
Seifermann-Harms, D. (1985). *Biochim. Biophys. Acta* **811**, 325–55.
Singh, R.K., Britton, G. and Goodwin, T.W. (1973). *Biochem. J.* **136**, 413–19.
Sokolovsky, V.Y., Kritsky, M.S., Belozerskaya, T.A. and Chernysheva, E.K. (1983). *Prikl. Biokhim. Mikrobiol.* **19**, 176–81.
Spurgeon, S.L. and Porter, J.W. (1980). In: *Biochemistry of Plants* (Ed. Stumpf, P.K.), vol. 4, pp. 419–84. Academic Press, New York.
Spurgeon, S.L. and Porter, J.W. (1983). In: *Biosynthesis of Isoprenoid Compounds* vol. 2 (Eds. Porter, J.W. and Spurgeon, S.L.), pp. 1–122, Wiley, New York.
Spurgeon, S.L., Turner, R.V. and Harding, R.W. (1979). *Arch. Biochem. Biophys.* **195**, 23–9.
Subden, R.E. and Goldie, A.H. (1973). *Genetica* **44**, 615–20.
Subden, R.E. and Threlkeld, S.F.H. (1970). *Genetics Res.* **15**, 139–46.
Subden, R.E. and Turian, G. (1970a). *Experientia* **26**, 935–7.
Subden, R.E. and Turian, G. (1970b). *Molec. Gen. Genetics* **108**, 358–64.
Swift, I.E. and Milborrow, B.V. (1981a). *Biochem. J.* **199**, 69–74.
Swift, I.E. and Milborrow, B.V. (1981b). *J. Biol. Chem.* **256**, 11609–11.
Swift, I.E., Milborrow, B.V. and Jeffrey, S.W. (1982). *Phytochemistry* **21**, 2859–64.
Tada, M. and Shiroishi, M. (1982a). *Plant Cell Physiol.* **23**, 541–8.
Tada, M. and Shiroishi, M. (1982b). *Plant Cell Physiol.* **23**, 549–56.
Tada, M. and Shiroishi, M. (1982c). *Plant Cell Physiol.* **23**, 567–73.
Tada, M. and Shiroishi, M. (1982d). *Plant Cell Physiol.* **23**, 615–21.
Tada, M., Shiroishi, M., Hasegawa, K., Suzuki, T. and Iwai, K. (1982). *Plant Cell Physiol.* **23**, 607–14.
Tanaka, A. and Tsuji, H. (1985). *Plant Cell Physiol.* **26**, 893–902.
Taylor, R.F. and Davies, B.H. (1973). *Biochem. Soc. Trans.* **1**, 1091–2.
Taylor, R.F. and Davies, B.H. (1974a). *Biochem. J.* **139**, 751–60.
Taylor, R.F. and Davies, B.H. (1974b). *Biochem. J.* **139**, 761–9.
Taylor, R.F. and Davies, B.H. (1976). *Biochem. J.* **153**, 233–9.
Taylor, R.F. and Davies, B.H. (1982). *Can. J. Biochem.* **60**, 675–83.
Torres-Martinez, S., Murillo, F.J. and Cerda-Olmedo, E. (1980). *Genetics Res.* **36**, 299–309.
Urbach, D., Suchanka, M. and Urbach, W. (1976). *Z. Naturforsch.* **31C**, 652–5.
Whitaker, B.D. and Shropshire, W. (1981). *Exp. Mycol.* **5**, 243–52.
Wild, A. (1979). *Ber. Deutsch. Bot. Ges.* **92**, 341–64.
Yakovleva, I.M., Vakulova, L.A., Feofilova, E.P., Bekhtereva, M.N. and Samokhvalov, G.I. (1980). *Mikrobiologiya* **49**, 274–8.
Yokoyama, H., Hsu, W.J., Poling, S.M. and Hayman, E. (1982). In: *Carotenoid Chemistry and Biochemistry* (Eds. Britton, G. and Goodwin, T.W.) pp. 371–385. Pergamon, Oxford.

4
The Function of Pigments in Chloroplasts

RICHARD COGDELL

Department of Botany, University of Glasgow, Glasgow G12 8QQ, Scotland

1 INTRODUCTION

It is well beyond the scope of this chapter to describe in detail the function of all the pigments located within the chloroplasts involved in the light reactions of photosynthesis. The author therefore intends to limit this chapter to a discussion of the function of the major light-absorbing pigments, that is, the chlorophylls, carotenoids and, in the lower algae and cyanobacteria, the bilipigments.

The secret of understanding the function of these pigments in the chloroplast is to realize that, with very few exceptions, they are bound to proteins (Thornber, 1986a). This may seem to be very obvious now, but only a few years ago it was by no means generally accepted. All of

PLANT PIGMENTS
ISBN 0-12-289847-8

the major light-absorbing pigments involved in photosynthesis are photochemically active molecules and can potentially participate in a range of photochemical reactions; which of these possible reactions is expressed *in vivo* is controlled by which protein the pigment binds to.

This can be clearly illustrated by considering the case of chlorophyll a. There is a division of labour among the chlorophyll a molecules present in the photosynthetic membrane. The majority serve a light-harvesting or antenna role and funnel absorbed radiant energy to a specialized few, which form photochemical reaction centres. Identical chlorophyll a molecules serve both these functions. The fate of each chlorophyll a molecule depends critically upon which type of protein it is bound to. If it is bound to an antenna apoprotein then it will function in light-harvesting, while if it is bound to a reaction centre subunit then it will participate in the primary redox reactions. It should already be clear to the reader that it is not possible to discuss the function of these pigments without also describing the structure of the pigment–protein complexes. This will be a constant theme dominating this chapter.

In general, the structure and function of the phycobiliproteins is much better understood than that of the chlorophyll/carotenoid pigment–protein complexes (Gantt, 1986; Glazer, 1985). This is because of basic differences in the properties of these two classes of pigment-protein complex. The chlorophylls and carotenoids are usually non-covalently bound to their apoproteins, while the bilipigments are covalently attached to theirs (Glazer, 1985; Gantt, 1986; Thornber, 1986a). The chlorophyll/carotenoid pigment–protein complexes are in the main hydrophobic, integral membrane proteins (Thornber, 1975). In contrast, the phycobiliproteins are only membrane surface-associated and are actually true water-soluble proteins (Gantt, 1981; Glazer, 1982). Before a pigment–protein complex can be characterized and the molecular details of its function clearly understood, it mut be isolated and purified. This is comparatively easy in the case of the phycobiliproteins. As they are only weakly membrane-associated they can be readily released from the thylakoids by "salt-washing" (Gantt, 1981). They are then water-soluble and coloured, and their subsequent isolation and purification are rather straightforward. Even if they are denatured they remain pigmented and so can be easily identified. This process is much more difficult for integral membrane proteins. Before a chlorophyll/carotenoid pigment–protein can be isolated and purified it must be solubilized by the addition of a suitable detergent (one which is strong enough to solubilise the complex under study yet mild enough not to cause too much denaturation; there is no perfect detergent and the choice is always a trade-off between these two factors). Still the most common detergent used to solubilise higher plant photosynthetic membranes is sodium dodecyl-sulphate (SDS) (Thornber, 1986a). Unfortunately, however, this detergent is also rather denaturing. This tends to introduce a great deal of variability into the composition and properties of these solubilized chlorophyll/carotenoid pigment–protein complexes, as they are prepared in the various different laboratories throughout the world. Slight changes in the exact solubilization procedures used causes marked changes in the properties of the isolated complex, and this has led to the accumulation of much apparently conflicting data. This problem has been further accentuated by the fact that most workers in this area of research still tend to use their own, usually rather inscrutable, nomenclature for the pigment–protein complexes that they isolate. It can be quite difficult, even for experts in this field, always to be sure which complex is being described. The author will present below (see Table 4.4, p. 197) a system of nomenclature for these pigment–protein complexes, outlined by Thornber (1986a, b) which introduces some order into this field and helps to simplify this situation.

Readers who are interested in the other chloroplast pigments involved in the slower electron transport reactions of the "Z" scheme (cytochromes plastocyanin, quinones, etc.) should consult *The Encyclopaedia of Plant Physiology*, vol. 19 (1986), which contains several excellent in-depth reviews on this topic.

2 THE PICTURE BEFORE 1976

In the few years leading up to 1975–6 a rather consistent picture of the organization and function of the light-absorbing pigments in photosynthesis emerged (Thornber, 1975; Thornber *et al.*, 1977). The photosynthetic apparatus was viewed as a tripartite structure [a term coined by Warren Butler (Butler and Kitajima, 1975) to describe a model of the photosynthetic apparatus developed to explain his fluorescence studies]. This tripartite model is illustrated in Fig. 4.1. The three components of the system are photosystem I, photosystem II and the chlorophyll a/b protein (the major light-harvesting complex in higher plants). In lower

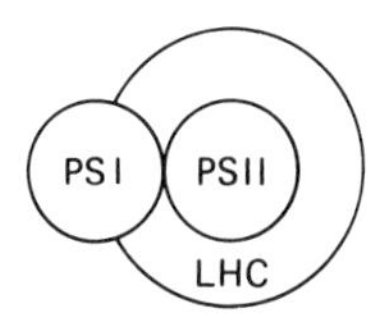

Fig. 4.1 Schematic representation of the tripartite model of photosynthesis. PSI = photosystem I; PSII = photosystem II; LHC = light-harvesting complex.

algae, which lack chlorophyll b, the chlorophyll a/b protein is replaced with a range of other types of antenna complexes. These are summarized in Table 4.1.

This type of model was supported by early studies which fractionated chloroplast membranes with SDS polyacrylamide gel electrophoresis (Fig. 4.2; Ogawa *et al.*, 1966; Thornber *et al.*, 1966). When the photosynthetic membranes were solubilised with SDS (typically Chl:SDS ratios of >1:10 w/w were used) and the resulting solution fractionated on a gel under non-denaturing conditions three distinct green bands were seen. These types of gels have now become known as "green gels". The top band was called CP1 (or the P700-Chl a protein) and represents photosystem I. The middle band was called CP2 (this is the chlorophyll a/b protein) and the bottom band was "free pigment", which came from denatured material. The top two bands only were associated with protein. Typically about 70% of the applied chlorophyll remained in these top two bands (Thornber, 1986a). Unfortunately photosystem II did not

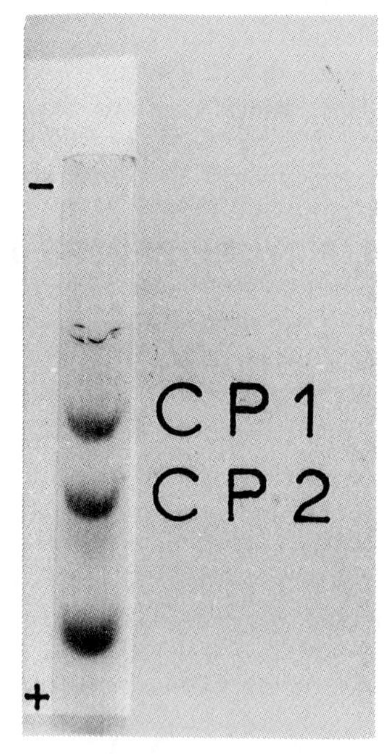

Fig. 4.2 Fractionation of the pigment–protein complexes from French bean chloroplasts by SDS polyacrylamide gel electrophoresis ("green gel").

Isolated chloroplasts were solubilized in 50 mM Tris HCl pH 8.0, 10% glycerol and 1% SDS (chlorophyll:SDS, 1:10 w/w) and run on a 7% polyacrylamide gel at 20°C for about 2 h. The three green (unstained) bands are CP1, CP2 and free pigment.

Table 4.1 The major light-harvesting pigment–protein complexes in photosynthetic organisms which do not contain chlorophyll b. Data from Siefermann-Harms (1985)

Group (representative species)	Major antenna component(s)	References
Cyanobacteria (e.g. *Anacystis nidulans*)	Phycobiliproteins[1]	Gantt (1981)
Rhodophyta (e.g. *Porphyridium cruentum*)	Phycobiliproteins[1]	Gantt (1981)
Dinophyta (e.g. *Glenodinium* spp.)	Chl a-peridinin complex[1,2]	Prezelin and Haxo (1976) Haxo *et al*. (1976)
	Chl a/c_2-xanthophyll complex[2]	Prezelin and Boczar (1981)
Bacillariophyta (e.g. *Phaeodactylum tricornutum*)	Chl a/c-fucoxanthin complex (several types)[2]	Friedman and Alberte (1984) Gugliemelli (1984)
Phaeophyta (e.g. *Acrocarpia paniculata*)	Chl a/c_2-fucoxanthin complex,[2] Chl a/c_1/c_2-violaxanthin-complex[2]	Barratt and Anderson (1980)
Crytophyta (e.g. *Chroomonas* spp.)	Chl a/c_2-xanthophyll complex[2]	Ingram and Hiller (1983)

[1]Hydrophilic membrane-associated complexes.
[2]Hydrophobic intrinsic membrane complexes.

survive this preparative procedure. Rather poorly resolved photosystem II particles could, however, be isolated from the chloroplast membrane by milder fractionation procedures, such as digitonin solubilization and differential centrifugation (Boardman and Anderson, 1964; Boardman, 1971).

At this time there was no firm evidence for complexes other than these three; however concern remained about what might have been destroyed in the SDS treatment, producing the large amount of free pigment. As will be seen below, attempts to reduce this level of denaturation to zero have led to the development of a much more detailed picture of the structure and composition of the photosynthetic apparatus (Thornber, 1986a).

The availability of partially purified but greatly enriched preparations of photosystems I and II allowed a much more detailed investigation of their photochemical reactions to begin. The enrichment factor was particularly important since it allowed for a significant increase in the signal to noise ratio for both optical and magnetic resonance studies. The P700-Chl a preparations were both more stable and more refined than those of photosystem II (Bearden and Malkin, 1977) and more fundamental progress was made with investigations of the structure and functions of photosystem I.

In the mid 1970s a range of photosystem I preparations were generally available. The properties of some of these are presented in Table 4.2. It seemed quite easy to reduce down to a preparation which contained about 30–40 chlorophyll a molecules per mole of P700, but the refinement of the preparation down to the level achieved with the purple photosynthetic bacteria, i.e. four chlorophylls per reaction centre, was not accomplished. Preparations with reduced chlorophyll content were reported (e.g. Ikegami and Katoh, 1975), however the functional integrity of these was unclear. The P700-Chl a complexes were shown to contain two prominent apoproteins in the 40–55 kD molecular weight range.

When a photon of light is absorbed by photosystem I it is very rapidly (within a few ps) transferred from the antenna chlorophylls to a specialised chlorophyll a molecule(s) which then undergoes an oxidation-reduction reaction. This specialized reactive site, where the excitation energy is "trapped", is called the reaction centre. In this primary redox reaction one or two molecules of chlorophyll a are oxidized. This is a one-electron oxidation reaction and results in a bleaching in the long wavelength absorption band of chlorophyll a, centred at 700 nm. This is why the reaction centre of photosystem I is called P700 (Kok, 1961). When P700 is oxidized by light a single electron is transferred to an electron acceptor. This is an extremely fast reaction and one which still occurs with high efficiency at cryogenic temperatures.

Table 4.2 Comparison of a range of different photosystem I preparations available in 1975

Preparation	Detergent used in solubilization	Ratio of P700:chlorophyll	References
D-144	Digitonin	1:200	Boardman (1971)
TSF-1	Triton X-100	1:100	Ke et al. (1975)
HP-700	Triton X-100	1:30	Vernon and Shaw (1971)
LDAO-PSI	LDAO	1:30	Malkin (1975)
P700-Chl a protein (higher plants)	Triton X-100	1:40	Shiozawa et al. (1974)
P700-Chl a protein (cyanobacteria)	SDS	1:40	Dietrich and Thornber (1971)
Digitonin-PSI	Digitonin	1:100	Bengis and Nelson (1975)
Di-ethyl ether extracted-PSI	—	1:10	Ikegami and Katoh (1975)

SDS = sodium dodecyl-sulphate; LDAO = lauryl-dimethylamine-N-oxide.

The properties of P700 have been characterized both optically and by electron spin resonance (ESR) spectroscopy (Commoner *et al.*, 1956, Katz and Norris, 1973; Norris *et al.*, 1971, 1975). P700 has an oxidation-reduction potential of about +500 mV (Bearden and Malkin, 1977; that is, slightly more oxidizing than the ferri-ferrocyanide redox couple). The generation of $P700^+$ not only causes changes in the optical spectrum, but also produces a free radical which can be detected by ESR (Commoner *et al.*, 1956). The ESR signal thought to represent $P700^+$ is called signal I. It has a g value of 2.0026 and line width (peak-to-peak) of 7.1 G. When an unpaired electron is localized over more than one molecule, the line-width of the ESR signal will narrow (see Katz and Norris, 1973). If that delocalization is symmetrical, then the narrowing will be proportional to $\sqrt{N}$, where N is the number of molecules which share the unpaired spin. The line-width of the ESR signal arising from monomeric $chlorophyll^+$ is 9.3 G. The ratio of these two line-widths is therefore nearly $\sqrt{2}$ and has led to the suggestion that P700 is a chlorophyll a dimer (Norris *et al.*, 1971; Katz and Norris 1973). The dimer hypothesis was also supported by ENDOR (electron nuclear double resonance spectroscopy) spectra of $P700^+$ and the monomeric chlorophyll cation (Norris *et al.*, 1974, 1975) where with $P700^+$ several proton hyperfine splittings were found to be half those of chlorophyll a^+.

Hiyama and Ke (1971) investigated the primary reactions in photosystem I particles and discovered a broad absorbance decrease centred at 430 nm. They called this change P430. The properties of this absorbance change led them to suggest that P430 represented the stable electron acceptor of photosystem I. The mid-point oxidation-reduction potential of P430 is in the range of −470 to −500 mV (Ke, 1973), i.e. it is very reducing. The reader should remember that the reducing side of photosystem I needs to produce a strong enough reductant to allow it to reduce NADP to $NADPH_2$. About the same time that P430 was being studied optically by flash photolysis, Malkin and Bearden (1971) were able to observe an ESR signal from the stable electron acceptor of photosystem I. If chloroplasts or P700-Chl a preparations were illuminated at 77K and then cooled to 25K, a rather stable free radical was generated whose ESR signal had resonances at $\boldsymbol{g}$ values of 1.86 and 2.05. Malkin and Bearden (1971) called the signal "bound ferredoxin" because of its similarities to soluble ferredoxin. In order to take account of these ESR experiments Ke (1973) suggested that P430 was an iron-sulphur protein. More detailed ESR studies on the "bound ferredoxin" revealed that in fact it represents two iron-sulphur centres (Evans *et al.*, 1974, 1976; Bearden & Malkin, 1977). These have g values at 1.86, 1.94 and 2.05, and 1.89, 1.92 and 2.05 respectively, and have become known as centre A and centre B.

At this time therefore the primary reactions in photosystem I could be summarized as follows:

$$\mathrm{Chl_{a(antenna)} + P700\ P430 \xrightarrow{h\nu} Chl^*_{a(antenna)} + P700\ P430}$$

$$\mathrm{Chl^*_{a(antenna)} + P700\ P430 \rightarrow Chl_{a(antenna)} + P700^*\ P430}$$

$$\mathrm{P700^*\ P430 \rightarrow P700^+\ P430^-}$$

Studies on the primary reactions of photosystem II were confined to using chloroplast membranes or the rather poorly refined particles, such as the digitonin D-10 particle (Boardman, 1971) and the TSF-2a particle of Vernon *et al.* (1971) and Vernon and Shaw (1971). The composition of these particles is summarized in Table 4.3. The primary donor (or reaction centre) of photosystem II is called P680 (again named after the wavelength of its maximum bleaching in the red most absorption bands of chlorophyll a; Doring *et al.*, 1969). P680 has proved rather difficult to study. The species $P680^+$ is very oxidizing, does not last for very long (less than a few μs under normal conditions) and photosystem II preparations are quite labile. The redox potential of P680 has not yet

Table 4.3 Some properties of two photosystem II preparations available in 1975

	Amount present (mol/100 mol Chl)	
Component	TSF-II[1]	D-10[2]
Chl a	67	69
Chl b	33	31
Chl a:Chl b	~2.0:1	2.27:1
β-Carotene	6	6
Lutein	16	11
Xanthophylls	5	8
Plastoquinone	2	Not determined

[1]Data taken from Vernon and Shaw (1971).
[2]Data taken from Boardman (1971).
P680 could not be quantified at this time, therefore data upon the P680:Chl ratio are not available.

been determined but it must be above +820mV since it can oxidize water (Diner, 1986). A transient ESR signal arising from $P680^+$ has been described (Malkin and Bearden, 1975; Van Gorkham *et al.*, 1975; Visser, 1975). It is centred at $\boldsymbol{g} = 2.002$ and has a line-width of 7–9 G (Malkin and Bearden, 1975). The reader might logically conclude therefore, using the same argument outlined above for $P700^+$, that P680 is a chlorophyll a dimer. However, views on this in the photosynthetic community oscillate, and it is not yet clear which—the monomer or dimer model—will prevail.

The stable electron acceptor of photosystem II was originally called Q (for quencher; Duysens and Sweers, 1983). This was because its presence was initially inferred from studies on the variable fluorescence emitted by chloroplasts at room temperature. The reader should note that most, if not all, room temperature fluorescence comes from photosystem II, not photosystem I (Butler, 1973). The phenomenon of variable fluorescence is illustrated in Fig. 4.3. When chloroplasts are in the dark the level of fluorescence is low. Upon illumination the fluorescence rises very rapidly to a level called F_o, and then continues to rise more slowly to a higher level called F_{max}. When the chloroplasts are in the state F_o, Q is oxidised, then as Q gets reduced photochemically in the light to Q^- the level of fluorescence rises. At F_{max} all of Q is in the state Q^-. The explanation why the fluorescence yield of chloroplasts should be sensitive to the redox state of Q is as follows. Consider the fate of singlet excited P680. $P680^*$ can decay back down to the ground state by a variety of competing pathways viz.

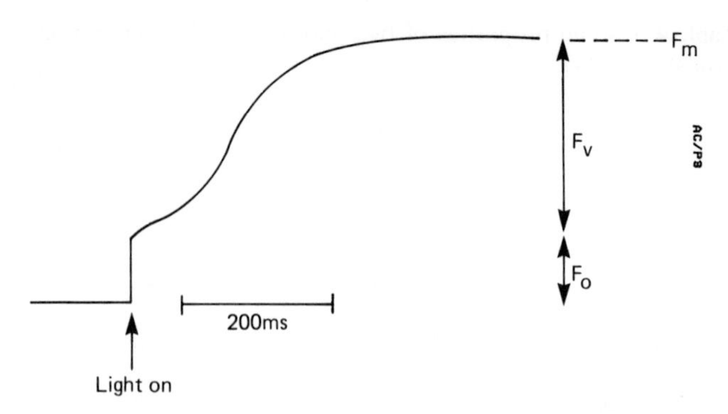

Fig. 4.3 A stylized representation of the kinetics of fluorescence induction seen on illuminating isolated chloroplasts.

In the dark the level of fluorescence is low. Upon illumination the level of fluorescence rises rapidly to F_o and then more slowly, over several hundred milliseconds, to F_m. The exact rise kinetics of the F_v depend on the intensity of the actinic illumination.

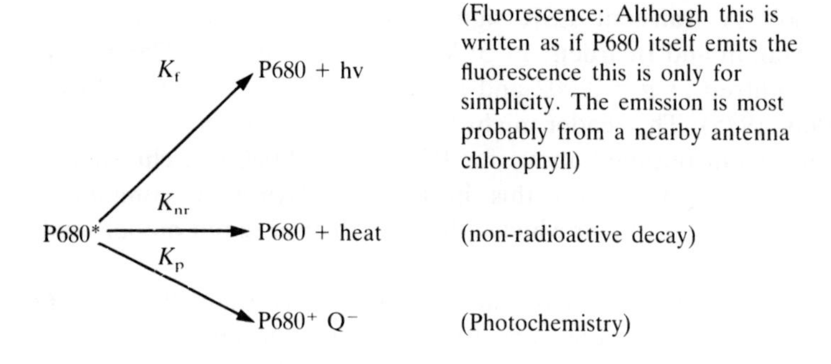

K refers to the respective rate constants for each of the competing processes. The quantum yield of fluorescence (ϕ_f) is then given by the following equation.

$$\phi_f = \frac{K_f}{K_f + K_{nr} + K_p Q}$$

When $Q = 1$, i.e. Q is 100% oxidized, ϕ_f is low because K_p is the dominant rate constant (i.e. the biggest) since the efficiency of photochemistry is high. On the other hand, when Q is reduced then $Q = 0$ and the yield of fluorescence will rise.

Stiehl and Witt (1969) discovered an absorbance change at 320 nm (they called it X320); its properties were found to correlate very strongly with these previously found for Q. The difference spectrum of X320 was very similar to that of the anionic plastosemiquinone, thus they suggested that Q is a bound plastoquinone molecule. Redox titrations of Q (determined by titrating the variable fluorescence of chloroplasts) have given confusing results. Two redox components of Q have been discovered; Q_L and Q_H. Q_L has a redox potential of −250 mV, while Q_H has a redox potential of −45 mV (Butler, 1973; Cramer and Butler, 1969; Ke *et al.*, 1976). The status of these apparently two different Qs was not clear at this time, so the primary reaction in photosystem II was simply viewed as the oxidation of P680 and the reduction of Q.

$$\mathrm{P680}\ Q \xrightarrow{h\nu} \mathrm{P680}^{*}\ Q$$
$$\mathrm{P680}^{*}\ Q \rightarrow \mathrm{P680}^{+}\ Q^{-}$$

The chlorophyll a/b protein was first isolated in 1966 (Ogawa *et al.*, 1966; Thornber *et al.*, 1966) and in the mid 1970s it was still thought that all the chlorophyll b found in the chloroplast was associated with this complex (Thornber, 1975). The presence of this antenna complex within the photosynthetic membrane is not essential for photosynthesis. Mutants lacking chlorophyll b [such as the now famous b-less barley mutant (Thornber and Highkin, 1974)] lack the chlorophyll a/b protein. These mutants do however grow photosynthetically.

When chloroplast membranes are fractionated with digitionin the chlorophyll a/b protein remains mainly associated with photosystem II (Boardman and Anderson, 1964; Boardman, 1971). Depending upon the plant and the conditions under which it has been grown, up to about 50% of the total chlorophyll present within the chloroplast can be found associated with this complex (Thornber, 1975). By the mid 1970s methods for the large-scale isolation of this antenna complex had been developed (Thornber, 1975). This then allowed experiments to be undertaken to study its structure and function *in vitro*. The absorption spectrum of the chlorophyll a/b complex isolated from *Chlamydomonas* chloroplasts is shown in Fig. 4.4 (Thornber *et al.*, 1977). The chlorophyll a/b protein has two absorption maxima in the red region of the spectrum. At room temperature these are located at 652 and 670 nm and represent the Q_y transitions of chlorophyll b and chlorophyll a respectively. Analysis of the polypeptide content of purifed samples of the chlorophyll a/b protein revealed the presence of an apoprotein with a molecular weight in the 27–35 kD range (Thornber *et al.*, 1977). The most widely quoted pigment content of this antenna complex was three molecules of chlorophyll a, three molecules of chlorophyll b and one molecule of carotenoid per single copy of the pigment-binding apoprotein. *In vitro* when a purified

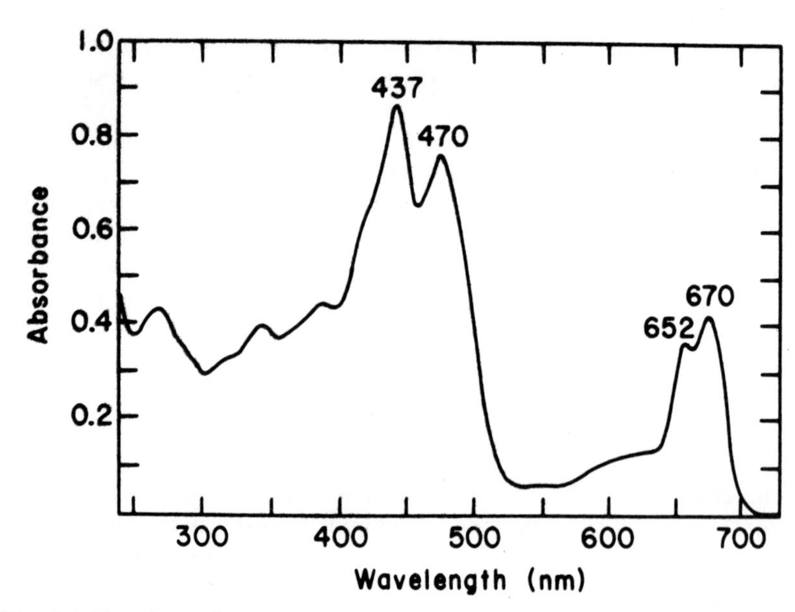

Fig. 4.4 The absorption spectrum of the light-harvesting chlorophyll a/b protein from *Chlamydomonas*. Reproduced with permission from Thornber *et al.* (1977).

sample of the chlorophyll a/b protein is illuminated no redox reactions are seen; most of the excitation energy is released either as fluorescence or through carotenoid triplets. The fluorescence emission spectrum shows a single peak centred at 685 nm at 77K (Thornber *et al.*, 1977).

Bogorad (1975) presented an excellent review on what was known at the time about the structure and function of the phycobiliproteins. These pigment–protein complexes were classified into three main types depending upon the position of their visible absorption bands and the chemical identity of their chromophores. The absorbance maxima of phycoerythrins range from 498 to 568 nm; that of phycocyanins is usually about 625 nm and most allophycocyanins absorb maximally around 650 nm, although they can range between 618 and 673 nm. In general though, as one progresses through the series PE, PC to APC the absorption maxima move to the red.

All the phycobiliproteins appear to be constructed on the same basic structural principle. There are usually two pigment-binding apoproteins (the α- and β-subunits). The exact molecular weights of these subunits vary, but they are normally in the range of 12–20 kD for the α-subunits and 15–22 kD for the β-subunits (Bogorad, 1975). The intact phycobiliproteins usually exist in solutions as aggregates, such as $(\alpha\beta)_3$,

$(\alpha\beta)_6$ or $(\alpha\beta)_{12}$. The chromophores, which are present on both types of apoprotein, were shown to be covalently linked to the polypeptide backbone; however at this time the nature and position of these linkages had not been determined.

In vivo the phycobiliproteins are organized into highly ordered extra-membrane structures. These are called phycobilisomes and can be clearly seen under the electron microscope as disc-shaped structures closely applied to the photosynthetic membranes. They function as rather efficient accessory light-harvesting complexes, funnelling absorbed radiant energy to the chlorophyll molecules located in the photosynthetic membrane. At this time however, very little was known about either the details of phycobilisome structure or how it functioned in light capture and energy transfer.

The phycobilisome content of cyanobacteria is regulated by light. Three patterns of regulation have been described (Bogorad, 1975). In some species only the number of phycobilisomes per cell is regulated by light intensity. The higher the levels of incident radiation, the lower the number of phycobilisomes per cell and vice versa. Two other more subtle types of regulation of the actual composition of the phycobilisomes, as well as their numbers per cell, are also seen. In these cases the spectral quality of the light is important. In the first of these two cases the content of phycocyanin in the phycobilisome remains constant but the level of phycoerythrin varies. Phycoerythrin synthesis is induced by green light and repressed by red light. In the second case both phycocyanin and phycoerythrin synthesis are regulated by the spectral quality of the incident radiation. Green light induces the synthesis of phycoerythrin and the repression of phycocyanin, while in red light these changes are reversed. This final type of regulation is called complementary chromatic adaptation, since when these cells are grown in red light they appear green, while when they are grown in green light the cells appear red. This phenomenon is readily demonstrated in the laboratory, but its functional significance for organisms growing in the wild is not at all clear (Bogorad, 1975).

3 THE RECENT PICTURE

Simultaneous advances in several areas have combined to make the past few years a very exciting time for researchers interested in understanding, at the molecular level, how chloroplast pigments participate in the primary reactions in photosynthesis. Biochemical techniques for isolating, purifying and characterizing membrane proteins have steadily improved. This led to a much more detailed understanding of the composition of the chloroplast membrane (Thornber, 1986). The powerful techniques of modern molecular biology have begun to be applied to photosynthesis research (Barber and Marder, 1986). Indeed, the entire chloroplast DNA from a liverwort, *Marchantia polymorpha*, has been sequenced (Ohyama *et al.*, 1986) and the primary structures of many of the apoproteins of the major photosynthetic pigment–protein complexes have now been determined (Dyer, 1985; Barber and Marder, 1986). Rapid kinetic techniques (ps and ns laser flash photolysis) have been used to investigate the very early photochemical events that occur within the primary reactions. X-ray crystallography has been used to produce high-resolution three-dimensional structures of several important photosynthetic pigment–protein complexes (for example, see Deisenhofer *et al.*, 1985, 1986; Schirmer *et al.*, 1985, 1986). Also, it has proved extremely helpful to apply to plants the detailed knowledge on the mechanism of the primary reactions gained from studying the much better resolved system of the purple photosynthetic bacteria (Parson, 1982; Cogdell, 1983; Rutherford, 1985). The bacterial system has laid the foundation for our present detailed understanding of the primary reactions in plants.

Thornber (1986a, b) presented a simplified scheme for naming the various chloroplast pigment–protein complexes. This scheme (Table 4.4) is a significant improvement for two main reasons: firstly, it is simple, and secondly, the names used are genuinely informative rather than confusing. The need for such a rational system of nomenclature had become quite acute. Recent studies using milder (non-denaturing) solubilization methods to isolate the photosynthetic pigment–protein complexes had revealed the presence of more types of complex, and the lack of a sensible system of nomenclature was causing a good deal of confusion. A general picture of the structure of each photosystem has now emerged and the Thornber system of nomenclature is based upon this. Each photosystem is viewed to consist of a "core" complex, called CCI or CCII, together with its associated antenna complexes, called LHCI and LHCII respectively (LHCI is associated with CCI and LHCII with CCII). If more than one type of antenna complex is found to be associated with CCI, for example, they would be called $LHCI_\alpha$ and $LHCI_\beta$ etc.

A CCI

CCI runs on "green gels" with an apparent molecular weight on ~ 110 kD, though this is certainly an underestimate of its true size (Lundell *et al.*, 1985; Thornber, 1986a). In this form it contains 40–60 chlorophyll a

Table 4.4 A simple system of naming plant chlorophyll–protein complexes. After Thornber (1986a,b)

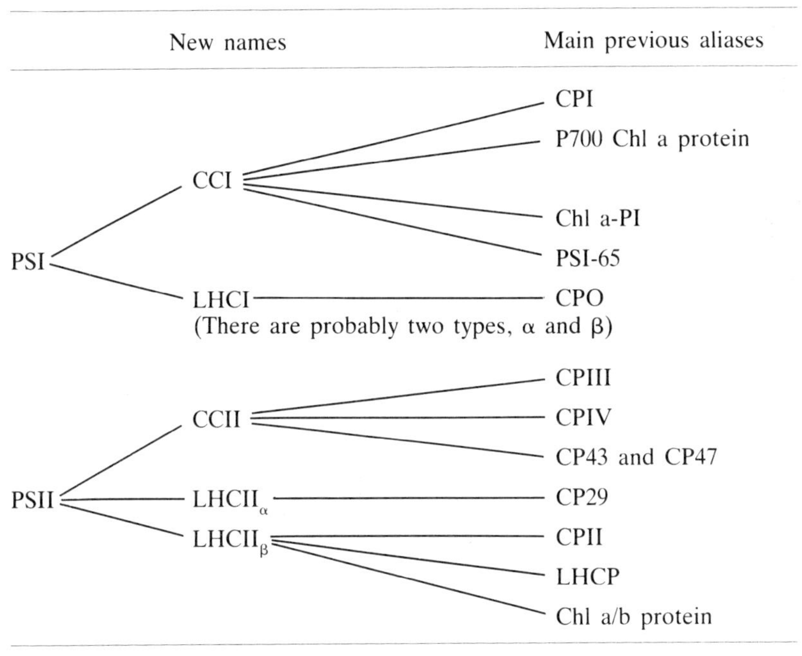

	New names	Main previous aliases
PSI	CCI	CPI
		P700 Chl a protein
		Chl a-PI
		PSI-65
	LHCI	CPO
	(There are probably two types, α and β)	
PSII	CCII	CPIII
		CPIV
		CP43 and CP47
	$LHCII_{\alpha}$	CP29
	$LHCII_{\beta}$	CPII
		LHCP
		Chl a/b protein

molecules and one or two molecules of β-carotene per molecule of P700. The subunit composition of CCI is complex and varies somewhat depending upon the species, especially in the case of the cyanobacteria (Bengis and Nelson, 1975, 1977; Nechushtai and Nelson, 1981; Nechushtai *et al.*, 1983; Lundell *et al.*, 1985). Table 4.5 compares the subunit composition of two types of CCI. P700 is generally assumed to be located on one or both of the two apoproteins found in the 60–70 kD region. There has been a good deal of argument as to whether these two apoproteins are identical or not (see Thornber, 1986a). However, recent evidence from molecular biological studies, where two non-identical genes coding for these apoproteins have been located on the chloroplast genome (Fish *et al.*, 1985), suggests that they are probably non-identical. The other smaller subunits of CCI are thought mainly to represent the various iron-sulphur proteins that form the chain of electron acceptors on the reducing side of photosystem I. The exact identities of these smaller subunits have not yet been unequivocally assigned, and it is more than likely that this will not happen until the genes which code for them have been cloned and sequenced (Setif and Mathis, 1986; Wollman, 1986). A model of CCI is presented in Fig. 4.5.

Table 4.5 Comparison of the typical polypeptide composition of CCI complexes isolated from higher plants and cyanobacteria. Data from Bricker *et al.* (1986) and Ort (1986)

Size (kD)			
Higher plants	Cyanobacteria	Function	Location of the structural gene
70'	64–70'	P700	chloroplast
25	16	unknown	nuclear
20		interaction with plastocyanin	nuclear
18	18,17.5	Fe/S	nuclear
16	11.5	Fe/S	nuclear
8	10	Fe/S	nuclear

'There are two distinct polypeptides in this molecular weight region.
Fe/S = iron-sulphur protein.

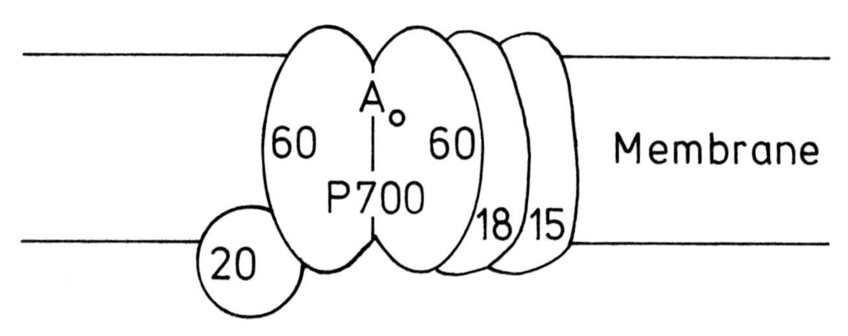

Fig. 4.5 A model of the subunit arrangement of CCI in the thylakoid membrane. The numbers refer to the apparent molecular weights of the major subunits of CCI in kD; see text for details.

B LHCI

The presence of LHCI, a chlorophyll a/b protein associated with photosystem I, was first really demonstrated by Mullet *et al.* (1980). These workers isolated a photosystem I preparation from pea chloroplasts, following solubilization with triton X-100. This preparation contained ~110 chlorophyll molecules per P700 and had at least 11 polypeptide subunits. Treatment with higher concentrations of triton X-100 reduced the chlorophyll content to ~65 chlorophyll molecules per P700. Removal of this chlorophyll was associated with the loss of three or four polypeptides in the 20–26 kD molecular weight range. It has since been shown that the higher triton treatment removed two antenna complexes associated with CCI (Haworth *et al.*, 1983; Lam *et al.*, 1984a, b). These have been called $LHCI_{\alpha}$ and $LHCI_{\beta}$. They are both chlorophyll a/b proteins. They have a chlorophyll a:chlorophyll b ratio of about 3.5:1 and their respective apoproteins are immunologically distinct (Lam *et al.*, 1984a).

C CCII

Tang and Satoh (1985) have described a digitonin extraction method for the isolation and purification of CCII from pea chloroplasts. CCII isolated in this way has quite a reproducible composition and shows photochemical activity. CCII contains five major protein subunits with apparent molecular weights of 47, 43, 34, 32 and 9-10 kD (see Table 4.6). A sixth subunit of rather lower molecular weight is also usually present, though it is often difficult to detect on gels because it is hard to stain with coomassie blue (Metz *et al.*, 1983; Widger *et al.*, 1984). Sometimes another subunit with an apparent molecular weight of ~29 kD is found as well [called CP29 by Camm and Green (1980, 1983)] but this has been shown to represent an antenna complex. CCII contains about 50 chlorophyll molecules and ~7β-carotenes for each pair of cytochrome b_{559} molecules (Ohno *et al.*, 1986). The two smallest subunits of CCII are the cytochrome b_{559} apoproteins. Their primary structures have been determined (Herrman *et al.*, 1984; Westhoff *et al.*, 1985), but the function of this cytochrome is not yet known.

Up until about 2 or 3 years ago it was always stated that P680 and *Q* were located on the 47 or the 43 kD subunits (Cogdell, 1983). It was then that the molecular biologists intervened. The genes for all five of

Table 4.6 The typical polypeptide composition of CCII. Data from Arntzen and Pakrasi (1986), Ort (1986) and Thornber (1986a)

Size of polypeptides (kD)	Function	Location of structural genes
47	Tightly bound	chloroplast
43	core antenna	chloroplast
34	P680, Q_a	chloroplast
32	Q_b	chloroplast
10	cytochrome-*b*559	chloroplast
6		chloroplast

the protein subunits of CCII have been cloned and sequenced (Alt *et al.*, 1984; Herrman *et al.*, 1984; Holschuh *et al.*, 1984; Morris and Herrman, 1984; Rassmussen *et al.*, 1984; Rochaix *et al.*, 1984; Westhoff *et al.*, 1985). They are all encoded on the chloroplast genome (see Table 4.6). The 47 and 43 kD polypeptides [true size 56.2 and 44.8 kD (Arntzen and Pakrasi, 1986; Barber and Marder, 1986)] have quite similar primary structures but are not completely homologous. But it is on the 34 and 32 kD subunits that most of the attention has now focused (Hearst, 1986). The 34 kD protein is also often called the D2 protein while the 32 kD protein is often referred to as the DI protein or the Q_B protein. When the primary structure of a protein is first determined it is usual to search through the database of all other known protein amino acid sequences to see if any sequence homologies can be discovered. This is done by computer. In *Chlamydomonas reinhardii* the overall sequence homology between the 34 and 32 kD proteins is only 27%; however there are several domains in the sequence where the degree of homology rises to between 33 and 58%. The 32 kD protein has been shown by numerous studies to bind the secondary electron accepting plastoquinone in CCII (Q_B) and it is in the region of the putative quinone-binding region that the highest level of homology is seen. Furthermore a similar, very striking homology is seen between the 34 kD protein and the "M" subunit of the reaction centre from the photosynthetic bacterium *Rhodobacter capsulatus*, and the 32 kD protein and the "L" subunit of the *R. capsulatus* reaction centre, when they are compared in the regions of their quinone-binding sites (Youvan *et al.*, 1984). This has naturally led to the suggestion that P680 and Q are, by analogy to the bacterial reaction centre, located on the 34 and 32 kD subunits of CCII (Hearst and Sauer, 1984; Hearst, 1986; Michel and Deisenhofer, 1986). The reactions of photosystem II have been studied for more than 20 years and it is interesting to think that until about 2 years ago nobody really knew, in biochemical terms, where P680 was located. In the past year the idea that P680 and Q are located on the 32 and 34 kD proteins has received strong biochemical support. The groups of Satoh and Barber (unpublished communications) have been able to isolate a CCII complex which lacks both the 47 and 43 kD subunits. This minimal complex still retains a measure of the normal photochemical reactions (see discussion below in section G) and appears to contain about four molecules of chlorophyll a, two molecules of pheophytin a and two molecules of plastoquinone per cytochrome b_{559}. A model of CCII is presented in Fig. 4.6.

At this point, because of the strong structural and functional similarity between the purple bacterial reaction centre and photosystem II, it is necessary to digress to discuss what is known about the structure and function of the bacterial reaction centre.

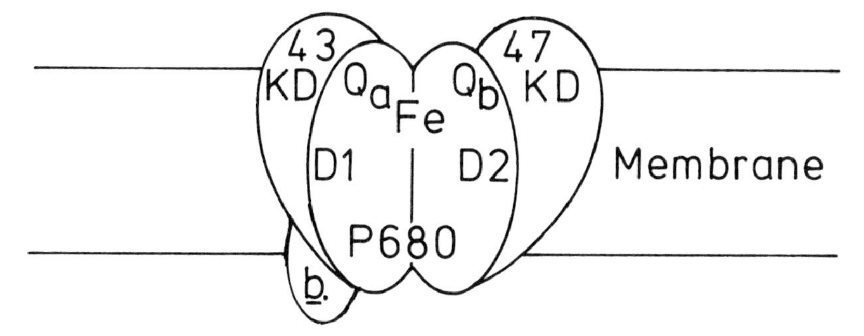

Fig. 4.6 A model of the subunit arrangement of CCI in the thylakoid membrane.

Reaction centres isolated from the bacteriochlorophyll *b*-containing purple, non-sulphur photosynthetic bacterium *Rhodopseudomonas viridis* have been crystallized (Michel, 1982). These crystals (for example, see Fig. 4.7A) have proved to be ideally suited for structural analysis by X-ray crystallography. As a result of this, in a brilliant piece of work, the group at the Max Planck Institute of Biochemistry in Munich have determined the complete three-dimensional structure of this reaction centre to a resolution of 3 Å (Diesenhofer *et al.*, 1984, 1985). Their structure is shown in Fig. 4.7B. The *R. viridis* reaction centre consists of four molecules of bacteriochlorophyll b, two molecules of bacteriopheophytin b, one molecule of menaquinone, one atom of ferrous iron and four protein subunits. The subunits are the L, M and H polypeptides and a bound, 4-heme, one polypeptide, *c*-type cytochrome. As part of their structural studies the Munich group also sequenced all four of their protein subunits (Michel *et al.*, 1985). On the lefthand side of Fig. 4.7B the full reaction centre structure is shown while on the righthand side the protein subunits have been removed to show the arrangement of the pigments. In plants there does not seem to be a direct equivalent of the bound cytochrome so details of its structure and function will not be presented here.

All the reaction centre pigments are bound to the L and M subunits. The most prominent feature in the spatial arrangement of the pigments is the approximate twofold axis of symmetry, which runs through the centre of the structure. The cytochrome lies down this line of symmetry. Two of the bacteriochlorophyll b molecules lie on this axis of symmetry and form a closely interacting dimer. They are stacked with pyrrole rings I on top of each other. This pair of bacteriochlorophyll b molecules forms the primary electron donor, called P960, and this is the analogue of P680 in CCII. The remaining reaction centre pigments are arranged in two parallel arms on either side of the twofold axis. The lefthand branch in

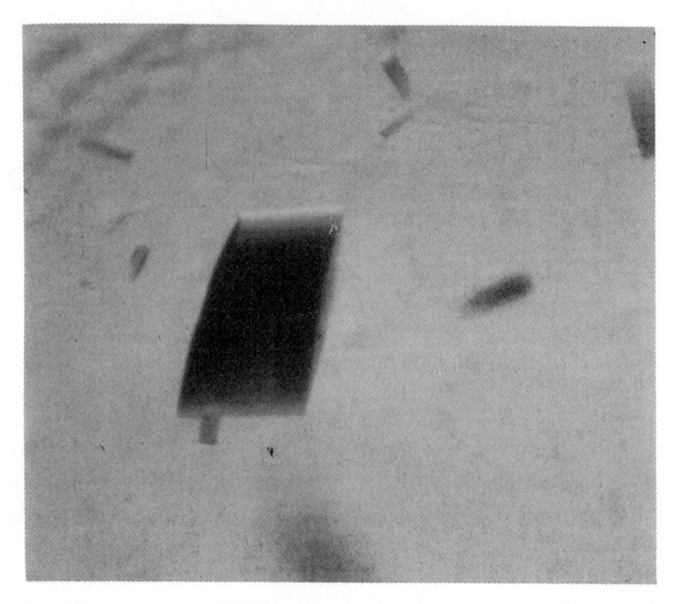

Fig. 4.7A. Crystals of reaction centres from *Rhodopseudomonas viridis*.
The pictures for Fig. 4.7 A and B were kindly provided by Drs. Michel and Deisenhofer. Reproduced in colour on the frontispiece.

Fig. 4.7B is called the M branch and the righthand side is called the L branch. These branches are named after the reaction centre subunit with which they are mainly associated, though it must be emphasized that nearly all the pigment binding sites are formed from amino acid residues from both the L and M subunits. Spectroscopic measurements on reaction centres, both in solution and in the crystalline state, have shown that under normal circumstances the electron transfer events associated with the primary reactions take place predominantly down the L branch (Zinth *et al.*, 1983). The discovery of the mirror-image M branch was one of the major surprises when the reaction centre structure was presented and its function is not yet clear. The "H" subunit of the reaction centre does not bind pigment and its function is also, as yet, not clear.

When P960 is excited by a photon of light an electron is transferred from the first excited singlet state of P960 down the L branch of the reaction centre. Initially the electron is transferred to bacteriopheophytin b_L. This reaction takes 2–4 ps (Breton *et al.*, 1986). Then subsequently the electron moves on to reduce the menaquinone in about 200–400 ps (Breton *et al.*, 1986; Parson and Holton, 1986). There is a monomeric

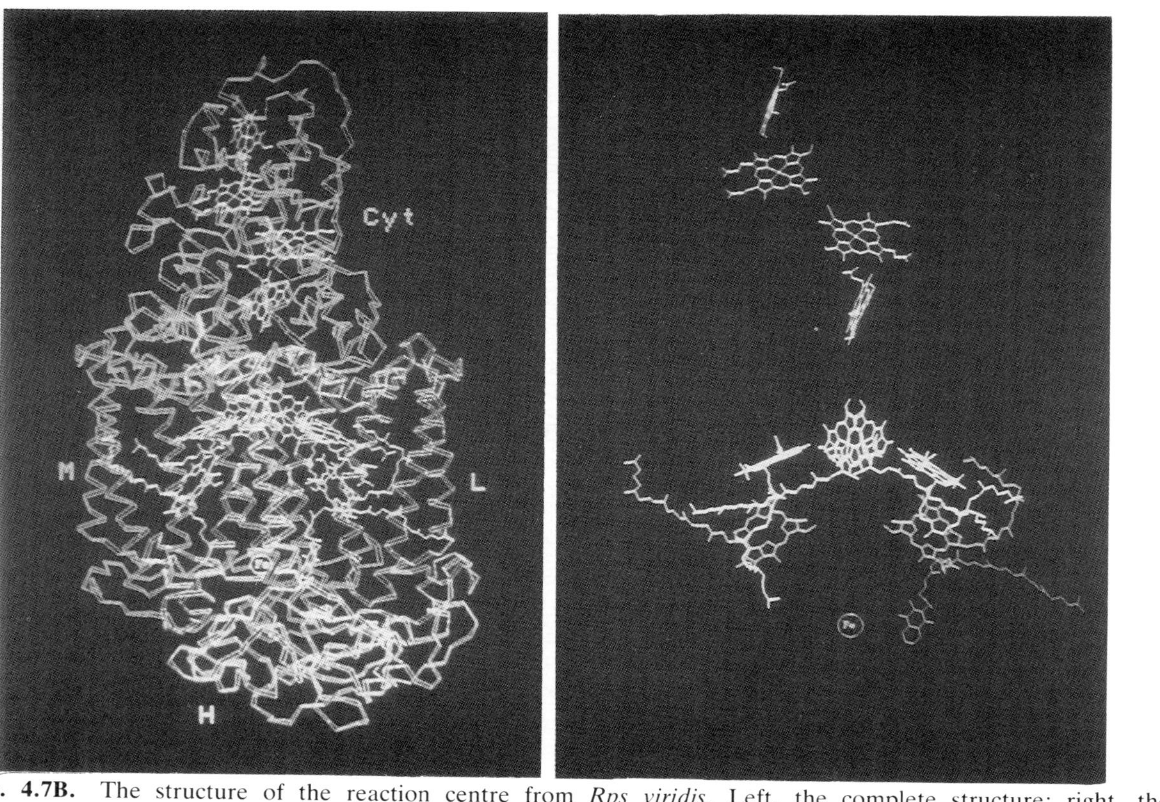

Fig. 4.7B. The structure of the reaction centre from *Rps viridis*. Left, the complete structure; right, the arrangement of the chromophores alone (orange = hemes; yellow = BChl b; blue = Bpheo b). Reproduced in colour on the frontispiece.

bacteriochlorophyll b molecule in the structure, which is located between P960 and bacteriopheophytin b. However, there is at present no convincing evidence that the electron can reside on this bacteriochlorophyll b during the primary charge separation. The role of this pigment has yet to be determined. The primary reactions taking place within the bacterial reaction centre can therefore be summarised as follows:

$$(\text{BChl-BChl})_{\text{P960}}\text{BChl}_{\text{L}}\text{Bpheo}_{\text{L}}Q \xrightarrow{h\nu} (\text{BChl-BChl})^{*}_{\text{P960}}\text{BChl}_{\text{L}}\text{Bpheo}_{\text{L}}Q$$

$$(\text{BChl-BChl})^{*}_{\text{P960}}\text{BChl}_{\text{L}}\text{Bpheo}_{L}Q \xrightarrow{2\text{–}4ps} (\text{BChl-BChl})^{+}_{\text{P960}}\text{BChl}_{\text{L}}\text{Bpheo}^{-}_{\text{L}}Q$$

$$(\text{BChl-BChl})^{+}_{\text{P960}}\text{BChl}_{\text{L}}\text{Bpheo}^{-}_{\text{L}}Q \xrightarrow{200\text{-}400ps} (\text{BChl-BChl})^{+}_{\text{P960}}\text{BChl}_{\text{L}}\text{Bpheo}_{\text{L}}Q^{-}$$

In vitro, in the absence of other redox components, there is a back reaction between $P960^{+}$ and Q^{-} which takes place over a few tens of ms (Shopes and Wraight, 1986). *In vivo* the electron from Q is transferred in μs to a secondary quinone and then on to the redox carriers in the bacterial cyclic electron transport system (Tiede, 1986).

In *R. viridis* reaction centres there is a space on the M branch in the site equivalent to the Q-binding site on the L branch. Quite recently the position of the pigments in reaction centres from *Rhodobacter sphaeroides* has also been determined by X-ray crystallography (Chang *et al.*, 1986). In this case the reaction centres contain two quinone molecules; the second quinone is bound to the site in the M branch which is the mirror-image of the menaquinone site in the *viridis* structure. This is therefore assumed to be the secondary quinone and probably corresponds to the Q_B site on the 32 kD protein in CCII.

D LHCII$_\beta$

Since the mid 1970s much experimental effort has been expended in studying the structure and function of LHCII$_\beta$, cf. Thornber (1986). This has not only been due to the fact that it is a major constituent of the chloroplast membrane but also due to the relative ease with which it can be prepared in large amounts (cf. Bennett, 1983).

This complex is not only the major antenna complex in chlorophyll b-containing organisms but it is also thought to be important in stabilizing the "stacked" regions of the thylakoid membranes (Barber and Marder, 1986).

The LHCII$_\beta$ apoproteins are coded for on the nuclear genome and form a multigene family (Dunsmuir *et al.*, 1983; Cashmore, 1984; Tobin *et al.*, 1984). They are synthesized on cytoplasmic ribosomes as precursor proteins. In this form they enter the chloroplast, are processed proteolytically (3–4 kD is removed from the N-terminus) and then assembled into the thylakoid membrane (Bennett, 1983; Karlin-Neumann and Tobin, 1986). It is not yet clear, however, at which stage in this process the chlorophylls are inserted into the apoprotein.

Most of the chloroplast's chlorophyll b is located in $LHCII_\beta$ and usual chlorophyll a:b ratios are in the range of 1:1 to 1.4:1. Most workers in this area now feel that this means that 4 chlorophyll a molecules and 3 chlorophyll b molecules and 1–2 xanthophylls are bound per apoprotein (Thornber *et al.*, 1979; Lichtenthaler *et al.*, 1982; Siefermann-Harms, 1985). As is usual, though, in this area of pigment–protein complexes where the pigments are non-covalently bound to their apoproteins, there are some dissenters from this view. Burke *et al.* (1978) favour 13 chlorophyll molecules per apoprotein. Knox and Van Metter (1979) investigated the spectral properties of this complex and proposed a model for the arrangement of its pigments within a single apoprotein. Their model places the three chlorophyll b molecules in a central position where they interact excitonically, and arranges the chlorophyll a molecules around the periphery.

When $LHCII_\beta$ is prepared and its polypeptide composition is analysed on SDS polyacrylamide gels, multiple apoproteins in the 24–27 kD molecular weight region are seen (Bennett, 1983). Different plants express different numbers of the apoproteins of this multigene family. In *Lemna gibba* there is a family of about 12 nuclear genes coding for $LHCII_\beta$ apoproteins and at least three different, but similar, apoproteins are expressed (Tobin *et al.*, 1984). The functional significance of these different apoproteins is entirely unclear. It is also unclear how these apoproteins associate to form the *in vivo* aggregated structure. Do homoaggregates or heteroaggregates form? Is there a heterologous mixture of aggregates? At present nobody knows.

The primary structure of an $LHCII_\beta$ apoprotein has been determined by sequencing of its structural gene in *Lemna* (Tobin *et al.*, 1984) and in pea (Cashmore, 1984). Once the primary structure of a membrane protein has been obtained it has become fashionable to try to predict how it will fold up and to try and pinpoint those regions which might be located in the hydrophobic part of the membrane (Kyte and Doolittle, 1982). The logic of these calculations is as follows. A window of between eight and 12 amino acids is moved, one amino acid at a time along the sequence, and at each position the section of the sequence in the window is given an index. The index is a measure of how hydrophobic or hydrophilic that segment of the sequence is. The value of the index is then plotted against the amino acid sequence and a so-called "hydropathy plot" is produced. Hydrophobic stretches are then good candidates to be located within the membrane and hydrophilic stretches are probably better positioned in the aqueous phases at the surface of the membrane (cf. Karlin-Neumann *et al.*, 1985). It has also become rather a dogma in this area that the most likely form for the membrane-spanning regions is an α-helix. Based on the primary structure of an $LHCII_{\beta}$ apoprotein from *Lemna*, Karlin-Neumann *et al.* (1985) have proposed a model for the folding of the polypeptide chain. The model is shown in Fig. 4.8. It has several interesting features. Firstly, it predicts that the $LHCII_{\beta}$ apoprotein is a transmembrane protein, then that there are three membrane-spanning α-helices and finally, that a large part of the structure (the stromal side in Fig. 4.8) is rather globular and hydrophilic. However, the model gives no information on where the chlorophylls are located. There is good evidence in the case of the antenna complexes from the bacteriochlorophyll-containing purple bacteria that the chlorophylls are ligated to histidine

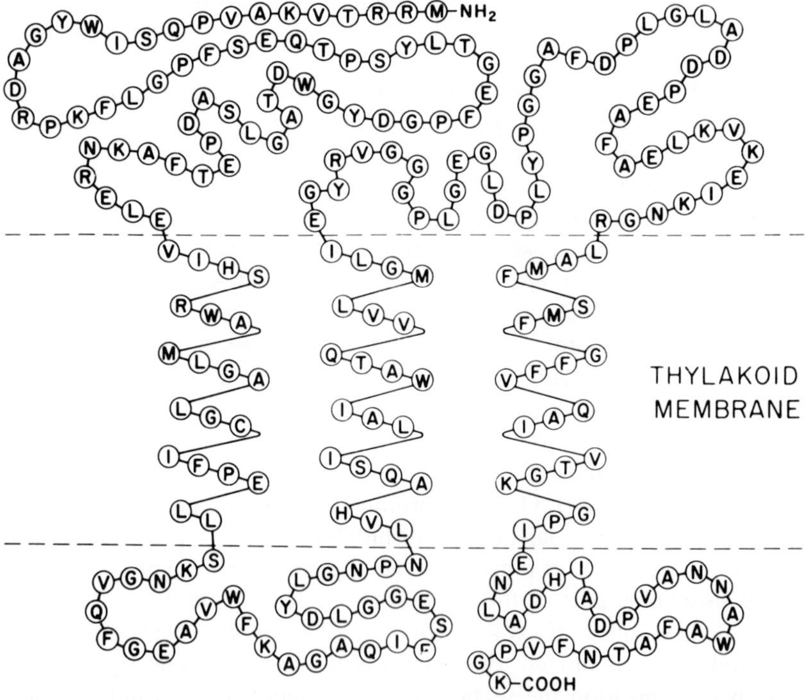

Fig. 4.8 A model for the folding of the $LHCII_{\beta}$ apoprotein and its association with the thylakoid membrane. The individual amino acids are represented by the single letter code. Note the presence of three membrane-spanning α-helices and the large polar domain on the stromal side of the membrane. Produced with permission from Karlin-Neumann *et al.* (1985).

residues. There are not enough histidine residues, however, in the $LHCII_{\beta}$ apoprotein for all the chlorophylls to be bound to histidines.

$LHCII_{\beta}$ has been crystallized in the form of ordered two-dimensional arrays (Li and Hollingshead, 1982; Kühlbrandt *et al.*, 1983; Kühlbrandt, 1984; Li 1985; Fig. 4.9). These two-dimensional crystals have been examined by electron microscopy and image reconstruction. A low-resolution structure of $LHCII_{\beta}$ has been determined, depicted in Fig. 4.9; it shows the details of the outline of the shape of the molecule. At this resolution the individual pigment molecules are not visible. $LHCII_{\beta}$ seems to be a trimer. This unit is ~27Å wide and about 60–65 Å long. These crystalline arrays are produced by reconstitution into artificial membranes and in these membranes they adopt a transmembrane orientation. Their membrane location is excentric, with the protein protruding some 20 Å on one side of the bilayer but only ~7Å on the other. In this respect these direct structural studies fit rather well with the model described above, based solely upon the primary structure of the $LHCII_{\beta}$ apoprotein. $LHCII_{\beta}$ has now been crystallized in the form of thin three-dimensional crystals (Kühlbrandt, personal communication). As yet however, these have not proved suitable for a high-resolution structural analysis by X-ray crystallography, but hopefully this will soon be achieved.

The Karlin–Neumann *et al.* (1985) model also predicts a rather substantial content of α-helix. In the absence of a high-resolution structure of a protein it is still possible to estimate its α-helical content. This can be achieved by investigating the circular dichroism spectrum of the protein in the far ultraviolet, in the spectral region where the peptide bond absorbs (Chen *et al.*, 1978). In the 190–250 nm region the shape of a protein's CD spectrum is sensitive to its secondary structure, i.e. its content of α-helix, β-pleated sheet etc. Nebedryk *et al.* (1984) have used this method together with infrared spectroscopy to show that $LHCII_{\beta}$ has an α-helical content of ~44% and that the helices lie nearly normal to the place of the membrane.

Although $LHCII_{\beta}$ is mainly associated with photosystem II it does play a role in controlling the distribution of absorbed radiant energy between the two photosystems (Allen, 1983; Bennett, 1983; Horton, 1983). It is now generally accepted that the two photosystems are arranged in series in an electron transport sequence called the "Z" scheme (Fig. 4.10). The efficient operation of the overall light reactions requires the activity of the two photosystems to be balanced. However, because the pigments associated with each photosystem absorb light differently at different wavelengths, the system can be unbalanced if light of certain wavelengths is used to excite photosynthesis (Myers, 1971). Investigation of the effect of inducing photosynthesis with light that is preferentially absorbed by

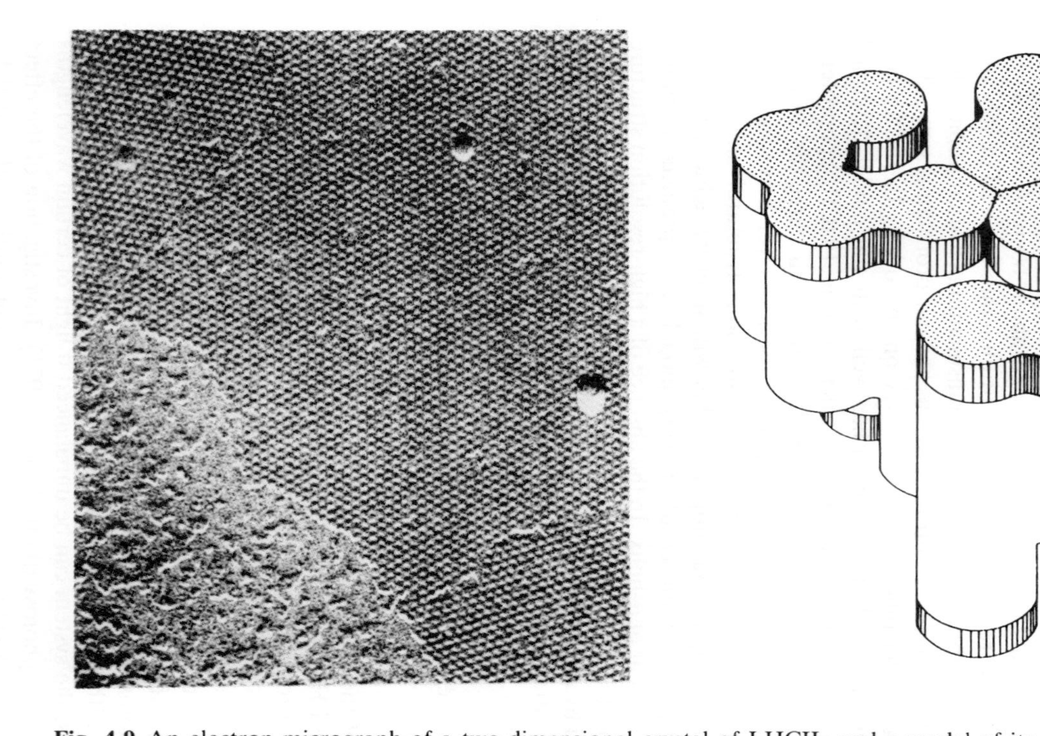

Fig. 4.9 An electron micrograph of a two-dimensional crystal of $LHCII_{\beta}$ and a model of its subunit organization deduced by image analysis of these crystals. Left, the two-dimensional crystalline array of $LHCII_{\beta}$ complexes. Right, the proposed model of the subunit arrangement. It is proposed that $LHCII_{\beta}$ is a trimer composed of three structurally equivalent subunits. The shaded portions show the part of the complex which would protrude from the bilayer (the bilayer is 45 Å thick). Adapted with permission from data in Kühlbrandt (1984).

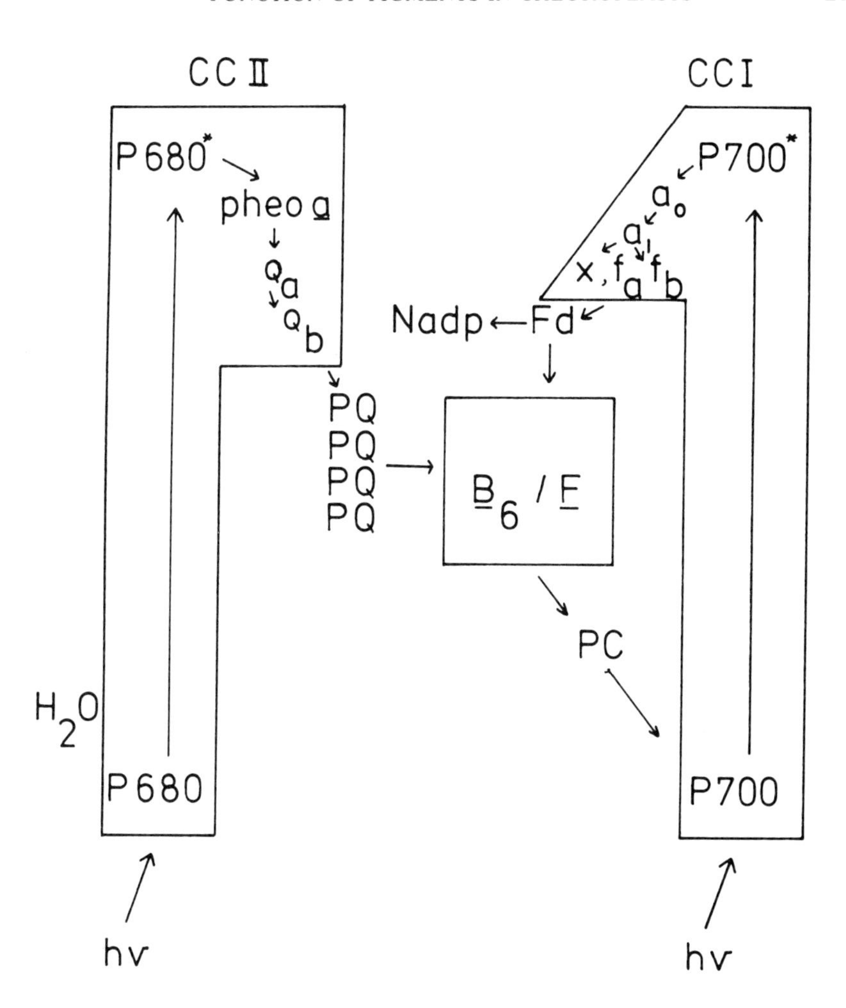

Fig. 4.10 A model of the "Z" scheme of photosynthetic electron transport, emphasizing the role of the three major, supramolecular membrane-bound complexes. In this representation the three major pigment–protein complexes (CCI, CCII and the cytochrome b_6/f complex, shown enclosed in boxes) are connected by mobile electron transport carriers (PQ = plastoquinone; PC = plastocyanin; Fd = ferredoxin; N_{adp} = nicotinamide adenine dinucleotide phosphate). The various components thought to be involved in the redox reactions within CCI and CCII are shown within their respective boxes; see text for details. $P680^*$ and $P700^*$ represent the first excited singlet states of P680 and P700 respectively.

photosystem I or photosystem II has uncovered a phenomenon called "state" changes (Bonaventura and Myers, 1969; Myers, 1971; Williams, 1977). If chloroplasts are illuminated with light preferentially absorbed by photosystem II (light 2) they are put into state 2. In this state more of the absorbed energy is diverted to photosystem I and the overall reaction moves back into balance. If light 2 is turned off, and light 1 (light preferentially absorbed by photosystem I) is now turned on the chloroplasts are put into state 1. In state 1 the absorbed radiant energy is now preferentially redirected to photosystem II. It has been proposed that $LHCII_\beta$ is responsible for these "state" changes and that the whole process is controlled by protein phosphorylation (Bennett, 1983).

When chloroplasts are exposed to light in the presence of adenosine triphosphate many thylakoid proteins become phosphorylated. This reaction is catalysed by a protein kinase(s). It has been well documented that $LHCII_\beta$ apoproteins are phosphorylated in the light (Allen, 1983). Phosphorylation occurs on a threonyl residue(s) in a segment of the $LHCII_\beta$ apoprotein which is surface exposed (Bennett, 1979, 1980). This is an N-terminal segment and it can be removed from the membrane-bound $LHCII_\beta$ by treatment with trypsin. Evidence has been presented to suggest that the phosphorylation is controlled by the activity of a membrane-bound protein kinase, whose activity is regulated by the redox state of the plastoquinone pool (Allen *et al.*, 1981; Horton and Black, 1981). The main features of this model are presented in Fig. 4.11. If photosystem II is overexcited the PQ pool will be reduced, the kinase switched on and $LHCII_\beta$ will be phosphorylated. On the other hand if photosystem I is overexcited the PQ pool will be oxidized and the kinase activity will be turned off. A protein phosphatase is present and always active. This means that in light 1, where the kinase is inactive $LHCII_\beta$ will become dephosphorylated. The model is completed by proposing that when LHCII is phosphorylated it directs more excitation energy to photosystem I and when it is dephosphorylated more excitation energy is distributed to photosystem II. It has been suggested that this energy redistribution is a result of the $LHCII_\beta$ being mobile and shuttling between the two types of photosystem under the influence of its phosphorylation state (Barber, 1983; Kyle *et al.*, 1983, 1984; Larsson *et al.*, 1983; Staehelin and Arntzen, 1983). This process will be discussed further in section 4 below.

E A general model of reaction centre function

It is now possible to discuss the mechanisms involved in the primary reactions in photosynthetic reaction centres in the term of a general

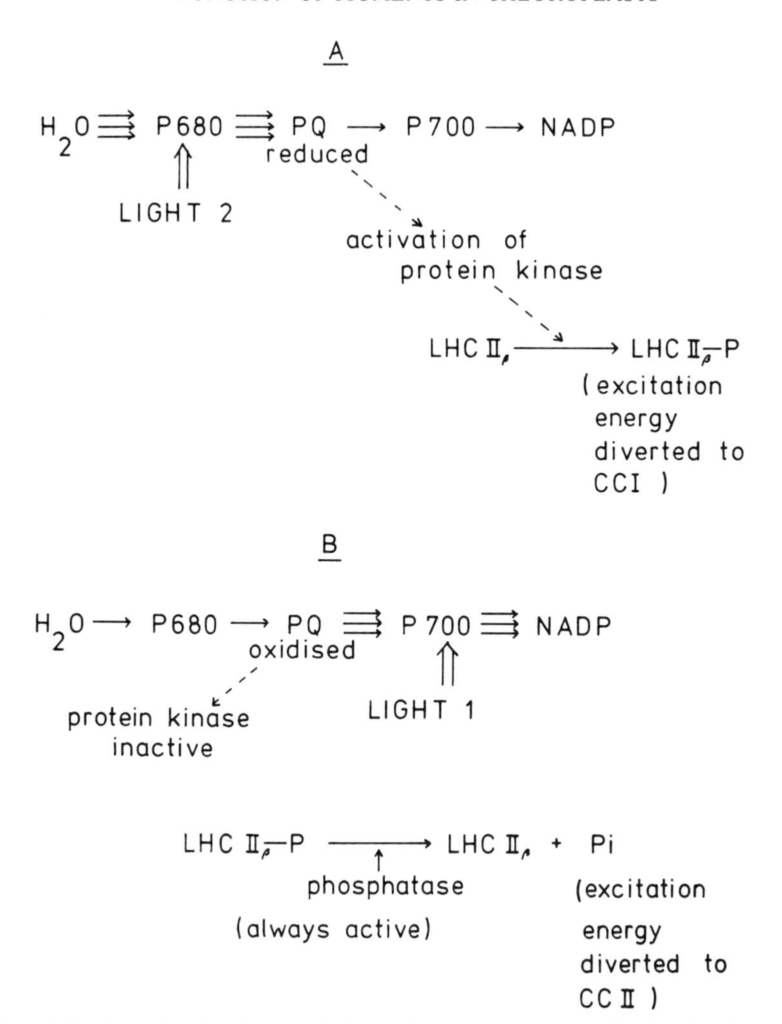

Fig. 4.11 A model for the regulation of excitation energy distribution between the two photosystems, controlled by the redox state of the plastoquinone pool and the phosphorylation of $LHCII_{\beta}$. **A** Light 2 (light preferentially absorbed by PSII) causes the over-reduction of the plastoquinone pool. This activates the protein kinase(s) and $LHCII_{\beta}$ is phosphorylated. In its phosphorylated state $LHCII_{\beta}$ directs absorbed excitation energy to photosystem I and so compensates for the imbalance between the activity of the two photosystems induced by light 2. **B** Light 1 (light preferentially absorbed by photosystem I) causes oxidation of the plastoquinone pool. This switches off the protein kinase(s). The phosphatase(s) is always active and so in this situation any $LHCII_{\beta}$ which is phosphorylated will be slowly dephosphorylated. Excitation energy is now redirected back to PSII and again the imbalance in the activities of the two photosystems is corrected. The basis of this model was proposed by Allen *et al.* (1981) and Horton and Black (1980).

model. This model is largely based upon applying the knowledge gained in the bacterial system to plants.

$$P(I_1\text{-}I_N)X \xrightarrow{h\nu} P^*(I_1\text{-}I_N)X \rightarrow P^+(I_1\text{-}I_N)^- X \rightarrow P^+(I_1\text{-}I_N)X^-$$

When a reaction centre is excited the primary electron donor, P, is raised to its first excited singlet state, P^*. Then, in less than 1 ns, a series of electron transfer steps through intermediates (I_1-I_N) takes place, resulting in the oxidation of P to P^+ and the reduction of an electron acceptor X to X^- (Cogdell, 1983; Parson and Holton, 1986). The state X^- is usually stable for microseconds or longer and used to be called the reduced primary electron acceptor (Cogdell, 1983). The series of electron acceptors is now recognised as an important feature of reaction centre function. At each step in the electron acceptor chain the forward reaction is fast. However with each reaction in the series the rate of the back reaction slows down by at least two orders of magnitude. This effectively stabilizes the initial charge separation and prevents the potential energy-wasting back reactions from occurring.

F The pattern of the primary reactions in photosystem I

As has already been discussed, in the 1970s it was widely accepted that P700 was a dimer of chlorophyll a. However, recently both these assumptions have been challenged. It has not only been questioned whether P700 is indeed a dimer, but also whether in fact it is really chlorophyll a (cf. Setif and Mathis, 1986). A different type of chlorophyll, called Chl-RCI, has been isolated and found to be stoichiometric with the PSI reaction centre (Dornemann and Senger, 1982). Scheer *et al.* (1983) have suggested that the Chl-RCI is a 13^2-hydroxy-20-chloro-chlorophyll a-derivative. However it still remains to be demonstrated, even if Chl-RCI is present in a stoichiometric ratio with P700, that it actually forms P700. The problem of whether P700 should be viewed as a monomer or a dimer is probably rather academic. It seems quite likely that it will depend upon which electronic state of P700 is being examined as to whether it displays monomeric or dimeric characteristics (Setif and Mathis, 1986). If, for example, $P700^+$ is examined by magnetic resonance techniques on the time scale of 10^{-7} s or less, then the "hole" appears to be localized on a single chlorophyll molecule (Wasielewski *et al.*, 1981). However, if the difference spectrum between triplet excited P700 and ground state P700 is determined (den Blanken and Hoff, 1983), then the dimer model gives a better fit with the data. This ambiguity will probably persist until a high-resolution, three-dimensional structure is available for CCI.

The reducing side of photosystem I has been investigated (Ke, 1973; Sauer *et al.*, 1978, 1979; Fenton *et al.*, 1979; Shuvalov *et al.*, 1979 a,b,c) by a combination of fast optical spectroscopy and slower, low-temperature ESR spectroscopy (Malkin and Bearden, 1971, 1978; Evans *et al.*, 1974, 1976; McIntosh and Bolton 1976; Heathcoate *et al.*, 1978; Malkin 1982; Setif and Mathis, 1986). These two approaches have yielded a great deal of information, but it is still extremely difficult fully to reconcile all the available data in one consistent picture.

If photosystem I preparations are excited by a short laser flash then P700 is oxidized and P430 is reduced. In the absence of any added electron donors or acceptors the state $P700^+$ $P430^-$ then decays at room temperature by a back reaction (Ke, 1973). The half-time for this decay is ~45 ms. If the redox potential is poised such that P430 is chemically reduced prior to the exciting flash, P700 will still be oxidized, following an actinic laser pulse, but now the half-time for the back reaction is faster at ~250 μs (Sauer *et al.*, 1978, 1979). This result has been interpreted to mean that there must be another electron acceptor interposed between P700 and P430. In this case charge separation could then occur between P700 and this intermediate electron acceptor. Sauer *et al.* (1978) called this acceptor A_2. These workers extended this approach and investigated what happened when the photosystem I preparation was poised so that A_2 was also chemically reduced in the dark. Laser excitation still induced the formation of $P700^+$ but the back reaction was even faster, with a half-time of ~7 μs. This led to the postulation of yet another intermediate electron acceptor, this time between P700 and A_2. This acceptor is now called A_1.

ESR experiments on photosystem I preparations were also proceeding in parallel with the optical studies (for example, see Malkin, 1982). When preparations were chemically reduced so that they were in the state $P430^-$, then frozen to liquid nitrogen temperature, illumination produced a new electron acceptor signal (McIntosh and Bolton, 1976). This signal was called X and had ***g*** values at 1.78, 1.88 and 2.08. It was proposed that this represented yet another iron-sulphur acceptor and it is now thought to be equivalent to A_2. More detailed ESR studies have also been performed on triton photosystem I preparations under conditions where A_2 has also been chemically reduced (Bonnerjea and Evans, 1982; Gast *et al.*, 1983). These experiments have been interpreted to show the presence of two further electron acceptors between P700 and A_2, called A_1 and A_0. There is now ESR and optical data which suggest that A_1

could be a vitamin K-like quinone and that A_0 might be a monomeric chlorophyll a species (Mansfield and Evans, 1986; Thurnauer *et al.*, 1986). These results, however, must at present be viewed as preliminary.

The difficulty in combining all these data into a single, consistent model arises from two basic problems. The first comes from a fundamental difference in the nature of the data obtained from optical experiments as opposed to those carried out using ESR. The optical studies have very good time resolution, but most of the signals studied have rather poorly defined difference spectra and so their identification is difficult. The ESR studies, on the other hand, have a low time resolution but are most useful in identifying the chemical nature of the electron-accepting species involved. It is therefore difficult to be sure that the optical and ESR studies are actually looking at the same things. Notwithstanding this, it is the second problem that is probably even more difficult to resolve. All the experiments that have just been discussed above involved blocking the normal electron transport reactions. When this is done it is always a worry that artifactual side-reactions will now be favoured. It is really rather difficult to prove, in the blocked system, that any intermediate that becomes photochemically reduced is in fact a member of the normal photochemical pathway that operates in "open" reaction centres. What is required is a kinetic experiment on "open" reaction centres in which the path of the electron along the chain of acceptors can be followed as it "hops" from one to the next. Some ps experiments have been carried out on "open" photosystem I preparations (Nuijs *et al.*, 1986); however more work needs to be done before definite conclusions can be drawn. Setif and Mathis (1986) have tried to present a unified picture of the electron transport reactions occurring within CCI and this is summarized in Fig. 4.12.

G The pattern of the primary reactions in photosystem II

The similiarity of the acceptor side of photosystem II to the acceptor side of the bacterial reaction centre has stimulated the search for a transient

$$P700 \rightarrow A_0(\text{chl}\,a?) \rightarrow A_1(\text{vit.k?}) \rightarrow X(A_2:\text{Fe/S})$$

$$P430(F_a,F_b:\text{Fe/S})$$

Fig. 4.12 The pathway of the electron on the acceptor side of CII. See text for details. The chemical identity of electron acceptors A_0 and A_1 is not yet certain; this uncertainty is represented by the question marks.

intermediate electron acceptor in photosystem II analogous to the Bpheo in the bacteria. Many of the types of experiments initially carried out on bacterial reaction centres have now been repeated with CCII. In bacterial reaction centres illumination at low redox potential allows the accumulation of $Bpheo^-$ (Tiede *et al.*, 1976; Shuvalov and Klimov, 1976; Okamura *et al.*, 1979). Under reducing conditions in the light, charge separation between the primary donor, P, and Bpheo takes place. In continuous light this state is continually being formed, decaying by a back reaction and the being formed again. Occasionally an added electron donor can fill the "hole" on P and trap Bpheo in its reduced, anionic form. $Bpheo^-$ will now persist long enough for its spectroscopic properties to be studied. Klimov *et al.* (1977) illuminated DT-20 photosystem II particles which had been poised at low redox potential. They were able to detect absorbance changes which were consistent with the reduction of pheophytin a. The ESR and ENDOR spectra of this trapped state have also been recorded and are consistent with it being the anionic radical of monomeric pheophytin a (Fajer *et al.*, 1980; Klimov *et al.*, 1980a; Rutherford *et al.*, 1981a). It is the production of this pheophytin radical anion that has been used as the assay for photosystem II activity in the CCII preparations which lack the 47 and 43 kD subunits. Under certain experimental conditions a very typical doublet ESR signal is observed for the state Pheo a^- $Q_A{}^-$ (Rutherford, 1985). An exactly analogous doublet ESR signal has been seen with bacterial reaction centres and shown to arise from an interaction of $Bpheo^-$ with the semiquinone-Fe complex (Tiede *et al.*, 1976a,b; Okamura *et al.*, 1979). This split ESR signal is therefore very diagnostic for the native primary reactions.

If photosystem II preparations are poised at low redox potential and excited by a 30-ps laser pulse the state $P680^+$ Pheo a^- is formed very rapidly and then decays with a lifetime of ~4 ns (Ke and Dolan, 1980; Klimov *et al.*, 1980b). Some of this decay, just as with bacterial reaction centres, yields triplet states which have been detected both optically and by ESR (Rutherford *et al.*, 1981b). The ESR signal of the triplet state is very strongly spin-polarized, which is consistent with it being formed by a back reaction of a biradical species. The redox potential of Pheo a^- in photosystem II has been determined by titrating the extent of this light-induced triplet state (Rutherford *et al.*, 1981a). It has an E_m of −604 mV. There has also been a report suggesting that there could be another intermediate electron acceptor between P680 and the Pheo a (Rutherford, 1981). Illumination of photosystem II particles in which Pheo a has already been reduced induces a new ESR signal. This signal grows in slowly upon illumination and may be due to chlorophyll a^-. This however remains to be confirmed. A simple model of the photochemical pathway in photosystem II is presented in Fig. 4.13.

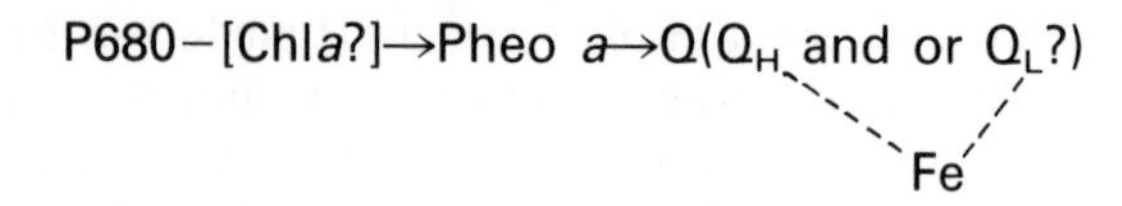

Fig. 4.13 The pathway of the electron on the acceptor side of CCII. See text for details. Note that the quinone acceptor(s) is magnetically coupled to the atom of ferrous iron.

Just as with the bacterial reaction centres, the presence of a ferrous ion atom in photosystem II, which is close enough to interact magnetically with reduced Q_A, has been demonstrated. The ESR signal which arises from the semiquinone-iron complex has been detected (Nugent *et al.*, 1981) and Mössbauer spectroscopy indicates that ferrous ion is high spin (Petrouleas and Diner, 1982).

There is a degree of heterogeneity in photosystem II. This was first noted in the case of the electron acceptor Q_A. As discussed above, redox titrations of Q_A showed the presence of two components, called Q_H and Q_L. It is still not clear whether a single photosystem II complex contains both Q_H and Q_L, or whether there are two types of photosystem II centres, one with Q_H and one with Q_L. There is a further type of heterogeneity that has been described for photosystem II. This is the so-called α- and β-centres. The thylakoid membranes are not uniform structures and as will be seen, the composition of the membrane regions which form the grana stacks is quite different from those of the stromal lamellae (Staehelin, 1986). The α-centres seem to be found in the granal regions and contain Q_H and Q_L. The β-centres appear to be located in the stromal membranes and have a quencher which has been called Q_β (see the discussion in Diner, 1986). The α- and β-centres are characterized by having different fluorescent properties. The origin and the function of this heterogeneity is, as yet, entirely unclear.

H The structure and function of the phycobilisomes

Research in this area of photosynthesis has been very fruitful over the past few years. The primary structure of a large number of phycobiliproteins has been described (Zuber, 1986) and the sites of bilin attachment have been determined (Glazer, 1985; Zuber, 1986). Colourless linker-polypeptides have been discovered in the phycobilisomes and shown to control assembly of, and energy distribution, within the phycobilisomes (Tandeau de

Marsac, 1983; Glazer, 1984, 1985). Much is known about the dynamics of energy transfer within the phycobilisome. Finally, in the last 2 years, the complete three-dimensional structure of two phycobiliproteins has been elegantly determined by X-ray crystallography (Schirmer *et al.*, 1985, 1986).

The bilin chromophores are linked to their apoproteins by way of one or two rather stable thioether bonds to cysteine residues. The tight-binding of the chromophore in the correct conformation is very important for its functioning in light-harvesting (Scheer, 1982, 1986; Glazer, 1985). When the biliproteins are denatured and the bilin is essentially "free", or with bilins alone in solution, the chromophores are extremely flexible. In this state they only have moderate absorption extinction coefficients (Brandlmeier *et al.*, 1981) and they have rather low ($\sim 10^{-3}$) fluorescence quantum yields (Scheer, 1982; Braslavsky *et al.*, 1983). Low fluorescence quantum yields means very fast decay from the first excited singlet state, a property which would make singlet–singlet energy transfer from one pigment molecule to another (i.e. the antenna function) very unfavourable. When the chromophores are correctly bound in rather rigid conformation, in the native phycobiliprotein the rapid deactivation of the first excited singlet state is severely hindered (Braslavsky *et al.*, 1983; Scheer, 1986). Now the absorption extinction coefficients are increased by a factor of about five and the fluorescence quantum yield rises to about 50% (Grabowski and Gantt, 1978). Both these changes greatly enhance the capacity of the chromophores to act as efficient light-harvesters.

Although numerous studies have been presented in which the excited state dynamics within phycobilisomes have been investigated on the ps time domain, still perhaps the easiest to understand is the study described by Searle *et al.* (1978). These workers, using phycobilisomes from *Porphyridium cruentum*, made use of the fact that the different phycobiliprotein types absorb and fluoresce at different wavelengths (see Table 4.7). Using ps excitation they were able to put light energy specifically into phycoerythrin and look for energy transfer by monitoring the fluorescence emitted by phycoerythrin itself, then from phycocyanin, allophycocyanin and finally from chlorophyll a. Searle *et al.* (1978) were thus able to see the sequential flow of excitation energy from PE→PC→APC→chlorophyll a, where PE = phycoerythrin, PC = phycocyanin and APC = allophycocyanin core. If PC is excited, energy is transferred to APC, not back to PE. The phycobilisome acts like a funnel where any absorbed light energy is transferred inwards and downwards to the chlorophyll-containing thylakoid membrane (see Fig. 4.14). Usually more than 90% of the absorbed incident radiation reaches the chlorophyll in the thylakoid membrane.

Table 4.7 Typical phycobilisome components, their spectral and molecular characteristics. Data from Bogorad (1975), Glazer (1977), Scheer (1980), Cohen-Bazaire and Bryant (1981), Gantt (1981)

Coloured apoproteins	Room temp. absorption peaks (nm)	Room temp. fl. emission peak (nm)	M_r (kD)	Subunit structure	Chromophore content
R-PE	498,542,565	578	260	$(\alpha,\beta)_6\ \gamma$	—
B-PE	498,545,563	575	260	$(\alpha,\beta)_6\ \gamma$	$2PEB_1$ 4PEB 2PEB + 2PUB
C-PE	565	578	230	$(\alpha,\beta)_6\ \gamma$	2PEB, 4PEB
R-PC	553,615	640	130	$(\alpha,\beta)_3$	1PCB, 1PCB + 1PEB
C-PC	620	650	120	$(\alpha,\beta)_3$	1PCB, 2PCB
APC-B	618,673	680	98	$(\alpha,\beta)_3$	1PCB, 1PCB
APC-I	654	680	145	$(\alpha,\beta)_3\ \gamma$	—
APC-II	650	660	105	$(\alpha,\beta)_3$	1PCB, 1PCB

Uncoloured (linker) poly-M_r (kD)	Cyanobacterial phycobilisomes e.g. *Synechococcus* 6301	Red algal phycobilisomes e.g. *Porphyridium cruentum*
95	—	present
75	present	—
45–75	—	present
35–45	present	present
30–35	present	present
25–30	present	present
10–25	present	present

R = Rhodophytan; B = Bangiophycean; C = Cyanophycean.
PEB = phycoerythrobilin; PCB = phycocyanobilin; PUB = phycourobilin.

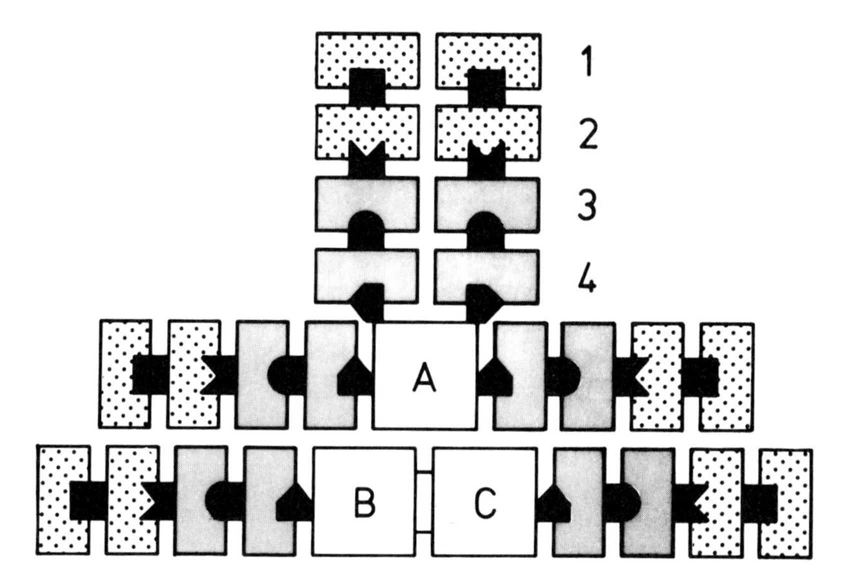

Fig. 4.14 Schematic representation of the *Synechocystis 6701* phycobilisome. The phycobilisome is constructed of "rods" connected to three "core" cylinders (represented by three squares in Fig. 4.14). A rod (dotted rectangle) is made up of four hexameric biliprotein complexes, each of which is attached through its specific linker polypeptide (shaded rectangle) to the component adjacent to it in the structure. In the following equations, abbreviations are: AP = allophycocyanin; PC = phycocyanin; PE = phycoerythrin; α^{AP}, β^{AP} etc. = α- and β-polypeptides of allophycocyanin; L = linker polypeptide (subscript denotes its location in the phycobilisome, i.e. R = in the rod; RC = rod/core junction; C = core; CM = core/membrane junction; the superscript denotes the apparent size in kD.

1: $(\alpha^{PE}\ \beta^{PE})_6 L_R^{30.5}$; **2**: $(\alpha^{PE}\ \beta^{PE})_6 L_R^{31.5}$; **3**: $(\alpha^{PC}\beta^{PC})_6 L_R^{33.5}$; **4**: $(\alpha^{PC}\beta^{PC})_6 L_{RC}^{27}$.

A: $((\alpha^{AP}\beta^{AP})_3)_2 + ((\alpha^{AP}\beta^{AP})_3 L_C^{10})_2$; **B** and **C**: $(\alpha^{AP}\beta^{AP})_3 + (\alpha^{AP}\beta^{AP})_3 L_C^{10} + (\alpha^{AP}\beta^{AP})_2 \beta^{18.5} L_{cm}^{99} + (\alpha_1^{APB}\alpha_2^{AP}\beta_3^{AP}) L_c^{10}$ Each core component is a tetramer of trimers. It is not yet known how the core cylinders are held together. The linker polypeptides hold the whole phycobilisome structure together and modulate the absorption and fluorescent properties of the pigments so as to promote excitation energy transfer down through the structure towards the chlorophyll-containing membrane. Adapted from data in Glazer (1985).

The organization of the phycobiliproteins within the phycobilisome and how phycobilisomes are assembled are now quite well understood (Glazer, 1985; Gantt, 1986). Fig. 4.14 illustrates the typical structure of a phycobilisome from the cyanobacterium *Synechocystis 6701* (Glazer, 1985). The allophycocyanin core is formed from three units. Each unit basically consists of (αβ) hexamers, where the differences between hexamers is the presence of different colourless linker (L) polypeptides. Rod-like structures are joined on to the core. These rods consist of phycoerythrin and phycocyanin hexamers stacked one on top of the other. The rods also contain linker polypeptides which are thought to be responsible for ensuring the correct interactions within the rods. Assembly of phycobilisomes can be achieved in solution from its individual components, but only if the correct linker polypeptides are present (Glazer, 1985). Several linker polypeptides have now been sequenced. They probably interact with the phycobiliproteins by way of ionic bonds.

The crystal structure of two phycocyanin molecules has now been determined to a resolution of 3Å, C-phycocyanin from *Mastigocladus laminosus* (a trimer) and C-phycocyanin from *Agmenellum quadruplicatum* (a hexamer; Schirmer *et al.*, 1985, 1986). The structure of the C-phycocyanin trimer is shown in Fig. 4.15.

4 THYLAKOID STACKING AND LATERAL HETEROGENEITY

In a normal, healthy chloroplast isolated from a higher plant the thylakoid membranes can be divided into stacked regions (called grana) and unstacked regions (called stromal lamellae). A considerable body of evidence has been accumulated (for example, see Barber, 1986) to suggest that membrane stacking requires a low surface charge density. Stacking can be reversed by putting chloroplasts into media at low ionic strength and induced by the addition of divalent cations, such as magnesium.

One of the consequences of stacking is that there is a lateral redistribution of charged species, and this appears to be at least one of the driving forces which cause the thylakoid membrane to show the phenomenon of lateral heterogeneity (for recent reviews see Staehelin, 1986 and Barber, 1986). If spinach chloroplasts, for example, are mechanically sheared, the resulting membrane vesicles can be separated into two distinct populations by phase partitioning (Andersson, 1978; Andersson and Akerlund, 1978; Albertsson *et al.*, 1982). As a result of extensive work it has become clear that these two membrane vesicle fractions are derived from different regions of the thylakoid system. One

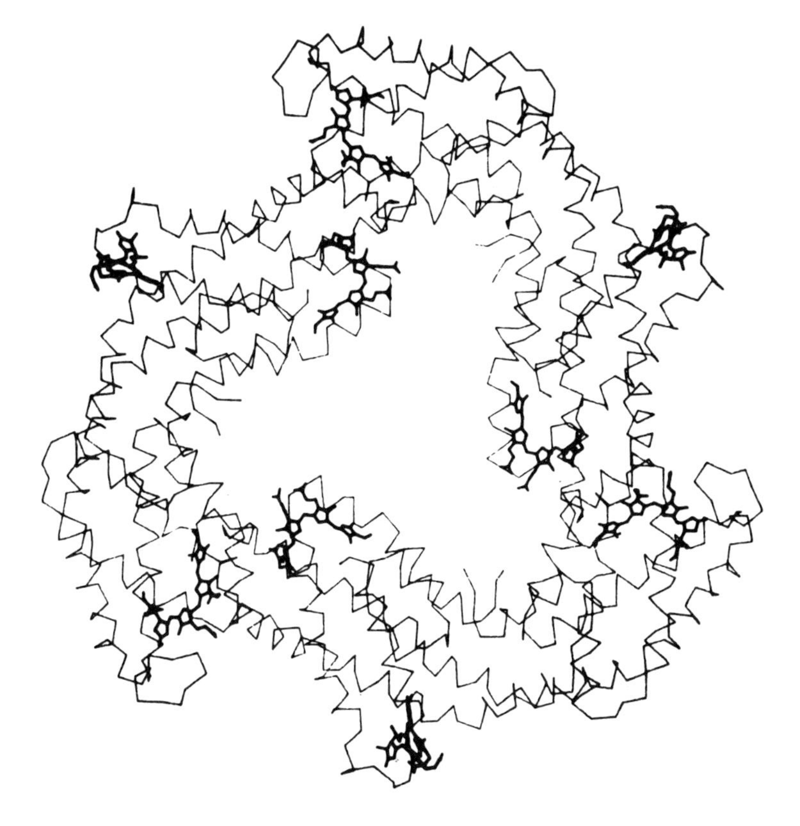

Fig. 4.15 The structure of the C-phycocyanin $(\alpha\beta)_3$ trimer isolated from *Mastigocladus laminosus*. Refined crystal structure: fainter lines represent the C_α-backbone of the apoproteins. Bold lines show the position and conformation of the chromophores. Courtesy of Prof. R. Huber.

fraction comes from the stacked or appressed regions and the other comes from the non-appressed stromal membranes. When the composition of these two fractions was investigated it was discovered that they are strikingly different (Andersson and Anderson, 1980; Anderson, 1982; Anderson and Melis, 1983). This is summarized in Fig. 4.16. The appressed regions are enriched in photosystem II while the non-appressed regions are enriched in photosystem I and the coupling factor (the membrane ATP synthetase). The cytochrome b_6/f complex is thought to be concentrated at the boundaries between the appressed and non-appressed regions.

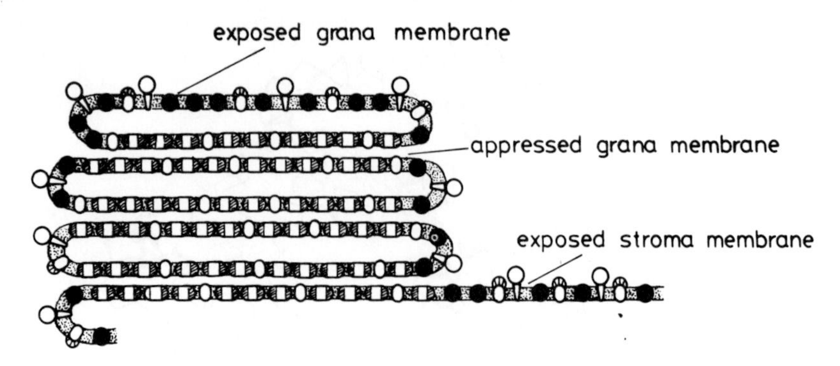

Fi. 4.16 Lateral separation of the proteins of the chloroplast membrane between the appressed in the non-appressed regions. This diagram illustrates current thinking about where the different major integral membrane proteins of the thylakoid membrane are located. Reductase means the ferredoxin-NADP reductase. Courtesy of Prof. J. Barber.

When $LHCII_\beta$ is phosphorylated it becomes strongly negatively charged. It has been proposed that this could lead to both unstacking and migration of the $LHCII_\beta$ out of the appressed regions into the stromal lamellae (for recent review see Barber, 1986). This is where the photosystem I complexes are located and this may be the explanation of the energy redistribution between the two photosystems seen in the "state" changes.

5 THE FUNCTION OF CAROTENOIDS IN PHOTOSYNTHESIS

Carotenoids have two major functions in photosynthesis. They act as photoprotective agents, preventing photooxidative damage, and as accessory light-harvesting pigments (Cogdell, 1978, 1985; Krinsky, 1979). The first of these two functions is essential. It must be quite clearly stated that without carotenoids there would be no photosynthesis in the presence of oxygen. The second function is more in the way of a bonus for the photosynthetic organisms, as it allows them to utilize light over a wider spectral range.

If a solution of chlorophyll is illuminated in the presence of oxygen it is very quickly and irreversibly photooxidized. In this reaction chlorophyll sensitizes its own destruction. The following series of reactions is responsible for this harmful process:

$$Chl + \xrightarrow{h\nu} Chl^{*} \qquad \text{(singlet excited Chl)}$$

$$Chl^{*} \rightarrow {}^{3}Chl^{*} \qquad \text{(triplet excited Chl)}$$

$${}^{3}Chl^{*} + O_2 \rightarrow Chl + {}^{1}\Delta_g\, O_2 \qquad \text{(singlet oxygen)}$$

When chlorophyll is excited by light a proportion of the molecules undergo intersystem crossing to the first excited triplet state. In contrast to singlet excited chlorophyll, triplet excited chlorophyll lasts for tens of microseconds or longer and is therefore able to interact with molecular oxygen in a bimolecular collision reaction. Oxygen in its ground state is a triplet. When it reacts with triplet chlorophyll one of the products formed is singlet oxygen (Foote, 1976). Singlet oxygen is a very powerful oxidizing agent. It can oxidize chlorophyll, lipids, proteins and indeed nucleic acids. Singlet oxygen is therefore very toxic for living cells and will rapidly kill them. Carotenoids prevent this photooxidative killing. All wild-type photosynthetic organisms contain carotenoids. Lack of carotenoids is a lethal condition. A good illustration of this is seen with plants treated with herbicides such as SAN-9785 (Moreland, 1980). This herbicide mainly kills plants by blocking the biosynthesis of their carotenoids and of course this killing reaction requires light.

In principal carotenoids can quench these harmful reactions with singlet oxygen in two ways. Carotenoids can react directly with singlet oxygen to detoxify it, or they can quench the $^{3}Chl^{*}$ sensitizer and so prevent singlet oxygen production (Foote, 1976). Carotenoids will react with singlet oxygen, at diffusion controlled rates, to produce carotenoid triplets.

$${}^{1}\Delta_g O_2 + Car \rightarrow {}^{3}Car^{*} + O_2$$

$${}^{3}Car^{*} \rightarrow Car + \text{heat}$$

These carotenoid triplets then decay harmlessly, producing heat rather than any toxic products. The energy level of singlet oxygen is 94 kJ/mol, and any carotenoid whose triplet state energy is less than this will be effective in quenching singlet oxygen: in essence, this means carotenoids with more then seven conjugated double bands.

When chlorophyll is excited by light, triplets are produced. Carotenoids are able to react with these chlorophyll triplets to produce carotenoid triplets and this effectively prevents the sensitized generation of single oxygen. When chloroplasts are overexcited with light, carotenoid triplets are produced (Wolff and Witt, 1969; Mathis, 1970). It is now quite clear that these arise from a triplet–triplet exchange reaction with triplet excited chlorophyll (Mathis, 1970).

$$\text{Chl} + \xrightarrow{h\nu} {}^{1}\text{Chl}^{*}$$

$${}^{1}\text{Chl}^{*} \rightarrow {}^{3}\text{Chl}^{*} \text{ (intersystem crossing)}$$

$${}^{3}\text{Chl}^{*} + \text{Car} \rightarrow {}^{3}\text{Car}^{*} + \text{Chl}$$

In vivo carotenoid triplets are produced, following laser flash excitation, in a few nanoseconds (Wolff and Witt, 1969; Mathis, 1970). This reduces the lifetime of ^{3}Chl* by more than three orders of magnitude and is undoubtedly the major photoprotective mechanism (Cogdell, 1985). This very rapid temperature-independent triplet–triplet exchange reaction requires the carotenoid and chlorophyll molecules to be arranged precisely in very close proximity to each other (Dirks *et al.*, 1980; Moore *et al.*, 1980). Studies on model systems with carotenoporphyrins have shown that rapid carotenoid triplet formation, following excitation of the porphyrins (the analogue of the chlorophyll), requires that the carotenoid and the porphyrin triplet donor must almost be touching (Dirks *et al.*, 1980; Moore *et al.*, 1980). Once again the vital role of the proteins in organizing the pigments so that the required reactions are promoted must be emphasized.

If the photoprotective role of carotenoids is a manifestation of triplet–triplet energy transfer then their light-harvesting function reflects singlet–singlet energy transfer. A good example of this is the peridinin-Chl a complex from dinoflagellates. If this complex is excited by light in the region of the carotenoid's absorption, then sensitized fluorescence is emitted by the chlorophyll (Song *et al.*, 1976). The efficiency of this singlet–singlet energy transfer process has been shown to be very nearly 100%. Though this energy transfer reaction is easy to demonstrate experimentally, the exact mechanism by which it occurs remains to be established. It certainly does not take place by the well known Förster dipole–dipole exchange mechanism. Carotenoids are essentially non-fluorescent and cannot be directly excited to their triplet state (Cogdell, 1985). The lifetime of the ^{1}Bu first excited singlet state is only a few ps (Wasielewski and Kispert, 1986; Wasielewski *et al.*, 1986b). This means that there is effectively not time for the energy transfer to take place by the long-range weak interaction, Förster dipole–dipole mechanism. In model systems it has been shown that the distance between carotenoid and the chlorophyll must be of the order of 2Å or less if efficient singlet–singlet energy transfer from the carotenoid to the chlorophyll is to be maintained (Wasielewski *et al.*, 1986a).

ACKNOWLEDGEMENTS

I would like to thank the various people who kindly supplied me with figures for use in this review, and my colleague Dr P. Dominy for reading the manuscript and giving me helpful criticism.

REFERENCES

Albertsson, P.-A., Andersson, B., Larsson, C. and Akerlund, H.-E. (1982). *Meth. Biochem. Anal.* **28**, 115–50.
Allen, J.A. (1983). *TIBS,* **8**, 369–73.
Allen, J.F., Bennett, J., Steinback, K.E. and Arntzen, C.J. (1981). *Nature* **291**, 25–9.
Alt, J., Morris, J., Westhalf, P. and Herrmann, R.G. (1984). *Curr. Genet.* **8**, 597–606.
Anderson, J.M. (1982). *FEBS Lett.* **138**, 62–6.
Anderson, J.M. (1986). *Ann. Rev. Pl. Physiol.* **37**, 93–136.
Anderson, J.M. and Barrett, J. (1986). *Encl. Pl. Physiol.* **19**, 269–85.
Anderson, J.M. and Levine, R.P. (1974). *Biochim. Biophys. Acta* **333**, 378–87.
Anderson, J.M. and Melis, A. (1983). *Proc. Natl. Acad. Sci. USA* **80**, 745–9.
Andersson, B. (1978). *Thesis.* University of Lund, Sweden.
Andersson, B. and Akerlund, H.-E. (1978). *Biochim. Biophys. Acta* **503**, 427–40.
Andersson, B. and Anderson, J.M. (1986). *Encl. Pl. Physiol.* **19**, 457–67.
Arntzen, C.J. and Pakrasi, H.B. (1986). *Encl. Pl. Physiol.* **19**, 457–67.
Barber, J. (1983). *Pl. Cell Environ.* **6**, 311–22.
Barber, J. (1986). *Encl. Pl. Physiol.* **19**, 651–64.
Barber, J. and Marder, J.B. (1986) *Biotech. Gen. Eng. Revs.* **4**, 355–404.
Barratt, J. and Anderson, J.M. (1980). *Biochim. Biophys. Acta* **590**, 309–23.
Bearden, A.J. and Malkin, R. (1977). *Brookhaven Symp. Biol.* **28**, 247–66.
Bengis, C. and Nelson, N. (1975). *J. Biol. Chem.* **250**, 2783–8.
Bengis, C. and Nelson, N. (1977). *J. Biol. Chem.* **252**, 4564–9.
Bennett, J. (1979). *Eur. J. Biochem.* **99**, 133–7.
Bennett, J. (1980). *Eur. J. Biochem.* **104**, 85–9.
Bennett, J. (1983). *Biochem. J.* **212**, 1–13.
Boardman, N.K. (1971). *Methods Enzymol.* **23**, 268–76.
Boardman, N.K. and Anderson, J.M. (1964). *Nature* **293**, 166–7.
Bogorad, L. (1975). *Ann. Rev. Pl. Physiol.* **26**, 369–401.
Bonaventura, C. and Myers, J. (1969). *Biochim. Biophys. Acta* **189**, 366–89.
Bonnerjea, J. and Evans, M.C.W. (1982). *FEBS Lett.* **148**, 313–16.
Borg, D.C., Fajer, J., Felton, R.H. and Dolphin, D. (1970). *Proc. Natl. Acad. Sci. USA* **67**, 813–20.
Brandlmeir, T., Scheer, H. and Rudiger, W. (1981). *Naturforsch.* **36c**, 431–9.
Braslavsky, S.E., Nolewarth, A.R. and Schaffner, K. (1983). *Angew. Chem.* **22**, 670–89.
Breton, J., Martin, J.-L., Migus, A., Antonetti, A. and Orszag, A. (1986). *Photochem. Photobiol.* **47**, 745.
Bricker, T.M., Guikema, J.A., Pakrasi, H.B. and Sherman, L.A. (1986). *Encl. Pl. Physiol.* **19**, 640–52.

Burke, J.J., Ditto, C.L. and Arntzen, C.J. (1978). *Arch. Biochem. Biophys.* **187**, 252–63.
Butler, W.L. (1973). *Acc. Chem. Res.* **6**, 177–83.
Butler, W.L. and Kitajima, M. (1975). *Biochim. Biophys. Acta* **396**, 72–85.
Camm, E.L. and Green, B.R. (1980). *Plant Physiol.* **66**, 428–32.
Camm, E.L. and Green, B.R. (1983). *Biochim. Biophys. Acta* **724**, 291–3.
Cashmore, A.R. (1984). *Proc. Natl. Acad. Sci. USA* **81**, 2960–4.
Chang, C.-H., Tiede, D.M., Tang, J., Smith, U., Norris, J.R. and Schifter, M. (1986). *FEBS Lett.* **205**, 82–6.
Chen, Y.-H., Yang, J.T. and Martinez, H.M. (1978). *Biochem.* **11**, 4120–31.
Cogdell, R.J. (1978). *Phil. Trans. Roy. Soc. Lond. B* **284**, 569–79.
Cogdell, R.J. (1983). *Ann. Rev. Pl. Physiol.* **34**, 21–45.
Cogdell, R.J. (1985). *Pure Appl. Chem.* **57**, 723–8.
Cohen-Bazire, G. and Bryant, D.A. (1981). In: *The Biology of the Cyanobacteria* (Eds. N. Carr and B. Whitton) pp. 143–190. Blackwell, New York.
Commoner, B., Heise, J.J. and Townsend, J. (1956). *Proc. Natl. Acad. Sci. USA* **42**, 710–18.
Cramer, W.A. and Butler, W.L. (1969). *Biochim. Biophys. Acta* **172**, 503–10.
den Blanken, H.J. and Hoff, A.J. (1983). *Biochim. Biophys. Acta* **724**, 52–61.
Deisenhofer, J., Epp. O., Miki, K., Huber, R. and Michel, H. (1984). *J. Mol. Biol.* **180**, 385–98.
Deisenhofer, J., Epp. O., Miki, K., Huber, R. and Michel, H. (1985). *Nature* **318**, 618–24.
Dietrich, W.E. and Thornber, J.P. (1971). *Biochim. Biophys. Acta* **245**, 482–493.
Diner, B.A. (1986). *Encl. Pl. Physiol.* **19**, 422–36.
Dirks, G., Moore, A.L., Moore, T.A. and Gust, D. (1980). *Photochem. Photobiol.* **32**, 277–80.
Doring, G., Bailey, J.L., Kreutz, W., Wiekand, J. and Witt, H.T. (1968). *Naturwissenschaften* **55**, 219–20.
Doring, G., Renger G., Vater, J. and Witt, H.T. (1969). *Z. Naturforsch. Teil B.* **24**, 1139–43.
Dornemann, D. and Senger, U. (1982). *Photochem. Photobiol.* **35**, 821–6.
Dunsmuir, P., Smith, S.M. and Bedbrook, J. (1983). *J. Mol. Appl. Genet.* **2**, 285–300.
Duysens, L.N.M. and Sweers, H.E. (1983). In: *Microalgae and Photosynthetic Bacteria* pp. 353–372. University of Tokyo Press, Tokyo.
Dyer, T.A. (1985). *Oxford Surveys Pl. Mol. Cell Biol.* **2**, 147–77.
Evans, M.C.W., Reeves, S.G. and Cammack, R. (1974). *FEBS Lett.* **49**, 111–14.
Evans, M.C.W., Sihra, C.K. and Cammack, R. (1976). *Biochem. J.* **158**, 71–7.
Fajer, J., Davis, M.S., Forman, A., Klimev, V.V., Dolan, E. and Ke, B. (1980). *J. Am. Chem. Soc.* **102**, 7143–5.
Fenton, J.M., Pellin, M.J., Govindjee and Kaufmann, K.J. (1979). *FEBS Lett.* **100**, 1–4.
Fish, L.E., Kuck, U. and Bogorad, L. (1985). *J. Biol. Chem.* **260**, 1413–31.
Foote, C.S. (1976). In: *Free Radicals and Biological Systems* (Ed. W.A. Pryor) pp. 85–133. Academic Press, New York.
Friedman, A.L. and Alberte, R.S. (1984). *Plant Physiol.* **76**, 483–9.
Gantt, E. (1981). *Ann. Rev. Pl. Physiol.* **32**, 327–47.
Gantt, E. (1986). *Enc. Pl. Physiol.* **19**, 260–8.
Gast, P., Swarthoff, T., Ebskamp, F.C.R. and Hoff, A.J. (1983). *Biochim. Biophys. Acta* **722**, 163–75.

Glazer, A.N. (1977). *Mol. Cell. Biochem.* **18**, 135–40.
Glazer, A.N. (1982). *Ann. Rev. Microbiol.* **36**, 173–98.
Glazer, A.N. (1984). *Biochim. Biophys. Acta* **768**, 29–51.
Glazer, A.N. (1985). *Ann. Rev. Biophys. Biophys. Chem.* **14**, 47–77.
Gounaris, K. and Barber, J. (1985). *FEBS Lett.* **188**, 68–72.
Gounaris, K., Barber J. and Harwood, J.L. (1986). *Biochem. J.* **237**, 313–26.
Grabowski, J. and Gantt, E. (1978). *Photochem. Photobiol.* **28**, 39–45.
Gugliemelli, L.A. (1984). *Biochim. Biophys. Acta* **766**, 45–50.
Haworth, P., Watson, J.L. and Arntzen, C.J. (1983). *Biochim. Biophys. Acta* **724**, 151–8.
Haxo, F.T., Kycia, J.H., Somers, G.F., Bennett, A. and Siegelman, H.W. (1976). *Pl. Physiol.* **57**, 297–303.
Hearst, J.E. (1986). *Encl. Pl. Physiol.* **19**, 382–9.
Hearst, J.E. and Sauer, K. (1984). *Z. Naturforsch.* **39c**. 421–4.
Heathcote, P., Williams-Smith, D.L., Sihara, C.K. and Evans, M.C.W. (1978). *Biochim. Biophys. Acta* **503**, 333–42.
Herrmann, R.G., Alt, J., Schiller, B., Widger, W.R. and Cramer, W.A. (1984). *FEBS Lett.* **176**, 239–44.
Hiyama, T. and Ke, B. (1971). *Arch. Biochem. Biophys.* **147**, 99–108.
Holschuh, K., Bottomley, W. and Whitfield, P.R. (1984). *Nucleic Acids Res.* **12**, 8819–34.
Horton, P. (1983). *FEBS Lett.* **152**, 47–52.
Horton, P. and Black, M.T. (1981). *Biochim. Biophys. Acta* **635**, 53–62.
Ikegami, I. and Katoh, S. (1975). *Biochim. Biophys. Acta* **376**, 588–92.
Ingram, K. and Hiller, R.G. (1983). *Biochim. Biophys. Acta* **722**, 310–19.
Inoue, Y., Ogowa, T. and Shibata, K. (1973). *Biochim. Biophys. Acta* **305**, 483–7.
Karlin-Neumann, G.A. and Tobin, E.M. (1986). *EMBO J.* **5**, 9–13.
Karlin-Neumann, G.A., Kohorn, B.D., Thornber, J.P. and Tobin, E.M. (1985). *J. Mol. Appl. Genet.* **3**, 45–61.
Katz, J.J. and Norris, J.R. (1973). *Cur. Top. Bioenerg.* **5**, 41–75.
Ke, B. (1973). *Biochim. Biophys. Acta* **301**, 1–33.
Ke, B. and Dolan, E. (1980). *Biochim. Biophys. Acta* **590**, 401–6.
Ke, B., Sugahara, K. and Shaw, E.R. (1975). *Biochim. Biophys. Acta* **408**, 12–25.
Ke, B., Hawkridge, F.M. and Sahu, S. (1976). *Proc. Natl. Acad. Sci. USA* **73**, 2211–15.
Klimov, V.V., Klevanik, A.V., Suvalov, V.A. and Krasnovsky, A.V. (1977). *FEBS Lett.* **82**, 183–6.
Klimov, V.V., Dolan, E. and Ke, B. (1980a). *FEBS Lett.* **112**, 97–100.
Klimov, V.V., Ke, B. and Dolan, E. (1980b). *FEBS Lett.* **118**, 123–6.
Knox, R.S. and Van Metter, R.L. (1979). *Ciba Found. Symp.* **61**, 177–90.
Kok, B. (1961). *Biochim. Biophys. Acta* **48**, 527–33.
Krinsky, N.I. (1979). *Pure Appl. Chem.* **51**, 649–60.
Kühlbrandt, W. (1984). *Nature* **307**, 478–80.
Kühlbrandt, W., Tharler, T. and Whrli, E. (1983). *J. Cell. Biol.* **96**, 1414–24.
Kyle, D.J., Staehelin, L.A. and Arntzen, C.J. (1983). *Arch. Biochem. Biophys.* **222**, 527–41.
Kyle, D.J., Kuang, T-Y., Watson, J.L. and Arntzen, C.J. (1984). *Biochim. Biophys. Acta* **765**, 89–96.
Kyte, J. and Doolittle, R.F. (1982). *J. Mol. Biol.* **157**, 105–32.
Lam, E., Ortiz, W., Mayfield, S., Malkin, S. and Malkin, R. (1984a). *Pl. Physiol.*

74, 650–5.
Lam, E., Ortiz, W. and Malkin, R. (1984b). *FEBS Lett.* **168**, 10–14.
Larsson, U.K., Jergil, B. and Andersson, B. (1983). *Eur. J. Biochem.* **136**, 25–9.
Li, J. (1985). *Proc. Natl. Acad. Sci. USA* **82**, 386–90.
Li, J. and Hollingshead, C. (1982). *Biophys. J.* **37**, 363–70.
Lichtenthaler, H.K., Prenzel, U. and Kuhn, G. (1982). *Z. Naturforsch.* **37c**, 10–12.
Lundell, D.J., Glazer, A.N., Melis, A. and Malkin, R. (1985). *J. Biol. Chem.* **260**, 646–54.
Machold, O. (1975). *Biochim. Biophys. Acta* **382**, 494–505.
McIntosh, A.R. and Bolton, J.R. (1976). *Biochim. Biophys. Acta* **430**, 555–9.
Malkin, R. (1975). *Arch. Biochem. Biophys,* **169**, 77–83.
Malkin, R. (1982). In: *Electron Transport and Photophosphorylation* (Ed. J. Barber) pp 1–47. Elsevier Biomedical Press, Amsterdam.
Malkin, R. and Bearden,. A.J. (1971). *Proc. Natl. Acad. Sci. USA* **68**, 16–19.
Malkin, R. and Bearden, A.J. (1975). *Biochim. Biophys. Acta* **396**, 250–9.
Malkin, R. and Bearden, A.J. (1978). *Biochim. Biophys. Acta* **505**, 147–81.
Mansfield, R.W. and Evans, M.C.W. (1986). *Proc. VIIth Int. Congr. Photosyn. Res.* 304–86.
Mathis, P. and Vermeglio, A. (1975). *Biochim. Biophys. Acta* **369**, 371–81.
Metz, J.G., Ulmer, G., Bricker, T.M. and Miles, D. (1983). *Biochim. Biophys. Acta* **725**, 203–9.
Michel, H. (1982). *J. Mol. Biol.* **158**, 567–72.
Michel, H. and Deisenhofer, J. (1986). *Encl. Pl. Physiol.* **19**, 371–81.
Michel, H., Wyer, K.A., Gruenberg, H. and Loftspeich, F. (1985). *EMBO J.* **4**, 1667–72.
Milner, P.A., Marder, J.B., Gounaris, K. and Barber, J. (1986). *Biochim. Biophys. Acta* **852**, 30–7.
Moore, A.L., Dirks, G., Gust, D. and Moore, T.A. (1980). *Photochem. Photobiol.* **32**, 691–6.
Moreland, D.E. (1980). *Ann. Rev. Pl. Physiol.* **31**, 597–638.
Morris, J. and Herrmann, R.G. (1984). *Nucleic Acids Res.* **12**, 2837–50.
Mullet, J.E., Burke, J.J. and Arntzen, C.T. (1980). *Pl. Physiol.* **65**, 814–22.
Myers, J. (1971). *Ann. Rev. Pl. Physiol.* **22**, 189–312.
Nebedryk, E., Andriaambinintsoa, S. and Breton, J. (1984). *Biochim. Biophys. Acta* **765**, 380–7.
Nechushtai, R. and Nelson, N. (1981). *J. Cell Biol.* **256**, 11624–8.
Nechushtai, R., Muster, P., Binder, A., Liveannu, V. and Nelson, N. (1983). *Proc. Natl. Acad. Sci. USA* **80**, 1179–83.
Norris, J.R., Uphaus, R.A., Crespi, H.L. and Katz, J.J. (1971). *Proc. Natl. Acad. Sci. USA* **68**, 625–8.
Norris, J.R., Scheer, H., Druyan, M.E. and Katz, J.J. (1974). *Proc. Natl. Acad. Sci. USA* **71**, 4897–900.
Norris, J.R., Scheer, H. and Katz, J.J. (1975). *Ann. NY Acad. Sci.* **244**, 261–80.
Nugent, J.H.A., Diner, B.A. and Evans, M.C.W. (1981). *FEBS Lett.* **124**, 241–4.
Nuijs, A.M., Shuvalov, V.A., van Gorkom, H.J. and Duysens, L.N.M. (1986). *Proc. VIIth Int. Congr. Photosyn. Res.* 304–87.
Ogawa, T., Obata, F. and Shibata, K. (1966). *Biochim. Biophys. Acta* **112**, 223–34.
Ohno, T., Satoh, K. and Katoh, S. (1986). *Biochim. Biophys. Acta,* **852**, 1–8.
Ohyama, K., Fukuzawa, H., Kokchi, T. *et al.* (1986). *Nature* **322**, 572–4.
Okamura, M.Y., Issacon, R.A. and Feher, G. (1979). *Biochim. Biophys. Acta*

546, 394–417.
Okamura, M.Y., Feher, G. and Nelson, N. (1983). In: *Photosynthesis* (Ed. Govindjee) pp. 195–272. Academic Press, London.
Ort, D.R. (1986). *Encl. Pl. Physiol.* **19**, 143–96.
Parson, W.W. (1982). *Ann. Rev. Biophys. Bioeng.* **11**, 57–80.
Parson, W.W. and Holton, D. (1986). *Encl. Pl. Physiol.* **19**, 338–43.
Petrouleas, V. and Diner, B.A. (1982). *FEBS Lett.* **147**, 111–16.
Prezelin, B.B. and Boczar, B.A. (1981). In: *Proceedings of the 5th International Congress on Photosynthesis* (Ed. G. Akoyunoglou), vol. 3, pp. 417–426. Balaban International Science Services, Philadelphia, PA.
Prezelin, B.B. and Haxo, F.T. (1976). *Planta* **128**, 133–41.
Rassmussen, O.F., Brookjans, G., Stunmann, B.M. and Hennigsen, K.W. (1984). *Plant Mol. Biol.* **3**, 191–9.
Rochaix, J.-D., Dron, M., Rahire, M. and Malone, P. (1984). *Plant Mol. Biol.* **3**, 363–70.
Rutherford, A.W. (1981). *Biochem. Biophys. Res. Commun.* **102**, 1065–70.
Rutherford, A.W. (1985). *Biochem. Soc. Trans.* **14**, 15–17.
Rutherford, A.W., Mullet, J.E. and Crofts, A.R. (1981a). *FEBS Lett.* **123**, 235–7.
Rutherford, A.W., Paterson, D.R. and Mullet, J.E. (1981b). *Biochim. Biophys. Acta* **635**, 205–14.
Sauer, K., Mathis, P., Acker, S. and Van Best, J.A. (1979). *Biochim. Biophys. Acta* **503**, 120–34.
Sauer, K., Mathis, P., Acker, S. and Van Best, J.A. (1982). *Biochim. Biophys. Acta* **545**, 466–72.
Searle, G.F.W., Barber, J., Porter, G. and Treadwell, C.J. (1978). *Biochim. Biophys. Acta* **501**, 246–56.
Scheer, H. (1982). In: *Primary Processes in Photosynthesis* (Ed. F.K. Fong) pp. 7–45. Springer-Verlag, New York.
Scheer, H. (1986). *Encl. Pl. Physiol* **19**, 327–41.
Scheer, H., Wieschoff, H., Schaefer, W. *et al.* (1983). In: Proceedings of the 6th International Congress on Photosynthesis. (Ed. C.Sysbesma) vol. II, p. 81–84. Nijhoff/Junk, The Hague.
Schirmer, T., Bode, W., Huber, R., Sidler, W. and Zuber, H. (1985). *J. Mol. Biol.* **184**, 257–77.
Schirmer, T., Huber, R., Schneider, M., Bode, W., Miller, M. and Hacket, M.L. (1986). *J. Mol. Biol.* **188**, 651–76.
Setif, P. and Mathis, P. (1986). *Encl. Pl. Physiol.* **19**, 476–86.
Shiozawa, J.A., Alberte, R.S. and Thornber, J.P. (1974). *Arch. Biochem. Biophys.* **165**, 388–97.
Shopes, R.J. and Wraight, C.A. (1986). *Proc. VIIth Int. Cong. on Photosyn. Res.* 205–309.
Shuvalov, V.A. and Klimov, V.V. (1976). *Biochim. Biophys. Acta* **440**, 587–99.
Shuvalov, V.A., Dolan, E. and Ke, B. (1979a). *Proc. Natl. Acad. Sci. USA* **76**, 770–3.
Shuvalov, V.A., Ke, B. and Dolan, E. (1979b). *FEBS Lett.* **100**, 5–8.
Shuvalov, V.A., Klevanik, A.V., Sharkov, A.V., Kryulov, P.G. and Ke, B. (1979c). *FEBS Lett.* **107**, 313–16.
Siefermann-Harms, D. (1985). *Biochim. Biophys. Acta* **811**, 325–55.
Song, P.S., Koka, P., Prezelin, B.B. and Haxo, F.T. (1976). *Biochemistry* **15**, 442–7.
Staehelin, L.A. (1986). *Encl. Pl. Physiol.* **19**, 1–84.
Staehelin, L.A. and Arntzen, C.J. (1983). *J. Cell. Biol.* **97**, 1327–37.

Stiehl, H.H. and Witt, H.T. (1969). *Z. Naturforsch. Teil B* **24**, 1588–99.
Tandeau de Marsac, N. (1983). *Bull. Inst. Pasteur* **81**, 201–54.
Tang, X.-S. and Satoh, K. (1985). *FEBS Lett.* **179**, 60–4.
Thornber, J.P. (1975). *Ann. Rev. Pl. Physiol.* **26**, 127–58.
Thornber, J.P. (1986a). *Encl. Pl. Physiol.* **19**, 98–142.
Thornber, J.P. (1986b). *Photochem. Photobiol.* **43** (suppl.), 111.
Thornber, J.P. and Highkin, H.R. (1974). *Eur. J. Biochem.* **41**, 109–16.
Thornber, J.P., Smith, C.A. and Bailey, J.L.(1966). *Biochem. J.* **100**, 14–15.
Thornber, J.P., Gregory, R.P.F., Smith, C.A. and Bailey, J.L. (1967). *Biochemistry* , 391–5.
Thornber, J.P., Alberte, R.S., Hunter, F.A., Shozawa, J.A. and Kan, K.S. (1977). *Brookhaven Symp. Biol.* **28**, 132–48.
Thornber, J.P., Markwell, J.P. and Reinman, S. (1979). *Photochem. Photobiol.* **29**, 1205–16.
Thurnauer, M.C., Gast, P., Petersen, J. and Stehlik, D. (1986). *Proc. VIIth Int. Cong. Photosyn. Res* 304–84.
Tiede, D.M. (1986). *Encl. Pl. Physiol.* **19**, 344–52.
Tiede, D.M., Prince, R.C. and Dutton, P.L. (1976a). *Biochim. Biophys. Acta* **449**, 447–69.
Tiede, D.M., Prince, R.C., Reed, G.H. and Dutton, P.L. (1976b). *FEBS Lett.* **65**, 301–4.
Tobin, E.M., Wimpee, C.F., Silverthorne, J., Stiekema, W.I., Neumann, G.A. and Thornber, J.P. (1984). In: *Biosynthesis of the Photosynthetic Apparatus* (Eds. J.P. Thornber, L.A. Staehelin and R. Hallick), pp 325–334. Liss, New York.
Van Gorkam, H.J., Pulles, M.P.J. and Wessels, J.S.C. (1975). *Biochim. Biophys. Acta* **408**, 331–9.
Vernon, L.P. and Shaw, E.R. (1971). *Meth. Enzymol.* **23**, 277–89.
Vernon, L.P., Shaw, E.R., Ogawa, T. and Raveed, D. (1971). *Photochem. Photobiol.* **14**, 343–57.
Visser, J.W.M. (1975). *Ph. D. Thesis.* University of Leiden, The Netherlands.
Wasielewski, M.R. and Kispert, L.D. (1986). *Chem. Phys. Lett.* **128**, 238–43.
Wasielewski, M.R., Norris, J.R., Shipman, L.L., Lin, C.P. and Svec, W.A. (1981). *Proc. Natl. Acad. Sci. USA* **78**, 2957–61.
Wasielewski, M.R., Liddell, P.A., Barrett, D., Moore, T.A. and Gust, D. (1986a). *Nature* **322**, 570–2.
Wasielewski, M.R., Tiede, D.M. and Frank, H.A. (1986b). In: *Ultrafast Phenomena* (Eds. G.R. Fleming and A.R. Siegman) vol. V. Springer-Verlag, Berlin.
Westhoff, P., Alt, J., Widger, W.R., Cramer, W.A. and Herrmann, R.G. (1985). *Plant Mol. Biol.* **4**, 103–10.
Widger, W.R., Cramer, W.A., Hermodson, M., Meyer, D. and Gullifor, M. (1984). *J. Biol. Chem.* **259**, 3870–6.
Williams, W.P. (1977). In: *Primary Procession Photosynthesis* (Ed. J. Barber) pp. 94–147. Elsevier, Amsterdam.
Wolff, C. and Witt, H.T. (1969). *Z. Naturforsch.* **24b**, 1031–7.
Wollman, F.-A. (1986). *Encl. Pl. Physiol.* **19**, 487–95.
Youvan, D.C., Bylina, E.J., Alberti, M., Begusch, H. and Hearst, J.E. (1984). *Cell* **37**, 949–57.
Zinth, W., Kaiser, W. and Michel, H. (1983). *Biochim. Biophys. Acta* **723**, 128–31.
Zuber, H. (1986). *Encl. Pl. Physiol.* **19**, 238–51.

5

Functions of Carotenoids other than in Photosynthesis

W. RAU

Botanisches Institut, Universität München, Menzinger Str. 67, 8000 München 19, F.R.G.

1 INTRODUCTION

Among the major pigments occurring in the plant kingdom, in fungi and in bacteria, carotenoids are obviously the most widespread; not only do all photosynthetic organisms contain, besides chlorophylls and other pigments, carotenoids but also the colouring of many fungi and bacteria is caused by these pigments. Considering this widespread occurrence it might be expected that carotenoids have important functions in plant life.

PLANT PIGMENTS
ISBN 0-12-289847-8

This, indeed, is true although in some cases our insight into the role and mechanism of these functions is incomplete and in other cases our knowledge is very meagre.

Colouring by carotenoids of flowers, seeds and fruits as well as of some fungi points to a function in reproduction of the organism; colouring may help to attract animals needed for the distribution of both pollen and seeds in higher plants and spores in fungi. However, it is surprising what huge variety of different carotenoids together with their complicated biosynthetic pathways has been developed in various organisms only to bring forth a relatively small variation in colour from pale-yellow to deep-red. Moreover, it is difficult to imagine to what extent this great expenditure of energy has led to an evolutionary advantage. The distribution of carotenoids in the various organisms and tissues has been compiled *in extenso* by Goodwin (1980). In photosynthetic and non-photosynthetic bacteria, in fungi and in algae carotenoids are often used as chemosystematic markers; reviews on this topic have been presented by Goodwin (1980) and by Liaaen-Jensen (1979, 1985).

Recognition of and adaptation to various environmental conditions is an obvious prerequisite for the survival of all living beings. For plants one of the most important of such factors is light. As well as its function as a source of energy for photosynthesis in all chlorophyll-containing plants, it is also an important regulating factor in plant development and movement not only in green but also in non-photosynthetic organisms. Sensory pigments are needed for recognition of the light environment. In phototropism and phototaxis carotenoids have been considered to be possible photoreceptors. However, such an assumption has to be extended to all photoresponses which are mediated by the absorption of near-ultraviolet (UV) and blue light. Therefore, the role of carotenoids in photoreception will be discussed in some detail.

In addition to its positive effects in photosynthesis and photoregulation light causes harmful effects, and therefore organisms had to develop protective mechanisms. That carotenoids apparently are involved in the photoprotection of plants and bacteria has been recognized for more than half a century (see Krinsky, 1968). Although the function of carotenoids in photoprotection has been investigated from various points of view and for some decades, our insight into the mechanism of this reaction is still incomplete. The present state of knowledge will be presented in section 3 where earlier reports are summarized and more recent results discussed.

Finally, two less obvious functions of carotenoids will be summarized briefly: the walls of spores and pollen contain the so-called sporopollenins, substances which are highly resistant macromolecules considered to be carotenoid polymers; these substances are also constituents of the cell wall of some algae. Carotenoids also play a role in the reproduction of fungi but the extent of their implication is not yet clear.

2 THE ROLE OF CAROTENOIDS IN PHOTORECEPTION

As noted above, for recognition of the light environment organisms need sensory pigments and for all photoresponses which are mediated by near-UV and blue light carotenoids may be considered to be photoreceptors. One of these photoresponses in higher plants, algae and fungi is photomorphogenesis, that is, the development of the "normal" habit of a plant under the influence of light; usually, developmental processes inside the plant, e.g. biosynthesis of pigments, development of organelles etc., are also included. For regulation of photomorphogenesis the effects of both the red part of the visible spectrum mediated by phytochrome (see Chapter 6) and of the UV/blue part are of importance. Phototropism is triggered exclusively by UV/blue; although very few mature plants show any phototropic responses in the natural environment, much attention has been paid to this reaction. Since both UV/blue light-mediated photomorphogenesis and phototropism pose common problems, they will be considered in the same section.

The variety of the "blue light" responses, including movement and metabolism, is summarized in two books (Senger, 1980, 1984) to which the reader is referred for further details.

A Photomorphogenesis and phototropism

As possible photoreceptors for the UV/blue part of the spectrum two major candidates have been discussed, flavins and carotenoids. However, these compounds are considered only in cases when the action spectrum of the corresponding response follows a characteristic pattern showing peaks or shoulders around 420, 450 and 480 nm and more or less an additional peak between 360 to 380 nm. Photoreceptors of this type have been named "cryptochrome"* because of their as yet cryptic nature (Gressel, 1979); some authors, however, still prefer the term "UV/blue-photoreceptor". It should be emphasised that as yet no evidence exists that there is only *one* "cryptochrome", in analogy to the single photoreceptor "phytochrome".

*Cryptochrome is an unfortunate choice of name; it had already been given to a carotenoid by Karrer in 1946 (Ed.)

It must be assumed that there are other photoreceptors in the UV/blue region, e.g. a specific UV-B photoreceptor with a main peak of action around 300 nm (see Gressel and Rau, 1983; Wellmann, 1983); in these reactions carotenoids are not considered to be involved.

The controversy over flavins or carotenoids is a long-standing story and has been reviewed and critically discussed a number of times from various points of view by Song and Moore (1974), Presti and Delbrück (1978), Shropshire (1980), Horwitz and Gressel (1986), and Senger and Schmidt (1986), to mention only some of the most recent. It is therefore not the intention to give a comprehensive summary of all the results, conclusions and opinions on this topic but rather to point to arguments which are in favour of carotenoids or to indicate experimental evidence which does not allow exclusion of carotenoids from involvement in photoreception. It should be emphasized, however, that the majority of investigators clearly favour a flavoprotein at present.

1 ACTION SPECTROSCOPY

Some of the "typical" cryptochromal action spectra with peaks and shoulders in the blue and a marked peak around 370 nm are shown in Fig. 5.1. Starting from the earliest discussions it has been argued that the similarity with the absorption spectra of carotenoids in the blue are in their favour but that the lack of an absorption peak of all-trans-carotenoids in the UV rules out carotenoids. However, comparison of all known action spectra makes it clear that the shapes of the spectra vary quite considerably (see Presti and Delbrück, 1978; Senger, 1980, 1984; Senger and Schmidt 1986). The differences may be due to several reasons, e.g. differences in the reciprocity or fluence rate/response relationships, screening, etc.

A striking example which shows that the same method used for elaboration of action spectra can, under different conditions, lead to different shapes is the work of Lipson *et al.* (1984). These authors re-investigated the phototropic response of the sporangiophores of *Phycomyces* using the balance method. Sporangiophores were exposed to two opposed beams of light: a constant reference beam and a test beam with variable wavelength and intensity. Depending on the wavelength and intensity of the reference beam they received either a typical cryptochromal action spectrum or spectra with no fine structure in the blue or nearly no effect in the UV.

Three more action spectra which indicate no effect or only a minor one on the UV and therefore support the theory of the action of carotenoids are shown in Fig. 5.2. However, it has been argued that this

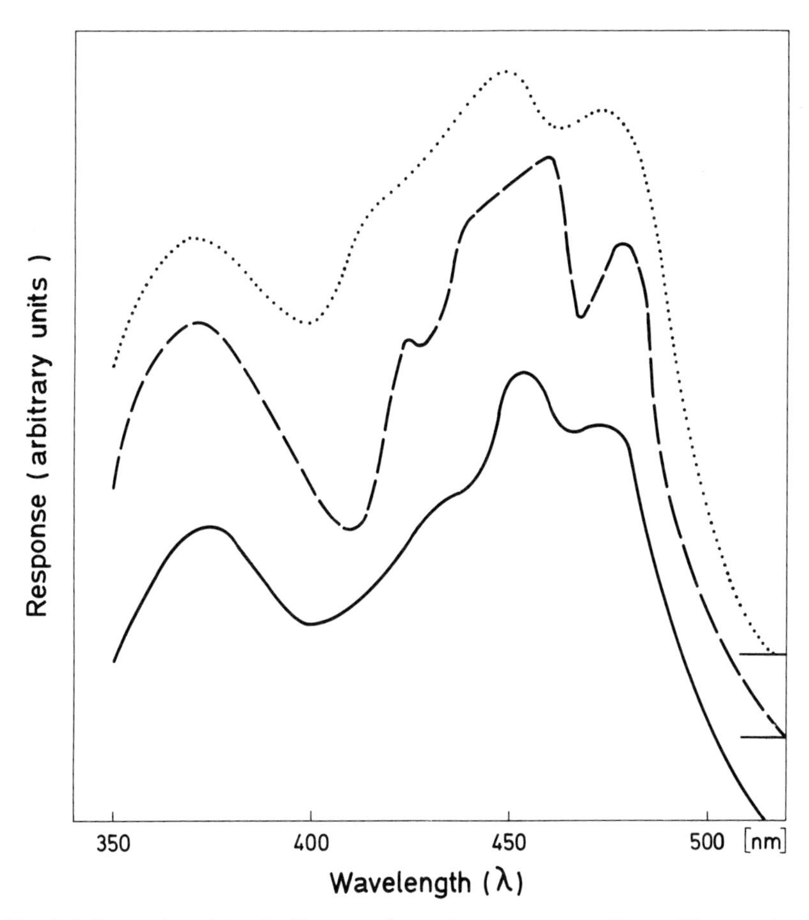

Fig. 5.1 Examples of "typical" cryptochromal action spectra. Dotted line = phototropism in *Avena* (Shropshire and Withrow, 1958); dashed line = perithecial formation in *Gelasinospora reticulispora* (Inoue and Watanabe, 1984); unbroken line = photoinduction of carotenoid biosynthesis in *Fusarium aquaeductuum* (Rau, 1967). The baselines have been shifted as indicated.

type of spectrum may be the result of shading by other UV-absorbing pigments. This is in fact possible, as has been shown for one of the photoresponses in Fig. 5.2, photoinhibition of phytochrome-controlled germination of spores in the fern *Pteris vittata*. In a re-investigation of the earlier action spectrum (Sugai, 1971) it was found that the spore coat has very low transmittance values in the UV-region. A corrected action spectrum based upon the transmission spectrum has an increased peak around 370 nm (Sugai *et al.*, 1984).

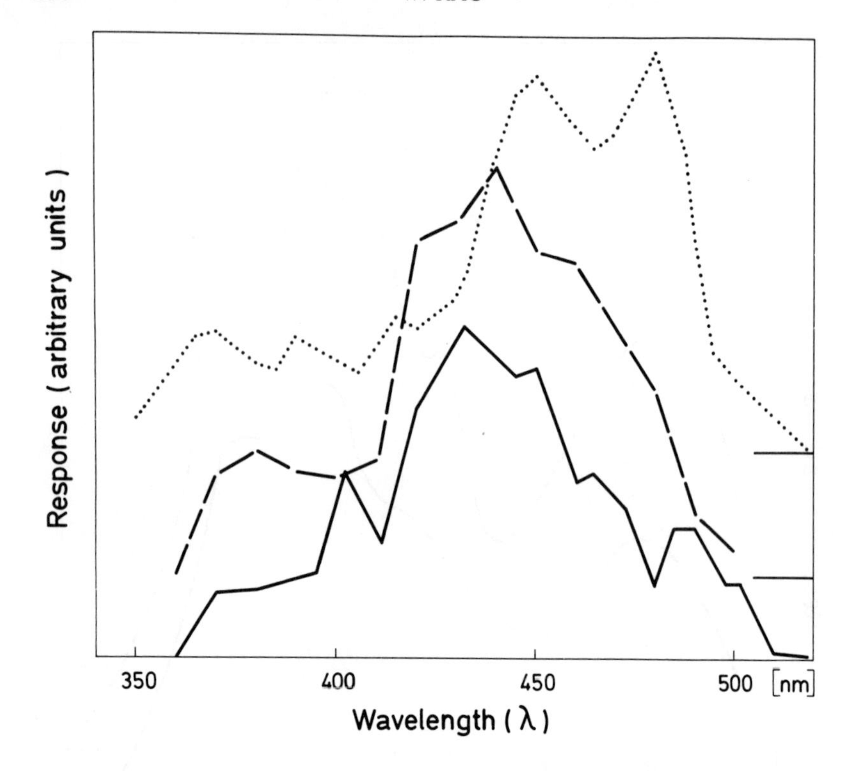

Fig. 5.2 Examples of UV/blue action spectra indicating minimal or no effect on UV. Dotted line = photoinduction of carotenoid biosynthesis in *Neurospora crassa* (DeFabo *et al.*, 1976); dashed line = photoinhibition of spore germination in *Pteris vittata* (Sugai, 1971); unbroken line = induction of two-dimensional growth in the brown alga *Scytosiphon lamentaria* (Dring and Lüning, 1975). The baselines have been shifted as indicated.

An additional factor which may lead to a distortion of an action spectrum is fluorescence; arguments in favour of carotenoids based on fluorescence have been summarized by DeFabo (1980) and are briefly repeated here. Many different types of compounds present in plants can be excited by UV and fluoresce in the blue. The blue quanta could then be re-absorbed by the carotenoid photoreceptor. The result of such a situation would be a higher response peak in the action spectrum coinciding with the peak of the exciting wavelength.

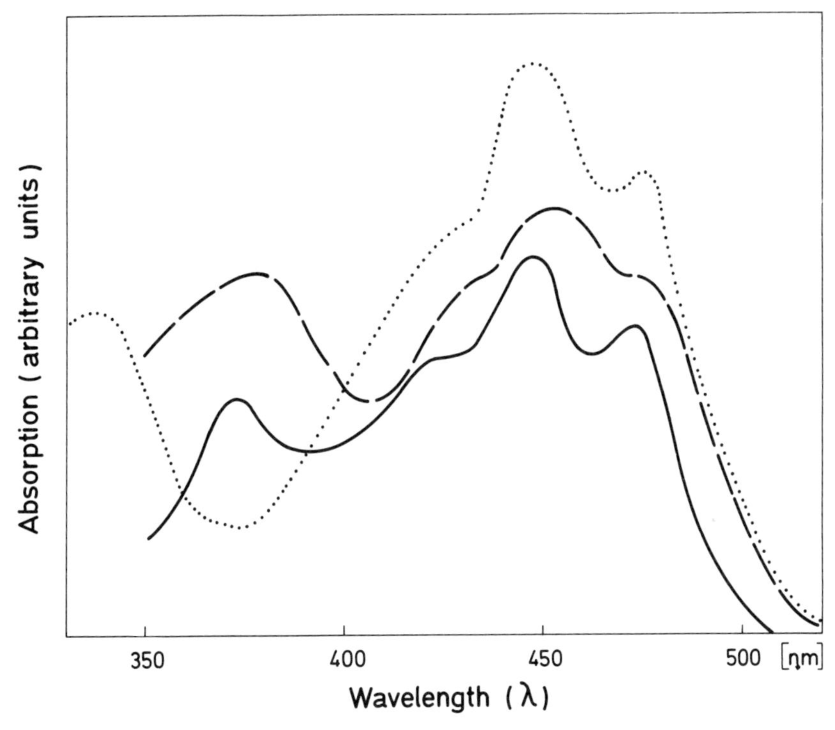

Fig. 5.3 Absorption spectra of some putative photoreceptors in the UV/blue region of the spectrum. Dotted line = 15,15′-*cis*-β-carotene (Zechmeister, 1962); dashed line = a flavoprotein (Robinson *et al.*, 1962); unbroken line = lutein in ethanol/H_2O mixture (Hager, 1970).

2 ABSORPTION SPECTRA OF PUTATIVE PHOTORECEPTORS

Absorption spectra of 15,15′-*cis*-β-carotene, of a flavoprotein and of a carotenoid in alcohol/water solvent are depicted in Fig. 5.3. Comparison of these absorption spectra with "typical" action spectra makes it quite clear that a *cis*-carotenoid is an unlikely candidate since the *cis*-peak of the absorption spectrum is at a much shorter wavelength than the UV-peak of the action spectra. All *trans*-carotenoids do not have a UV-peak in organic solvents. However, as Hager (1970) showed, a UV-peak appears when water is added to alcohol (polar) solution of carotenoids. He argued that a carotenoid complex which is bound to membranes or particles in the intact cell may have the same environment—a polar/nonpolar mixed phase—as in this solution. The absorption spectrum closely matches a "typical" action spectrum just as well as does the absorption spectrum of a flavoprotein.

3 PHOTOPHYSICAL AND PHOTOCHEMICAL ASPECTS

In a detailed analysis Song and Moore (1974) arrived at the conclusion that carotenoids were not suited as photoreceptors for cryptochrome-mediated responses. The main reason for this is the short lifetime due to efficient internal conversion from the lowest excited singlet state. However, from more recent results and calculations, summarized and discussed by Song (1980, 1984), the lowest excited state is expected to have a much longer lifetime than predicted. Under certain circumstances carotenoids can act as a photoreceptor but there is still no photochemical evidence that carotenoids are involved in cryptochrome reactions.

4 MUTANTS ALTERED IN CAROTENOID CONTENT

In different mutant strains of *Phycomyces* which varied in their ability to synthesize β-carotene, Galland and Russo (1979a) found that the irradiance threshold for photoinduction of sporangiophores was increased by a factor ranging from 100 to 2000 and that this was a direct function of the carotene concentration in the hyphae. They further observed (1979b) that the oxygen threshold required for the light-induced synthesis of β-carotene in these mutants also depends on the carotene content. Even though their data might support carotenoids as the photoreceptor they conclude that carotenes have a role in the dark reactions rather than as photoreceptor.

Contradictory results were reported by Presti *et al.* (1977). A double albino mutant of *Phycomyces* in which the β-carotene content of sporangiophores is reduced to 4×10^{-5} the level of that contained in the wild-type exhibits an unaltered phototropic response. However, these contrary results might not exclude carotenoids as photoreceptor if one keeps in mind that a calculation of Meissner and Delbrück (1968) led to the conclusion that a very low concentration of the photoreceptor is sufficient to cause the photoresponses.

5 INVOLVEMENT OF CAROTENOIDS IN LIGHT GRADIENTS

In phototropism where the *direction* of the incident light is the decisive factor for the response, an important problem is how a light gradient across the responsive organ of the plant can be established. Since the earlier work of Bünning (1937, 1938) a "two-pigment" or "shading" hypothesis has been discussed. This hypothesis is based on the assumption that one pigment is the photoreceptor (possibly a flavin) but its concentration is too low to establish a light gradient and, therefore, an additional pigment absorbing in the same spectral region is responsible for shading. Indeed, the tip of a coleoptile, which is the most sensitive part for the phototropic response, contains a high concentration of carotenoids. Whether or not these carotenoids are responsible for a shading effect or whether a light gradient might be caused by the high content of flavones in the vacuoles of the cells is still an open question.

An example of a possible role of carotenoids as shading pigments comes from the results of Hartmann and Schmidt (1980). In hypocotyl hooks of etiolated dwarf beans blue light causes a biopotential change; the amount of this change was dependent on the carotenoid content caused by an inhibition of carotenoid synthesis by a herbicide.

6 CAROTENOIDS AS ANTENNA PIGMENTS

The function of carotenoids as antenna pigments in photosynthesis is discussed in Chapter 4. Apart from this role, carotenoids may also act as antenna pigments in photoreception since an energy transfer from carotenoids to other pigments seems to be possible (see Mathis and Schenk, 1982; Cogdell, 1985).

B Photomovement of motile organisms

Like photomorphogenesis and phototropism (section A above), the phenomena and mechanisms of photomovement in motile organisms have been reviewed a number of times. Among these reviews, those from Goodwin (1980) and Burnett (1976) also give some historical background. More recent and detailed reviews have been compiled by Haupt and Feinleib (1979), Nultsch (1980), Häder (1984) and Haupt (1986), to mention only a few.

Photoregulation of movement of motile organisms comprises three types of reaction:

(i) Photokinesis is the dependence of the velocity of movement on the incident fluence rate, independently of the direction of light. Some organisms move only when they are illuminated but very bright light is inhibitory.

(ii) A photophobic response is caused by an abrupt change in light intensity. Both a step-up and step-down in light intensity may induce a response, i.e. a reversion or change of direction of movement when an organism enters or leaves a dark or bright light area. This type of reaction can be demonstrated by "light-trap" experiments.
(iii) Phototaxis is the orientation of the movement by light direction; it may be positive (towards the light source) or negative (away from the light source) depending on the light intensity.

1 PHOTOKINESIS

In purple bacteria, such as *Rhodospirillum*, and in blue-green algae photokinesis is caused by photosynthetic ATP-synthesis. Therefore, carotenoids are involved only in so far as they are antenna pigments in photosynthesis or constituents of the reaction centres.

2 PHOTOPHOBIC RESPONSE

In purple bacteria and in blue-green algae light inducing the phobic response is also absorbed by the photosynthetic pigments. In flagellates and in motile forms of algae a wide diversity of action spectra was found; in some organisms they even differed as regards the positive and negative response. In cases where the shape of the action spectrum is cryptochromal, such as in *Euglena* or in the diatom *Nitzschia communis*, the same controversy, flavins versus carotenoids, as in photomorphogenesis and phototropism arises and, therefore, need not be repeated here. A further example of this type of response is the so-called "avoidance reaction" in species of the true slime mould (myxomycete) *Physarum*.

In a number of organisms, however, action spectra exhibit maximal response between 400 and 550 nm—in some cases with a peak at about 500 nm—but lack a UV-peak at about 370 nm; action spectra of this type have been found, e.g. in some pyrrophyta (Halldal, 1958, 1961), in *Chlamydomonas* (Nultsch *et al.*, 1971) and in the red alga *Porphyridium* (Nultsch, 1980). This clearly excludes flavins as photoreceptors: carotenoids or carotenoproteins are assumed instead.

The archaebacterium *Halobacterium halobium* is bipolarly flagellated and uses bacteriorhodopsin inserted in purple patches in the cell membrane for ATP-synthesis via a proton pump. It also shows photophobic responses, a step-up as well as a step-down response, of which the action spectra are different. As a photoreceptor for these reactions a related pigment, the so-called slow-cycling or sensory rhodopsin, is very likely; the mechanism of signal transduction is, however, not yet completely understood (see Wagner, 1984).

3 PHOTOAXIS

The long-known and intensively investigated example of real orientation to the direction of light is *Euglena*. This single-cell flagellate rotates during swimming around its longitudinal axis. Its photoactic orientation is due to periodic shading of the photoreceptor structure—the paraflagellar body—by the so-called stigma or eyespot containing high amounts of carotenoids. The periodical shading causes a series of small photophobic responses as long as no further shading occurs.

A more complicated situation exists in *Volvox*, a colony of thousands of flagellates. The colony rotates while moving forward and a step-up illumination causes a flagellate to stop beating. Thus, just as in *Euglena*, photoaxis is brought about by a sum of photophobic responses not in a single cell but by a cooperation between all the flagellates. The action spectra for phototaxis and photophobic response exhibit activity from approximately 370 to nearly 600 nm, but lack a peak in the UV (Schletz, 1976). Temporal shading of the photoreceptor produced by light absorption by the stigma and by the chloroplast is due to carotenoids and chlorophylls. The shapes of the action spectra are interpreted as indicating a carotenoid as the photoreceptor pigment; in this respect a rhodopsin has also been considered.

C Intracellular photomovement

In some plant cells investigated in this respect protoplasmic streaming is regulated by light. A more interesting and important reaction is the ability of chloroplasts to change their position in response to light conditions, in particular to the light direction. The chloroplasts are redistributed in such a manner that under low and medium fluence rates they gather at those cell walls which are perpendicular to the incident light, i.e. parallel to the surface, whereas in high fluence rates (e.g. bright sunlight) they are located at the anticlinal walls. Light-dependent rearrangements of chloroplasts have been detected and investigated in many algae, mosses, pteridophyta and higher plants (Zurzycki, 1980).

All action spectra available for both protoplasmic streaming and chloroplast orientation resemble—with two special exceptions where phytochrome is involved—cryptochromic action spectra. Therefore, once more the controversy of flavin versus carotenoid arises. However, a screening effect is responsible for measuring the direction of light; carotenoids, as well as chlorophyll, are being considered for this effect.

D Photoisomerization in carotenoid biosynthesis

In many organisms biosynthesis of carotenoids is under photocontrol; a wide variety of photoreceptors has been found to be responsible for this photoregulation in various organisms (Rau, 1983, 1985). In respect to whether or not carotenoids can act as photoreceptors, an interesting photoreaction was reported by Schiff *et al.* (1982). A mutant strain of *Euglena gracilis* accumulates ζ-carotene in the *cis*-form in the dark. Light causes a photoisomerization to the *trans*-form and it was shown that the photoreceptor for this reaction is *cis*-ζ-carotene itself. The authors concluded that the isomerization from the *cis*- to the *trans*-form may be part of the normal biosynthetic pathway in *Euglena* and may be performed in wild-type cells in the dark; mutation causes a loss of this ability and, therefore, the organism is light-dependent for this step.

Some mutants of *Chlorella* and *Scenedesmus* synthesize only acyclic carotenes in the dark and need light for the formation of cyclic carotenes and xanthophylls; in the dark-grown cultures the acyclic pigments are present primarily, or at least to an appreciable extent, as *cis*-isomers (Rau, 1983). Since the production of cyclic carotenes starts immediately, i.e. without a lag phase, one might assume a reaction similar to that in the *Euglena* mutant. However, since action spectra are lacking or point to chlorophyll as the responsible photoreceptor, the assumption that a direct *cis* to *trans* photoisomerization is the rate-limiting step for the subsequent biosynthesis of more unsaturated carotenes and xanthophyll is highly speculative.

3 PHOTOPROTECTION

The role of carotenoids in protecting organisms against harmful effects of radiation has been summarized a number of times during the last years from various viewpoints; the mechanism has been reviewed comprehensively by Krinsky (1979, 1984), the occurrence in various tissues by Goodwin (1980), participation in photosynthesis by Mathis and Schenk (1982), and Cogdell (1985), and some recent results by Schrott (1985). Therefore, it is the intention of this section to summarize only the present state of knowledge briefly.

A Mechanisms of protection by carotenoids

Carotenoids absorb in a broad band in the blue part of the visible spectrum. In addition, in non-photosynthetic organisms absorption by carotenoids is extended to the near UV due to their content of less unsaturated compounds, i.e. ζ-carotene, phytofluene and phytoene. Protection by screening has not been considered to a great extent. However, bearing in mind that in particular high fluence rates of irradiation cause harmful effects, protection by screening should not be excluded completely.

The protective role of carotenoids so far discussed is that against deleterious effects caused by photosensitized oxidations. The photochemical basis for such reactions is the following: absorption of a quantum of radiation by a photosensitizing molecule(s) leads to formation of an excited photosensitizer in its singlet state (^1S) which has a very short lifetime (10^{-11} s). This singlet excited species may dissipate its energy either by emitting a photon in the form of fluorescence or may undergo an intrasystem crossing to form the triplet excited species (^3S). Triplet states can return to the ground state accompanied by phosphorescence. They have, however, a sufficiently long lifetime (10^{-4} s) to interact with other compounds and initiate photochemical reactions. The type of these subsequent reactions depends on the availability of oxygen and the nature of other potentially reactive molecules in the environment. These photochemical reactions of ^3S have been classified by Gollnick and Schenk (1967) as type I reactions without participation of O_2, and type II reactions with O_2. Type I reactions are redox reactions which may result in the formation of radical species, frequently involving hydrogen or electron abstraction depending on the nature of both the sensitizer and the reactant. In the type II reaction, ^3S interacts directly with groundstate O_2, regains its own ground state and forms singlet oxygen ($^1O_2$). This species of oxygen is very reactive and can rapidly react with many cellular components.

Both free radicals from type I reactions and singlet oxygen are capable of causing damage to cellular constituents or metabolic processes, e.g. destruction of membranes, inhibition of enzyme activity etc., which ultimately may lead to the death of the organisms; other reactions may be mutagenicity or carcinogenicity.

Numerous investigations have provided sufficient evidence that carotenoids are very effective agents for protecting systems against the harmful effects of photochemically induced oxidations (for details see the reviews cited above). This is due to their capacity to interact with all three photoreactions, as illustrated in Fig. 5.4.

Carotenoid pigments have the ability to quench the first potentially harmful intermediate, the triplet state of a sensitizer (^3S) by absorption of its energy, to form a triplet state carotenoid which then dissipates its energy to its surroundings in the form of harmless heat (reaction 1 in the Fig. 5.4). Carotenoids are also capable of quenching singlet oxygen ($^1O_2$);

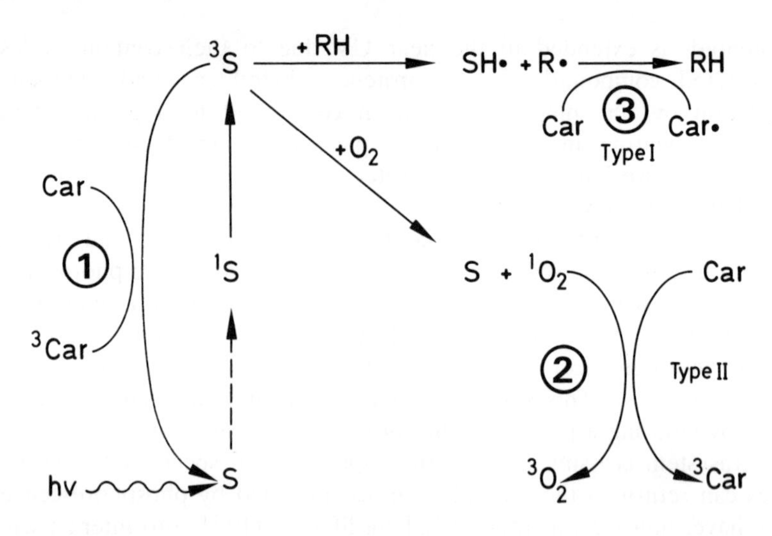

Fig. 5.4 Scheme illustrating interaction of carotenoids in photochemically induced oxidations (adapted from Krinksy, 1979). Car = carotenoid; S = sensitizer; R = reactant. For further explanation see text.

the exact mechanism of this reaction still needs final clarification (reaction 2). By an inadequately understood mechanism carotenoids can quench free radical intermediates (reaction 3).

B Photoprotection in bacteria and algae

For more than half a century it has been hypothesized that carotenoids play a role in protecting organisms against the deleterious effects of light. It was, however, not until 1956 that this hypothesis was confirmed by the crucial observations of Sistrom *et al.*, who investigated the physiology of *Rhodopseudomonas spheroides* wild-type and a blue-green mutant strain. From their observation that in the presence of air and light the mutant strain stopped growing whereas the wild-type did not, they suggested that the coloured carotenoid pigments are the protective compounds. Since then much work on photodamage in organisms lacking carotenoids has accumulated, dealing with photosynthetic bacteria as well as non-photosynthetic bacteria, and to a lesser extent with algae. The results of earlier work including some historical background were reviewed comprehensively by Krinsky (1968, 1979), by Burnett in the second edition of this book, and by Goodwin (1980); the main findings can be summarized as follows. Although in very few cases a protective role of carotenoids was not found—e.g. in *Micrococcus roseus* wild-type, colourless mutants and cultures rendered colourless with diphenylamine (Schwartzel and Cooney, 1974)—a number of investigations supported the suggested function of carotenoids in protecting organisms against destruction by photosensitized reactions.

One of the most important findings in this respect is the apparent relationship between the protective effectiveness and the number of conjugated double bonds in the carotenoid molecule. Only carotenoids with nine or more conjugated double bonds seem to be effective quenchers of singlet oxygen ($^1O_2$), those with shorter polyene chains are much less efficient, e.g. in a mutant strain of *M. luteus* containing only a carotenoid with eight conjugated double bonds, protection was three times less effective (Mathews-Roth *et al.*, 1974). In mutant strains of the alga *Chlorella*(Claes, 1961) carotenoid protection against the anaerobic photobleaching of chlorophyll, which presumably proceeds through the triplet excited state of chlorophyll, increases drastically between five and seven conjugated double bonds. This indicates that carotenoids containing seven or more conjugated double bonds may be effective quenchers of triplet state sensitizers.

Of special interest is the question whether or not carotenoids are capable of protecting organisms not only against visible light of high irradiance but also against radiation of shorter wavelengths, i.e. UV-radiation. This question is part of the problem of to what extent increased radiation in the UV-B part of the solar spectrum (280-320 nm) may cause harmful effects to organisms. The problem arose when it was recognized that some anthropogenic compounds, e.g. chlorofluoro-hydrocarbons, nitrogen oxides and others, when released into the atmosphere may lead to a gradual breakdown or at least a depletion of the natural stratospheric ozone layer. Ozone is responsible for absorbing the greater part of the extra-terrestrial solar UV-radiation and a reduction in the ozone layer would lead to an increase in the UV-B waveband reaching the earth's surface (Caldwell, 1981).

A completely different effect is caused by the shortest wavelengths of the UV, the so-called UV-C (<280 nm), which is not part of the solar spectral radiation reaching the earth's surface. This radiation has been extensively used in investigations on photokilling of bacteria ("germicidal lamp"). Damage by UV-C is mainly due to its absorption by DNA which causes the formation of pyrimidine dimers in the DNA molecule. It is well known that in bacteria several "repair" mechanisms exist, including the so-called "photoreactivation", i.e. a deleterious effect of short-wavelength irradiation can be overcome by a subsequent irradiation with longer wavelengths, e.g. blue light (Jagger, 1985). Protection by carotenoids against deleterious effects of UV-C was investigated, e.g. in *M. luteus* (Mathews and Sistrom, 1960); the results showed no protective effect by carotenoids.

Results on the involvement of carotenoids in the protection against so-called "near UV" (UV-B/UV-A; >280 nm) are available from investigations on two organisms. Buckley and Houghton (1976) found a strain of the blue-green alga *Gloeocapsa alpicola* which is more resistant against UV than the wild-type and contains higher amounts of carotenoids. Irradiation with UV reduced the pigment content of the wild-type but not that of the resistant strain. Furthermore, reduction of the carotenoid content reduced the survival rates on UV-irradiation. A somewhat different function of carotenoids was found in the archaebacterium *Halobacterium* by Sharma *et al.* (1984). This bacterium contains large quantities of carotenoids in addition to the retinal proteins. Halobacteria differing in their carotenoid content did not show a difference in the resistance to UV-B between coloured and pigment-free mutants if the cultures were kept in darkness after UV-irradiation. However, when the bacteria were illuminated with photoreactivating light after UV treatment, the pigmented cells were photoreactivated more effectively compared with the carotenoid-free cells. The authors suggested that the role of carotenoids in this case may be that of an accessory pigment to the photoreactivating enzyme which is assumed to contain a flavin.

C Photoprotection in fungi

Earlier investigations on photoprotection by carotenoids in fungi are mentioned by Burnett (1976) and by Goodwin (1980).

Concerning a possible protective role against UV-radiation, two earlier studies on *Neurospora crassa* are relevant. Morris and Subden (1974) found that both albino strains and those normal strains made colourless by treatment with the inhibitor of carotenoid biosynthesis, β-ionone, were much more sensitive to UV radiation than the wild-type strain. Estimating the survival rate of *N. crassa* conidia after different irradiation conditions (Blanc *et al.*, 1976) revealed that conidia from an albino strain are somewhat more sensitive to black light (300–400 nm) and to white light in the presence of the exogenous photosensitizer methylene blue than those of the wild-type; however, there is no difference in the survival rate between carotenoid-containing and carotenoid-less conidia after irradiation with short wavelength UV, i.e. UV-C.

Considering the problem of a possible enhancement of UV-radiation on the earth's surface caused by an ozone depletion, outlined in section 3B, we investigated the effect of UV-irradiation and the protective role of carotenoids using the fungal species *Fusarium aquaeductuum* and *N. crassa* (Huber and Schrott, 1980; Schrott, 1985; Huber-Willer *et al.*, 1987). The intention of these investigations was, unlike the studies on bacteria and fungal conidia in which the killing rate was mainly tested, to focus on harmful but sublethal effects of radiation on the metabolism of mycelia.

Mycelia of *Fusarium* and *Neurospora* seem to be well suited for this type of study for the following reason: biosynthesis of carotenoids is light-dependent in the mycelia; after a short pulse (in the order of minutes) of blue light pigment accumulation occurs within a few hours (Rau, 1980). Therefore, from physiologically and genetically homogenous cell populations aliquots can be obtained, containing no or different amounts of carotenoids. The results gathered so far show that irradiation with UV-A (320–400 nm) as well as UV-B (280–320 nm) of sufficient high fluences or fluence rates inhibits the respiration rate, the biosynthetic capacity for carotenoids, and the uptake and incorporation of radioactively labelled amino acids. Mycelia pre-illuminated with blue light and thus containing carotenoids at the time of UV-irradiation are less sensitive; moreover, the amount of carotenoids present seems to be correlated with the extent of protection.

In conclusion, although further work is needed to elucidate a protective role of carotenoids unequivocally, in particular the mechanism of such a protection, in view of the reactions compiled in Fig. 5.4 carotenoid photoprotection against UV seems possible if an intracellular sensitizer absorbs in the UV spectral region as well as visible light, thus leading to the same reactions as white light.

D Photoprotection in higher plants

Evidence for a protective role of carotenoids in higher green plants has come from studies with mutants which exhibit aberrations in carotenoid biosynthesis; such mutants have been found mainly in maize and in *Helianthus*. Since the results of investigations with these mutants have been reviewed extensively by Goodwin (1980), only the main findings are summarized here.

The mutants carry blocks in the biosynthetic pathway of carotenoids and therefore are unable to synthesize the normal set of carotenoids in leaves. Instead they contain either no C_{40}-compound or only more saturated carotenes but minimal or no β-carotene. Exposure to bright light causes photodestruction of chlorophylls in the chloroplasts, which is lethal to the plant; however, the mutants will grow in very dim light or in the dark heterotrophically. All observations on mutants from maize and *Helianthus* suggest that carotenoid synthesis is under nuclear control.

Photolabile plants have also been reported in variegated strains of *Oenothera* (Kandler and Schötz, 1956; Schötz and Bathelt, 1965). They are the result either of a plastid mutation or of hybridization leading to plants with an incompatibility between the genome and the plastome. During photobleaching of chlorophylls carotenoid content also decreases but since a certain amount is still present photosensitivity may not be due to deficiency of carotenoids in this case.

One manifestation of the protective ability of carotenoids in higher plants has become very important recently: various herbicides interfere with the normal pathway of carotenoid biosynthesis. Since the earliest reports (e.g. Bartels and Hyde, 1970; Burns *et al.*, 1971) it has become more and more obvious that such compounds are toxic because they inhibit the production of the normal set of leaf carotenoids, resulting in the accumulation of colourless polyenes. This effect, leading to a susceptibility to photodestruction of the treated plant, is now important not only from a scientific but also from an economic viewpoint.

Herbicide action on the carotenoid level was reviewed by Goodwin (1980) and comprehensively by Ridley (1982). According to this last review the essential characteristics of this herbicide action are as follows: in photosynthetic tissues of plants carotenoids are located in the chloroplasts. Although some of the herbicides are known to affect the biosynthetic pathway directly, others may interfere with other processes in the chloroplast and, therfore, have an indirect effect on carotenoid accumulation. Most of the herbicides that are believed to act by inhibiting carotenoid synthesis directly by interfering with the desaturation reactions between phytoene and lycopene. They can be divided into two groups, those that mainly cause accumulation of phytoene and those that mainly cause an accumulation of ζ-carotene. Among the first group are norfluorazon (San 9789), metflurazon (San 6706), difunone and fluridone; the second group is represented by dichlormate and some other compounds investigated by Ridley. For formulae of the compounds, detailed results and references the reader is referred to Ridley's (1982) review. For the compound Aminotriazole accumulation of ζ-carotene as well as that of lycopene was reported; accumulation of lycopene indicates that this substance may also interfere with lycopene cyclization. Some herbicides that block chloroplast electron transport, like dichlorophenyldimethylurea (DCMU), cause photodestruction not only of pre-existing chlorophyll but also of carotenoids which have already been synthesized.

Finally, it should be pointed out that the lethal effect of these herbicides due to inhibition of carotenoid biosynthesis and, as a consequence, photobleaching of chlorophyll, eventually leading to a disorientation of the chlorolast structure in most cases, proceeds only in bright and not dim light. Such a reaction is similar to that found in the mutants mentioned above.

Recent investigations in connection with the acute damage to forest trees (*Waldsterben*) revealed that treatment of spruce (*Picea abies*) with very high ozone concentrations (much more than are present in the atmosphere) together with bright light causes a decrease of β-carotene (but not of xanthophylls), followed by a decrease in the chlorophyll content (Senser *et al.*, 1987).

E The role of photoregulation of carotenoid biosynthesis

Photoregulation of carotenoid biosynthesis has been reported in angiosperms and in some fungi and non-photosynthetic bacteria, whereas in algae only some mutants show this phenomenon. The extent of increase of carotenoid synthesis caused by light differs from organism to organism. In many organisms, e.g. in all angiosperms, low amounts of carotenoids are synthesized in darkness and illumination only enhances the rate of synthesis. In some fungi and bacteria, however, illumination is obligatory for substantial biosynthesis.

Why do these organisms produce carotenoids only as a response to light? Has this photoregulation any ecological meaning? The main functions of carotenoids in plants, i.e. their role in photosynthesis and in photoprotection, are related only to growth of the plant in the light. Consequently, the bulk of carotenoids is needed only when, for example, the seedling emerges from the ground or fungal mycelia grow into illuminated areas. Therefore, photoregulation acts quite clearly as a saving mechanism to avoid waste of both material and energy.

Photoregulation of carotenoid biosynthesis and its characteristics has been reviewed (Harding and Shropshire, 1980; Rau, 1980, 1983, 1985). The main findings may be reported briefly as follows: seedlings of higher plants grown in the dark have some capacity for carotenoid production but only angiosperms show a photostimulation of carotenoid biosynthesis during the development of the young plant. This stimulation is only a part of the photomorphogenic transformations in the chloroplast and, therefore, is not entirely independent of them. An acting photoreceptor for the photoregulation of carotenoid accumulation is phytochrome; however, it is uncertain whether or not cryptochrome (see section 2A) may also be involved. The response type is either an "induction" mechanism via pulse illumination or the carotenoid formation is dependent on continuous illumination, indicating a different mechanism.

Few species of fungi and non-photosynthetic bacteria synthesize only trace amounts of carotenoids when grown in the dark. A brief exposure to light, e.g. for a few seconds, induces substantial pigment production, but higher fluences are required for maximum synthesis. Thus photoregulation in these organisms exhibits all of the features of an induction mechanism. The photoreceptor responsible for this induction is in some fungal species cryptochrome; in bacteria porphyrins have been considered. From pertinent results it has been assumed that as a consequence of photoinduction the carotenogenic enzymes are synthesized *de novo* by way of gene derepression.

4 REPRODUCTION OF FUNGI

In Zygomycetes, e.g. species of *Phycomyces*, *Mucor*, *Blakeslea trispora*, formation of zygospores is dependent on signals between + and − mating type. Earlier assumptions that carotenoids are implicated in this sexual reproduction have had to be corrected since the discovery and structural identification of trisporic acids acting as gamones. Since these compounds are probably derived from β-carotene, carotenoids are involved indirectly in these processes. Reviews on this topic are available from Bu'Lock (1973), Burnett (1976) and Goodwin (1980). More recent work on compounds acting as zygotropic pheromones can be taken from pertinent papers (Sutter, 1986).

Two further examples from a possible indirect role of carotenoids in fungal reproduction should be mentioned. As decribed in section 2A4 in a different connection, formation of sporangiophores in *Phycomyces blakesleeanus* is inhibited by oxygen deficiency. In mutant strains deficient in the biosynthesis of carotenoids to a different extent the irradiance threshold for photoinduction of sporangiophores is raised by a range of 100 to 2000 times. The increased threshold can be overcome by application of retinol and, therefore, it might be concluded that the role of β-carotene in the system is to provide a source of retinol (Russo, 1977; Galland and Russo, 1979).

It has been known for a long time that in *Neurospora crassa* biosynthesis of carotenoids is photoregulated in the mycelium (see section 3E) but not in the conidia. This situation indicates that the developmental step to the formation of conidia is accompanied by a *de novo* synthesis of carotenogenic enzymes without photoinduction. This is further indicated by results from a mutant strain, designated "white collar", which has lost the capacity for photoinduced pigment synthesis in its mycelium but still has pigmented conidia (Harding and Shropshire, 1980).

5 SPOROPOLLENIN

Sporopollenin is an integral constituent of the outer wall of pollen, the exine, and also of zygospores and ascospores in some fungal species. It consists of macromolecules extraordinarily resistant against chemical agents and against biological degradation. Therefore, spores are found in layers of peat after millions of years in practically unchanged structure; this allows conclusions on the history of vegetation by so-called pollen analysis.

Sporopollenin is assumed to be a biopolymer of cross-linked carotenoids or carotenoid esters (Brooks and Shaw, 1968; Brooks *et al.*, 1971; Gooday *et al.*, 1973). Evidence for this assumption comes mainly from investigations using fungal spores; the results revealed that radioactivity derived from labelled mevalonic acid or β-carotene was incorporated into sporopollenin. Furthermore, Shaw (1971) was able to synthesize a sporopollenin from carotenoids by oxidative polymerization. The properties of this synthetic compound are similar to those of a sporopollenin from *Mucor mucedo*. From Shaw's results he calculated a formula of *Mucor* sporopollenin: $C_{90}\ H_{130}\ O_{33}$. In some species of fungi a correlation between synthesis of carotenoids and that of sporopollenin was found; however, as Burnett (1976) pointed out "it is not certain that accumulation of carotenes is always related to sporopollenin production".

Sporopollenin was also detected in the cell wall of a few species of green algae (Chlorophyta) e.g. in *Chlorella* (Atkinson *et al.*, 1972; Good and Chapman, 1978; Burczyk and Hesse, 1981; Biedlingmaier *et al.*, 1987). In order to supply further evidence for the hypothesis of a biogenetic connection between the ability to synthesize carotenoids and sporopollenin Burczyk and Czygan (1983) investigated mutants of *Chlorella fusca* deficient in pigment synthesis. Whereas the wild-type strain synthesizes ketocarotenoids in cases of nitrogen deficiency and the cell walls contain ketocarotenoids as well as sporopollenin, none of the mutants have the capacity to form ketocarotenoids under all conditions tested; they also contain no sporopollenin in the cell walls. From these results the authors concluded that there was a correlation between the ability of ketocarotenoid and sporopollenin synthesis and a biosynthetic pathway of sporopollenin via ketocarotenoids.

In an investigation on the role of sporopollenin in protecting against the effect of detergents in the medium it was found that cells of several green algae are protected only when the morphological arrangement of sporopollenins in the cell wall is in a closed structure (Biedlingmaier *et al.*, 1987).

For pollen of higher plants there still exists an uncertainty about the metabolic pathway which produces sporopollenin. For instance, Heslop-Harrison and Dickenson (1969) were unable to detect carotenoids during the stages of pollen ripening when sporopollenin is intensively accumulated. In a recent study (Prahl *et al.*, 1985) on the effect of inhibitors of carotenoid biosynthesis on sporopollenin accumulation in pumpkin (*Cucurbita pepo*) it was found that only norfluorazon (San 9789; see section 3D) inhibited carotenoid synthesis in developing anthers or pollen, leading to an accumulation of phytoene and phytofluene, but did not result in a drastic inhibition of sporopollenin accumulation.

Sporopollenin was further detected in myxobacteria (Strohl *et al.*, 1977) and in phycobionts of lichens (König and Peveling, 1980; Honegger and Brunner, 1981).

REFERENCES

Atkinson, A.W., Gunning, B.E.S. and John, P.C.L. (1972). *Planta* **107**, 1–32.
Bartels, P.G. and Hyde, A. (1970). *Plant Physiol.* **45**, 807–10.
Biedlingmaier, S.,Wanner, G. and Schmidt, A. (1987). *Z. Naturforsch.* (in press).
Blanc, P.L., Tuveson, R.W. and Sargent, M.L. (1976). *J. Bacteriol.* **125**, 616–25.
Buckley, C.E. and Houghton, J.A. (1976). *Arch. Mikrobiol.* **107**, 93–7.
Bu'lock, J.D. (1973). *Pure Appl. Chem.* **34**, 435–61.
Bünning, E. (1937). *Planta* **26**, 719–36.
Bünning, E. (1938). *Planta* **27**, 148–58, 583–610.
Burczyk, J. and Czygan, F.-C. (1983). *Z. Pflanzenphysiol.* **111**, 169–74.
Burczyk, J. and Hesse, M. (1981). *Plant System. Evol.* **138**, 121–37.
Burnett, J.H. (1976). In: *Chemistry and Biochemistry of Plant Pigments* (Ed. Goodwin, T.W.) pp. 655–679. Academic Press, London.
Burns, E.R., Buchanan, G.A. and Carter, M.C. (1971). *Plant Physiol.* **47**, 144–8.
Brooks, J. and Shaw, G. (1968). *Nature* **219**, 523–4.
Brooks, J., Grant, P.R., Muir, M., van Gijzel, P. and Shaw, G. (Eds.) (1971). *Sporopollenin.* Academic Press, London.
Caldwell, M.M. (1981). In: *Encylopaedia of Plant Physiology*, vol. 12A, pp. 169–198. Springer-Verlag, Berlin.
Claes, H. (1961). *Z. Naturforsch.* **16b**, 445–54.
Cogdell, R.J. (1985). *Pure Appl. Chem.* **57**, 723–8.
DeFabo, E. (1980). In: *The Blue Light Syndrome* (Ed. Senger, H.) pp. 187–197. Springer-Verlag, Berlin.

DeFabo, E.C., Harding, R.W. and Shropshire, W. Jr. (1976). *Plant Physiol.* **57**, 440–5.
Dring, M.J. and Lüning, K. (1975). *Z. Pflanzenphysiol.* **75**, 107–17.
Galland, P. and Russo, V.E.A. (1979a). *Photochem. Photobiol.* **29**, 1009–14.
Galland, P. and Russo, V.E.A. (1979b). *Planta* **146**, 257–62.
Gollnick, K. and Schenk, G.O. (1967). In: *1,4-Cycloaddition Reactions* (Ed. Hamer, J.) pp. 255–344. Academic Press, New York.
Good, B.H. and Chapman, R.L. (1978). *Am. J.Bot.* **65**, 27–33.
Gooday, G.W., Green, D., Fawcett, P. and Shaw, G. (1973). *J. Gen. Microbiol.* **74**, 233–9.
Goodwin, T.W. (1980). *The biochemistry of the carotenoids*, 2nd Edn., vol. I. Chapman and Hall, London.
Gressel, J. (1979). *Photochem. Photobiol.* **30**, 749–54.
Gressel, J. and Rau, W. (1983). In: *Encyclopedia of Plant Physiology*, vol. 163, pp. 603–639. Springer-Verlag, Berlin.
Häder, D.-P. (1984). In: *Blue Light Effects in Biological Systems* (Ed. Senger, H.), pp. 435–443. Springer-Verlag, Berlin.
Hager, A. (1970). *Planta* **91**, 38–54.
Halldal, P. (1958). *Physiol. Plant.* **11**, 118–53.
Halldal, P. (1961). *Physiol. Plant.* **14**, 133–9.
Harding, R.W. and Shropshire, W. (1980). *Ann. Rev. Plant Physiol.* **31**, 217–38.
Hartmann, E. and Schmidt, K. (1980). In: *The Blue Light Syndrome* (Ed. Senger, H.), pp. 221–237. Springer-Verlag, Berlin.
Haupt, W. (1986). In: *Photomorphogenesis in Plants* (Eds. Kendrick, R.E. and Kronenberg, G.H.M.), pp. 415–441. Martinus Nijhoff, Dordrecht.
Haupt, W. and Feinleib, M.F. (Eds.) (1979). *Encyclopedia of Plant Physiology*, vol. 7. Springer-Verlag, Berlin.
Heslop-Harrison, J. and Dickenson, H.G. (1969). *Planta* **84**, 199–214.
Honegger, R. and Brunner, U. (1981). *Can. J. Bot.* **59**, 2713–34.
Horwitz, B.A. and Gressel, J. (1986). In: *Photomorphogenesis in Plants*. (Eds. Kendrick, R.W. and Kronenberg, G.H.M.), pp. 159–183. Martinus Nijhoff, Dordrecht.
Huber, A. and Schrott, E.L. (1980). In: *The Blue Light Syndrome* (Ed. Senger, H.), pp. 299–308. Springer-Verlag, Berlin.
Huber-Willer, A., Mitzka-Schnabel, U., Rau, W. and Schrott, E.L. (1987). (in press).
Inoue, Y. and Watanabe, M. (1984). *Plant Cell Physiol.* **25**, 107–13.
Jagger, J. (1985). *Solar-UV action on living cells.* Praeger, New York.
Kandler, O. and Schötz, F. (1956). *Z. Naturforsch.* **11b**, 708–18.
König, J. and Peveling, E. (1980). *Z. Pflanzenphysiol.* **98**, 459–64.
Krinsky, N.I. (1968). In: *Photophysiology*, vol. 3 (Ed. Giese, A.C.), pp. 123–195. Academic Press, New York.
Krinsky, N.I. (1979). *Pure Appl. Chem.* **51**, 649–60.
Krinsky, N.I. (1984). In: *Oxygen Radicals in Chemistry and Biology* (Eds. Bors, W., Saran, M. and Tait, D.), pp. 453–473. Walter de Gruyter, Berlin.
Liaaen-Jensen, S. (1979). *Pure Appl. Chem.* **51**, 661–75.
Liaaen-Jensen, S. (1985). *Pure Appl. Chem.* **57**, 647–58.
Lipson, E.D., Galland, P. and J.A. Pollock (1984). In: *Blue Light Effects in Biological Systems* (Ed. Senger, H.), pp. 228–236. Springer-Verlag, Berlin.
Mathews, M.M. and Sistrom, W.R. (1960). *Arch. Mikrobiol.* **35**, 139–46.

Mathews-Roth, M.M., Wilson, T., Fujimori, E. and Krinsky, N.I. (1974) *Photochem. Photobiol.* **19**, 217–22.
Mathis, P. and Schenk, C.C. (1982). In: *Carotenoid Chemistry and Biochemistry* (Eds. Britton, G. and Goodwin, T.W.), pp. 339–351. Pergamon Press, Oxford.
Meissner, G. and Delbrück, M. (1968). *Plant Physiol.* **43**, 1279–83.
Morris, S.A.C. and Subden, R.W. (1974). *Mutation Res.* **22**, 105–9.
Nultsch, W. (1980). In: *The Blue Light Syndrome* (Ed. Senger, H.), pp. 38–49. Springer-Verlag, Berlin.
Nultsch, W., Throm, G. and Rimscha, I. (1971). *Arch. Mikrobiol.* **80**, 351–69.
Prahl, A.K., Springstubbe, H., Grumbach, K. and Wiermann, T. (1985). *Z. Naturforsch.* **40c**, 621–6.
Presti, D. and Delbrück, M. (1978). *Plant Cell Environ.* **1**, 81–100.
Presti, D., Hsu, W.-J. and Delbrück, M. (1977). *Photochem. Photobiol.* **26**, 403–5.
Rau, W. (1967). *Planta* **72**, 14–28.
Rau, W. (1980). In: *The Blue Light Syndrome* (Ed. H. Senger), pp. 283–298. Springer-Verlag, Berlin.
Rau, W. (1983). In: *Biosynthesis of Isoprenoid Compounds*. (Eds. Porter, J.W. and Spurgeon, S.L.), pp. 123–157. John Wiley, New York.
Rau, W. (1985). *Pure Appl. Chem.* **57**, 777–84.
Ridley, S.M. (1982). In: *Carotenoid Chemistry and Biochemistry* (Eds. Britton, G. and Goodwin, T.W.), pp. 353–369. Pergamon Press, Oxford.
Robinson, J.C., Keay, L., Molinari, R. and Sizer, I.W. (1962). *J. Biol. Chem.* **237**, 2001–10.
Russo, V.E.A. (1977). *Plant Sci. Lett.* **10**, 373–80.
Schiff, J.A., Cunningham, F.X. and Green, M.S. (1982). In: *Carotenoid Chemistry and Biochemistry* (Eds. Britton, G. and Goodwin, T.W.). Pergamon Press, Oxford.
Schletz, K. (1976). *Z. Pflanzenphysiol.* **77**, 189–211.
Schötz, F. and Bathelt, H. (1965). *Planta* **64**, 330–62.
Schrott, E.L. (1985). *Pure Appl. Chem.* **57**, 729–34.
Schwartzel, E.H. and Cooney, J.J. (1974). *Can. J. Microbiol.* **20**, 1015.
Senger, H. (ed.) (1980). *The Blue Light Syndrome*. Springer-Verlag, Berlin.
Senger, H. (ed.) (1984). *Blue Light Effects in Biological Systems*. Springer-Verlag, Berlin.
Senger, H. and Schmidt, W. (1986). In: *Photomorphogenesis in Plants* (Eds. Kendrick, R.E and Kronenberg, G.H.M.), pp. 137–158. Martinus Nijhoff, Dordrecht.
Senser, M., Höpker, K. and Peuker, A. (1987). *Allgem. Forstzeitung* (in press).
Sharma, N., Hepburn, D. and Fitt, P.S. (1984). *Biochim. Biophys. Acta* **799**, 135–42.
Shaw, G. (1971). In: *Sporopollenin* (Eds. Brooks, J., Grant, P.R., Muir, M., van Gijzel, P. and Shaw, G.) pp. 305–350. Academic Press, London.
Shropshire, W., Jr. (1980). In: *The Blue Light Syndrome* (Ed. H. Senger), pp. 172–186. Springer-Verlag, Berlin.
Shropshire, W., Jr. and Withrow, R.B. (1958). *Plant Physiol.* **33**, 360–5.
Sistrom, W.R., Griffiths, M. and Stanier, R.Y. (1956). *J. Cell. Comp. Physiol.* **48**, 473–515.
Song, P.-S. (1980). In: *The Blue Light Syndrome* (Ed. H. Senger), pp. 157–171. Springer-Verlag, Berlin.

Song, P.-S. (1984). In: *Blue Light Effects in Biological Systems* (Ed. Senger, H.), pp. 75–80. Springer-Verlag, Berlin.
Song, P.-S. and Moore, T.A. (1974). *Photochem. Photobiol.* **19**, 435–41.
Strohl, W.R., Larkin, J.M., Good, B.H. and Chapman, R.L. (1977). *Can. J. Microbiol.* **23**, 1080–3.
Sugai, M. (1971). *Plant Cell Physiol.* **12**, 103–9.
Sugai, M., Tomizawa, K., Watanabe, M. and Turuya, M. (1984). *Plant Cell Physiol.* **25**, 205–12.
Sutter, R.P. (1986). *Exper. Mycol.* **10**, 256–8.
Wagner, G. (1984). In: *Blue Light Effects in Biological Systems* (Ed. H. Senger), pp. 48–54. Springer-Verlag, Berlin.
Wellmann, E. (1983). In: *Encyclopedia of Plant Physiology* vol. 16B, pp. 745–756. Springer-Verlag, Berlin.
Zechmeister, L. (1962). *Cis-trans-isomeric Carotenoids, Vitamins A and Arylpolyenes.* Springer-Verlag, Vienna.
Zurzycki, J. (1980). In: *The Blue Light Syndrome* (Ed. Senger, H.), pp. 50–68. Springer-Verlag, Berlin.

6
Phytochrome

GARRY WHITELAM and HARRY SMITH

Department of Botany, University of Leicester, University Road, Leicester LE1 7RU, U.K.

1 INTRODUCTION

Phytochrome is the blue-green, photochromic pigment controlling a wide range of developmental and metabolic processes in green plants. Phytochrome, first isolated and detected *in vitro* in 1959 (Butler *et al.*, 1959) by virtue of its unique photoreversible absorbance changes, is a protein with a linear tetrapyrrole chromophore existing in two stable yet interconvertible forms: Pr, which absorbs maximally in the red region of the spectrum; and Pfr, which absorbs maximally in the far red. These reversible spectral changes form the basis of a spectrophotometric assay which facilitated the subsequent purification of the protein and physicochemical characterization of the molecule. Recently, fresh insights into the structure and properties of the phytochrome molecule have been derived from the application of immunochemical and molecular biological

PLANT PIGMENTS
ISBN 0-12-289847-8

techniques. The aim of this review is to summarize the progress being made towards characterizing the native phytochrome molecule, the ultimate aim of which is to gain an understanding of the biochemical action of the photoreceptor. We have not considered aspects of the physiology of phytochrome and photomorphogenesis in any detail largely owing to limitations of space and because of the enormous breadth of published information in this area. For instance, in the period since the beginning of 1983 to the time of writing some 1332 published articles have appeared in the area of photomorphogenesis, including 437 with the term "phytochrome" in the title [data obtained from the Current Awareness in Biological Science Database (CABS Pergamon-Infoline)]. For further information on this topic the reader is referred to the comprehensive treatises of Shropshire and Mohr (1983) and Kendrick and Kronenberg (1986).

2 THE PROTEIN MOIETY

A Size and terminology

Analysis of phytochrome *in vitro* requires that the chromoprotein be obtained in a highly purified, undenatured and undegraded state. A major problem of phytochrome purification is that phytochrome is susceptible to degradation by proteases present in crude plant extracts (Gardner *et al.*, 1971; Vierstra and Quail, 1982a,b). This problem is particularly serious for extracts of etiolated oat seedlings, the most frequently used starting material. Before 1973 virtually all phytochrome purification procedures yielded a monomeric protein with M_r *ca.* 60 kD, as judged by SDS-PAGE. It was subsequently shown that this 60 kD M_r species was in fact a relatively stable proteolytic fragment derived from a much larger species, with an apparent monomer size of 120 kD (Gardner *et al.*, 1971). With modifications to purification protocols designed to limit *in vitro* proteolysis, phytochrome of 120 kD M_r has been isolated from a range of plant species (Pratt, 1982) and its physicochemical properties have been extensively studied (for reviews see Pratt, 1982; W. O. Smith, 1983).

Recently, it was shown that the 120 kD species of phytochrome isolated from oat seedlings was composed of a mixture of proteolytic fragments of 118 and 114 kD M_r, derived from the intact molecule which has a monomeric weight of 124 kD (Vierstra and Quail, 1982b). This limited *in vitro* proteolysis was found to occur especially when the phytochrome was in the Pr form (Fig. 6.1), the more stable form of phytochrome *in*

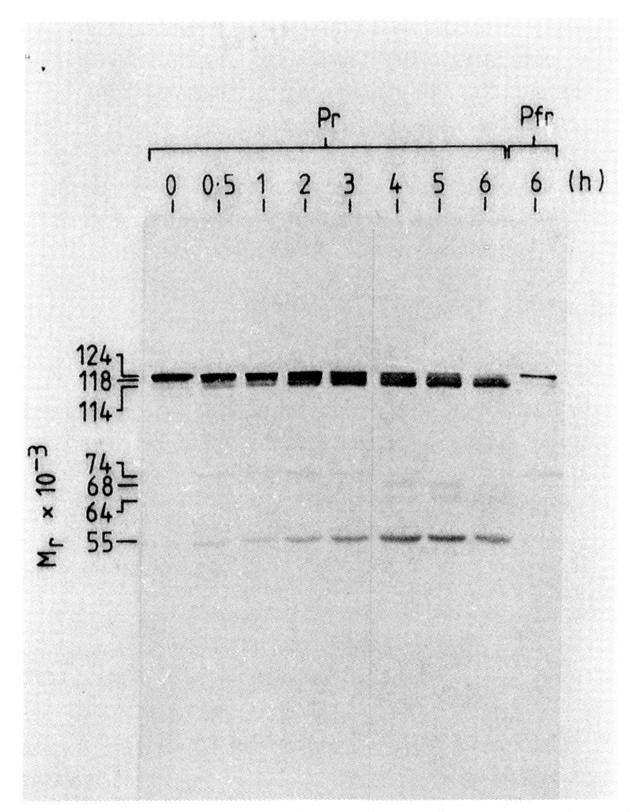

Fig. 6.1 Partial proteolysis of phytochrome in crude extracts. A crude extract of *Avena* shoots containing Pr, or mainly Pfr, was incubated at 20°C for up to 6 h and the peptides derived from phytochrome have been visualized by SDS-PAGE and immunoblotting. Data from M. L. Holdsworth.

vivo, and the form in which phytochrome was traditionally isolated (Pratt, 1982). This also accounts for the earlier observation that phytochrome extracted as Pfr exhibited a higher M_r than phytochrome extracted as Pr (Boeshore and Pratt, 1980). Partial proteolysis has now been shown to occur in extracts of several plant species, although there is variation in the size of the intact molecule and of the degradation products (Vierstra *et al.*, 1984). Several pieces of experimental evidence support the view that 124 kD genuinely represents the monomeric M_r of intact oat phytochrome. These include comparison of the sizes of purified phytochrome with the products of *in vitro* translation of phytochrome mRNA (Bolton and Quail, 1982), direct extraction of freeze-dried tissue into boiling SDS followed by rapid immunoprecipitation, thus precluding post-homogenization proteolysis (Vierstra and Quail, 1982b) and, most convincingly, comparison of the sizes of purified phytochrome with the size estimated from the amino acid sequence derived from the phytochrome cDNA sequence (Hershey *et al.*, 1985).

Generic terms used to describe the differently degraded states of the phytochrome molecule are rather confusing. For example, in order to distinguish between the 60 kD proteolytic fragment and the 120 kD, but still partially degraded species, the terms "small", or "degraded", and "large", or "undegraded", were widely adopted. The subsequent demonstration that "large/undegraded" phytochrome was also the product of proteolysis has led to the use of the terms "native", "intact" and "full-length" to describe the 124 kD species. The terminology and the sizes of the phytochrome species to which they refer are summarized in Table 6.1 for a number of plant species. For convenience, where necessary, we use the generic terms "intact", "large" and "small" to describe the appropriate species of phytochrome.

Table 6.1 Terminology used in describing various phytochrome species. Adapted from Vierstra and Quail (1986)

Operational term	Plant	Apparent monomeric M_r (kD)[1]
small, degraded	oat, rye, pea	60
large, undegraded, 120kD	oat	118/114
	maize	123
	rye	115
	pea	117/114
	Cucurbita	118/117/114/112
native, intact	oat	124
	maize	127
	rye	124
	pea	121
	Cucurbita	120

[1]From SDS-PAGE.

B Physicochemical properties

1 AMINO ACID COMPOSITION AND SEQUENCE

The amino acid sequence of intact oat phytochrome has been derived from the nucleotide sequence of the phytochrome gene (Hershey *et al.*,

1985). Although at least four phytochrome genes are expressed in etiolated oat seedlings, nucleotide sequence analysis of three of them revealed that they show greater than 98% conservation at the amino acid level (Hershey *et al.*, 1985). The full-length derived amino acid sequence establishes that the phytochrome polypeptide is composed of 1128 amino acids and has a monomeric M_r of 124.9 kD. Prior to this nucleotide sequence analysis only about 1% of the protein sequence was known, namely an 11 amino acid sequence to which the phytochrome chromophore is attached (Lagarias and Rapoport, 1980). This undecapeptide has the following structure:

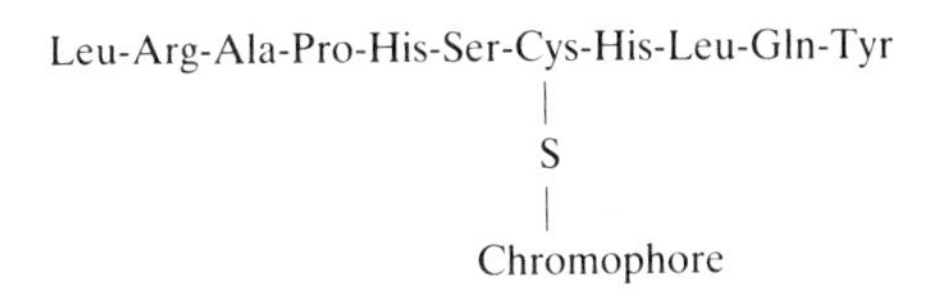

Since this unique structure, common to all chromophore-bearing peptides derived from proteolysis of oat phytochrome, occurs only once in the complete derived amino acid sequence, the conclusion of Lagarias and Rapoport (1980) that phytochrome has a single chromophore per monomer, is confirmed. The chromophore is attached in the NH_2-terminal half of the molecule at cysteine-321.

Complete, or partial, amino acid sequences have also been derived from the nucleotide sequences of the zucchini and pea phytochrome genes (Komeda *et al.*, 1986; Sharrock *et al.*, 1986). Comparisons of the sequences reveal regions of high homology around the chromophore attachment site and regions of divergence in the COOH-terminal portion of the molecule. However, evidence obtained from the cross-reactivities of monoclonal antibodies to phytochrome suggests that there is also a region of high homology located within the COOH-terminal portion of the molecule. An antibody to pea phytochrome, which maps to an epitope in the COOH-terminal portion of the phytochrome molecule, cross-reacts with SDS-denatured and electroblotted phytochrome from a wide range of plant species, including ferns and algae (Cordonnier *et al.*, 1986b). Similarly, an oat phytochrome monoclonal antibody raised to an epitope located in the same region shows a very wide cross-reactivity with denatured phytochrome from many plants (M. Holdsworth and G. C. Whitelam, unpublished observations).

The amino acid composition of oat phytochrome derived from the sequence data agrees well with that obtained from amino acid analysis of the purified 124 kD protein (Vierstra and Quail, 1983a) and is similar to that obtained by compositional analysis of predominantly intact rye and zucchini phytochromes (Cordonnier and Pratt, 1982; Vierstra and

Quail, 1986). From the available amino acid composition data, between 46 and 50% of the amino acids are polar; this is within the range observed for most soluble proteins (Capaldi and Vanderkooi, 1972).

The only known prosthetic group associated with phytochrome is the open chain tetrapyrrole chromophore. Analysis of phosphate content of "large" oat phytochrome indicates the presence of one phosphate per monomer (Hunt and Pratt, 1980). Phytochrome does not appear to contain carbohydrate residues (Pratt, 1982) despite earlier evidence to the contrary (Roux *et al.*, 1975). There are indications of another group, of an unknown nature, responsible for a blocked NH_2-terminus (Litts *et al.*, 1983; Vierstra and Quail, 1983a).

2 SECONDARY AND TERTIARY STRUCTURE

Ultraviolet circular dichroism (CD) spectroscopy has been used to make predictions about the secondary structure of phytochrome. Tobin and Briggs (1973) estimated "large" rye phytochrome to contain 20% α-helix, 30% β-structure and the remaining 50% random coil structure. Similar analyses for "large" oat phytochrome yielded estimates of 35% α-helix, 23% β-structure and 42% random coil (Hunt and Pratt, 1980). More recently, the availability of the complete derived amino acid sequence has allowed predictions of the secondary structure by Chou-Fasman analysis (Vierstra and Quail, 1986) as well as the construction of hydropathy profiles (Hershey *et al.*, 1985). From such analyses it is apparent that the chromophore is attached at a β-turn in a mildly hydrophilic region between two strongly hydrophobic regions. It has been suggested that the more extensive of these two hydrophobic regions, between residues 80 and 315, could provide a cavity in which the chromophore is housed (Hershey *et al.*, 1985). This is consistent with a number of studies indicating that the chromophore is located in the interior of the molecule, and so is relatively inaccessible to the hydrophilic external medium (Eilfeld and Rüdiger, 1984; Song, 1985; Rüdiger *et al.*, 1985). Furthermore, hydropathy profile analysis has revealed that the polypeptide segment at the NH_2-terminus of the protein, known to be important to the maintenance of stable protein–chromophore interaction (Vierstra and Quail, 1983a,b), is largely hydrophilic and therefore likely to be surface-located. This is consistent with suggestions that this segment shields the chromophore, as evidenced by the increased chemical reactivity of the chromophore in partially degraded, "large" phytochrome (Eilfeld and Rüdiger, 1984; Hahn *et al.*, 1984; Rüdiger *et al.*, 1985; Song, 1985).

The mean hydropathic index across a window of 19 amino acids does not exceed 1.06 for intact oat phytochrome (Hershey *et al.*, 1985). This value is consistent with the behaviour of phytochrome as a water-soluble globular protein and indicates that there are no regions of the polypeptide suitable for spanning cellular membranes.

3 QUARTERNARY STRUCTURE AND STRUCTURAL DOMAINS

Both intact and "large" phytochrome exist in solution as dimers of identical subunits with apparent molecular masses of 253 and 240 kD respectively (Pratt, 1982; Jones and Quail, 1986). Although for "large" phytochrome there are reports of possibly larger aggregates (Grombein and Rüdiger, 1976) it seems that at physiological pH and ionic strength the dimer represents the predominant species. Partial dissociation of the dimer at high salt concentrations, with essentially no effect on the spectral properties, as well as the observation that migration of phytochrome in non-denaturing gels or size exclusion columns is unaffected by the presence of reductant, indicates that the phytochrome dimer is held together by ionic rather than covalent interactions (Hunt and Pratt, 1980; Jones and Quail, 1986).

In solution, neither the monomer nor the dimer exhibits characteristics of an ideal spherical protein (Briggs and Rice, 1972; Jones and Quail, 1986). On size exclusion gels in Tris buffers intact phytochrome migrates with an apparent molecular mass of 350 kD for the dimer and 170 kD for the dissociated monomer (Jones and Quail, 1986). Size exclusion chromatography data have been used to provide estimates of the Stokes radius of the phytochrome molecule. Under conditions of minimal aggregation a value of 5.6 nm has been estimated for the dimer of intact oat phytochrome (Lagarias and Mercurio, 1985; Jones and Quail, 1986), whilst values of 6.5 and 6.1 nm were previously estimated for "large" rye and oat phytochromes (W. O. Smith, 1975; Litts, 1980). A much larger Stokes radius of 8.1 nm has been calculated for intact oat phytochrome from the diffusional coefficient determined by quasi-elastic light scattering (Sarkar *et al.*, 1984), but the reason for this discrepancy is not known. Using a Stokes radius of 5.6 nm, a frictional ratio (f/f_o) of 1.37 has been calculated for the dimer of intact oat phytochrome (Jones and Quail, 1986). The frictional ratio relates to the deviation of the molecular shape from that of a sphere and so provides evidence that dimer-intact phytochrome has an elongated shape. Similar analyses have been performed for the dissociated monomer of intact phytochrome, which has a Stokes radius of 3.3 nm and a frictional ratio of 1.34; i.e., the phytochrome monomer also has an elongated shape (Jones and Quail, 1986).

Studies on the proteolytic cleavage of intact phytochrome have provided information on the three-dimensional organization of the polypeptide into domains (Lagarias and Mercurio, 1985). This technique is greatly enhanced when used in combination with immunoblotting with domain-specific monoclonal antibodies (Daniels and Quail, 1984; Jones *et al.*, 1985; Jones and Quail, 1986). Fig. 6.2 summarizes the proteolytic cleavage data and shows the designation of monoclonal antibodies which have been mapped to the various domains. Proteolysis of Pfr by endogenous proteases, present in crude extracts, produces two main polypeptides. A polypeptide of approximately 74 kD M_r is released from the NH_2-terminal end of phytochrome with the concomitant production of an approximately 55 kD M_r COOH-terminal fragment (Jones *et al.*, 1985; Jones and Quail, 1986). The 74 kD polypeptide bears the chromophore and migrates as a spherical, globular monomer on gel filtration columns (Jones and Quail, 1986). Additionally, the spectral properties of the isolated 74 kD polypeptide are indistinguishable from those of the intact 124 kD molecule (Jones and Quail, 1986). These data suggest that about 60% of the intact phytochrome monomer represents a discrete globular domain which, apart from the obvious continuity of the polypeptide chain, lacks stable linkages with the COOH-terminal domain. Furthermore, the COOH-terminal domain does not appear to be involved in determining the conformation of the polypeptide segment or segments which interact with the chromophore. However, studies on the proteolytic cleavage of phytochrome with a number of commercial endoproteases have shown that cleavage sites within the 55 kD COOH-terminal domain are kinetically favoured in Pfr, indicating that there must be some form of interdomain interaction (Lagarias and Mercurio, 1985).

The isolated COOH-terminal 55 kD M_r domain migrates on gel filtration columns as a 160 kD species (Jones and Quail, 1986). This is thought to

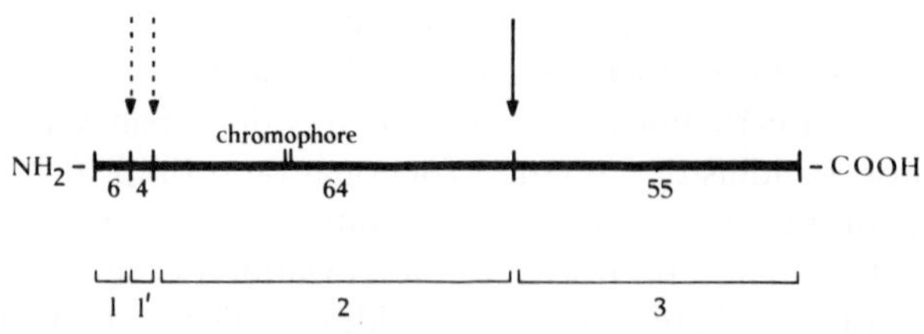

Fig. 6.2 Peptide map of 124 kD M_r *Avena* phytochrome. Bars indicate the positions of the major proteolytic cleavage sites, and the molecular masses of the derived peptides are given directly below. Solid arrows indicate cleavage sites which are preferred in Pfr, dashed arrows indicate Pr-preferred cleavage sites. The designation of monoclonal antibodies recognizing epitopes borne on the peptides is given below (1,1′,2,3).

represent the migration of a non-spherical dimer of polypeptides derived from the COOH-terminal end of the molecule. Further evidence that the dimerization site is located in this region of the molecule comes from the observation that tryptic peptides containing the COOH-terminal 42 kD also migrate as molecules two to three times larger than their monomeric molecular masses (Jones and Quail, 1986). The elongated nature of these COOH-terminal fragment dimers could reflect the dimerization of two essentially globular regions; alternatively, each individual COOH-terminal region could have an elongated shape. Support for the latter interpretation comes from the observation that the COOH-terminal domain is very vulnerable to further proteolytic cleavage, whilst the NH_2-terminal domain is rather protease-resistant. This has been interpreted to indicate that the COOH-terminal domain has an extended structure, whilst the NH_2-terminal domain has a more compact conformation (Lagarias and Mercurio, 1985; Jones and Quail, 1986).

Proteolysis of Pr by endogenous proteases reveals the presence of an important subdomain at the NH_2-terminus. Incubation of Pr in crude extracts leads to the rapid proteolytic removal of a 6–10 kD segment from the NH_2-terminus (Fig. 6.1) of 124 kD oat phytochrome to yield polypeptides of 118–114 kD M_r (i.e. "large" phytochrome). Similarly, the 74 kD domain is susceptible to the same proteolytic cleavages to yield 68–64 kD products (i.e. "small" phytochrome). Loss of the 6–10 kD NH_2-terminal segment leads to a blue-shift in the Pfr absorption maximum from 730 to 722 nm (Vierstra and Quail, 1982a, 1983a; Jones *et al.*, 1985). Other effects which accompany this partial proteolysis include an increase in the rate of non-photochemical dark reversion of Pfr to Pr (Vierstra and Quail, 1982a; Litts *et al.*, 1983), a decrease in the quantum yield for Pr to Pfr photoconversion (Vierstra and Quail, 1983b), an increase in the reactivity of the chromophore towards the oxidant tetranitromethane (Hahn *et al.*, 1984) and changes in the Pfr CD spectrum (Litts *et al.*, 1983; Vierstra *et al.*, 1987). The effects of proteolytic removal of the 6–10 kD NH_2-terminal fragment can, to some extent, be mimicked by the binding of monoclonal antibodies to the 6 kD NH_2-terminal fragment (Cordonnier *et al.*, 1985) and to the adjacent 4 kD fragment (Holdsworth and Whitelam, 1987). These data provide convincing evidence that the 6–10 kD NH_2-terminal subdomain plays a critical role in stable interaction between the polypeptide and the Pfr chromophore and suggest that the 6–10 kD segment functions to "protect" the chromophore, within its hydrophobic cavity, from the external hydrophilic environment.

Recently, evidence has been presented which indicates that the generation of 118 and 114 kD polypeptides by the action of endogenous proteases on Pr results from proteolytic cleavages at both the NH_2-

terminus and the COOH-terminus (Grimm *et al.*, 1987). These data apparently conflict with the assignment of the 6–10 kD fragment to the NH_2-terminus alone. However, cleavages at both the NH_2-terminus and the COOH-terminus are themselves not consistent with the observation that the sequential loss of 6 and 4 kD segments from the 124 kD polypeptide is recapitulated for the isolated 74 kD chromophore-containing domain, with identical effects on spectral properties (Vierstra and Quail, 1982a, 1983a; Jones *et al.*, 1985).

3 THE CHROMOPHORE AND ITS CONVERSIONS

The chromophore of phytochrome is an open chain tetrapyrrole (i.e. a bilin, or bile pigment). The two stable forms of phytochrome, Pr and Pfr, are mutually interconvertible by appropriate radiation; in other words, the pigment is photochromic. The absorption spectra of pure, native *Avena* phytochrome in the Pr (i.e. far-red-irradiated) and Pfr (i.e. red-irradiated) forms are shown in Fig. 6.3A. Red-irradiated phytochrome is an equilibrium mixture of 87% Pfr and 13% Pr; Fig. 6.3B shows the absorption spectrum (long wavelength region only) of Pfr isolated from Pr by immunoprecipation of Pr by a monoclonal antibody which discriminates absolutely against Pfr (Holdsworth and Whitelam, 1987). The differences in spectral properties between Pr and Pfr reflect chemical differences both within the chromophore and within the apoprotein. The data pertaining to the chemistry of the Pr and Pfr chromophores and their phototransformations have been comprehensively reviewed in recent years (Rüdiger, 1983, 1986; Rüdiger and Scheer, 1983; Rüdiger *et al.*, 1985; Song, 1985); this section is intended as a brief summary of the current position. The coverage extends only to etiolated-tissue phytochrome, and the majority of the data has been acquired from studies of "small" and "large", as opposed to intact, phytochrome.

A The Pr chromophore

Long before the term "phytochrome" had been coined, the Hendricks and Borthwick group at Beltsville had predicted, on the basis of similarity between the action spectra for photomorphogenetic responses and the absorption spectra of phycocyanin, that the responsible photoreceptor had a chromophore similar to that of phycocyanin (Parker *et al.*, 1950).

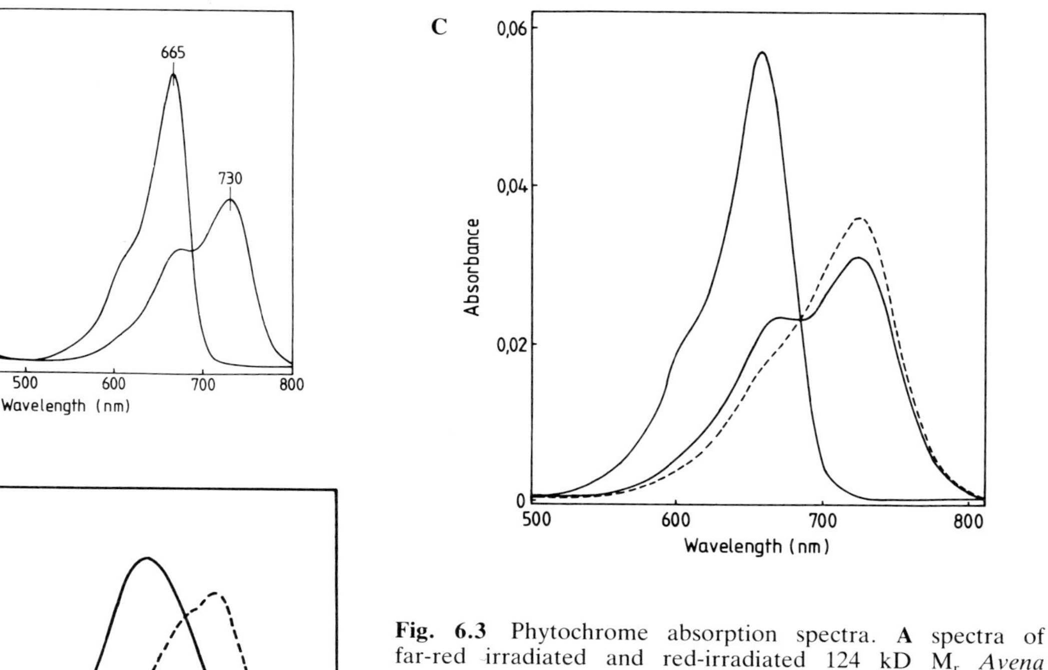

Fig. 6.3 Phytochrome absorption spectra. **A** spectra of far-red irradiated and red-irradiated 124 kD M_r *Avena* phytochrome; **B** spectra of Pr, red-irradiated phytochrome, and Pfr isolated from Pr using a Pr-specific monoclonal antibody (dashed line); data from M. L. Holdworth and G. C. Whitelam); **C** spectra of a Pfr-chromopeptide (unbroken line) and a Pr-chromopeptide (dashed line) (redrawn with permission from Rüdiger, 1986).

Throughout the subsequent investigations of the chemistry of the phytochrome chromophore, the biliproteins phycocyanin, phycoerythrin, and allophycocyanin have been employed as model compounds. Proof that the phytochrome chromophore is a bilin (i.e. "phytochromobilin"), however, awaited the application of the technique of chromic acid oxidation by Rüdiger and Correll in 1969. In this powerful and sensitive approach, biliprotein chromophores are oxidized *in situ*, yielding defined oxidation products (e.g. maleimides and succinamides) with unaltered β-pyrrolic substituents, capable of resolution and identification by thin-layer chromatography and appropriate staining methods (Rüdiger, 1969). Degradation studies can only provide definitive evidence on chromophore side chains, but the technique has provided enough data to construct a hypothetical structure for phytochromobilin which differs from that for phycocyanobilin only by the exchange of a vinyl group for an ethyl group at C-3 in ring D (see Fig. 6.4).

Further evidence for this structure, and for the linkage between ring A and the apoprotein, has been derived from comparison of spectral properties. In addition to a variable band in the near ultraviolet (UV), free bile pigments characteristically have a single broad band of absorbance in the visible, the wavelength maximum of which may be shifted by derivatization (i.e. cation or zinc complex formation). These properties may be used to characterize unknowns by comparison with the range of available bile pigments; data collected and reviewed by Rüdiger and Scheer (1983) show that phytochromobilin fits into the series of fully conjugated bilins. Comparison between free bile pigments and biliproteins is not useful because interactions between the chromophore and the apoprotein drastically modify the chromophore absorption properties. After unfolding of the polypeptide chain, however, the chromophore behaves similarly to a free bile pigment, allowing meaningful comparisons to be made. When the spectral properties of phytochrome (unfolded in 6 M guanidinium chloride; Grombein *et al.*, 1975) were compared with those reported for phytochromobilin (extracted from the isolated chromoprotein; Siegelman *et al.*, 1966), a spectral shift was seen which was attributed to the presence of an ethylidene group at ring A in phytochromobilin, but absent from unfolded phytochrome. A similar spectral shift is seen between phycocyanobilin and phycocyanin, and is taken as evidence that the ethylidene groups are the site of linkage to the polypeptide chain, thus only contributing to the spectral properties after the linkage is cleaved.

Siegelman *et al.* (1966) attempted cleavage of the chromophore from the apoprotein using boiling methanol, but were able to achieve only very low yields. The most effective cleavage technique for other biliproteins is treatment with HBr in trifluoroacetic acid, giving, for example, 100%

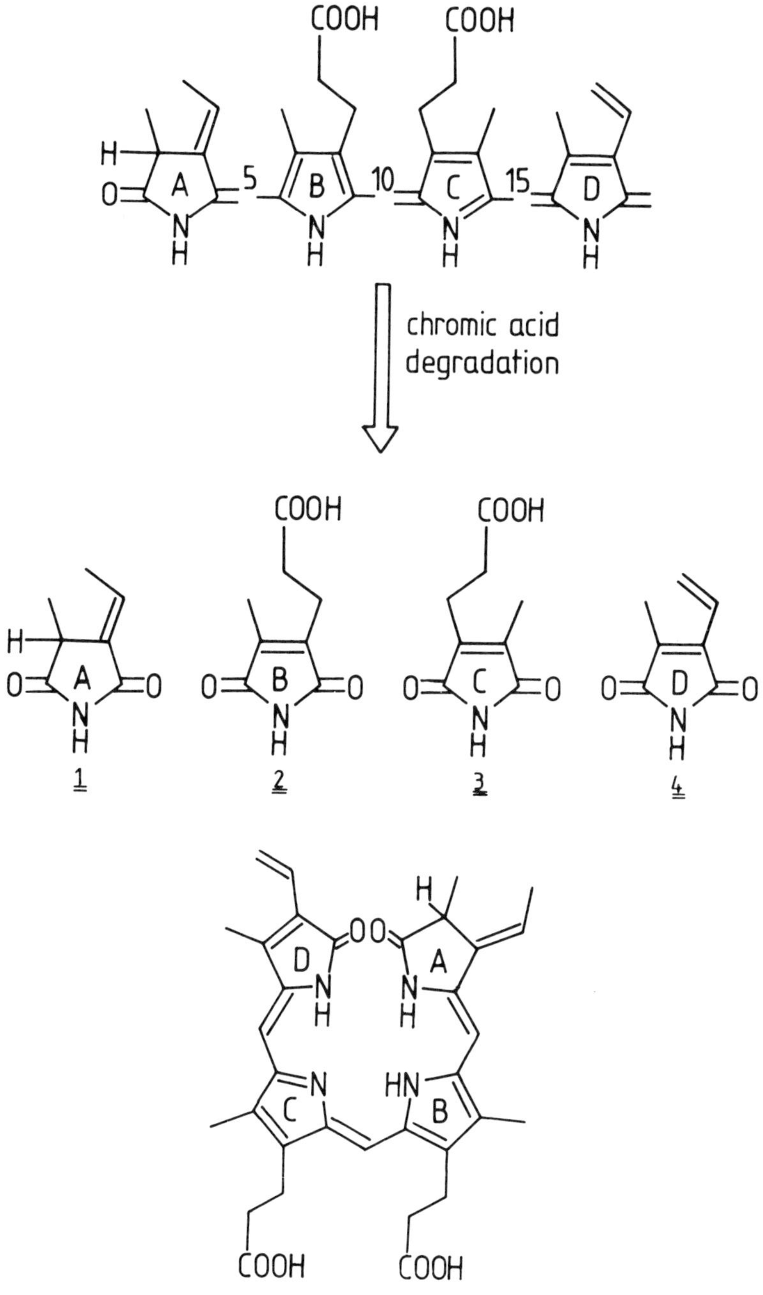

Fig. 6.4 Chromic acid/ammonia degradation of phytochromobilin. Intermediate steps have been omitted (redrawn with permission from Rüdiger, 1986).

release of phycocyanobilin from phycocyanin (Schram and Kroes, 1971). Even this approach was unsuccessful with phytochrome (Kroes, 1970), and successful cleavage was not achieved until proteolytically derived chromopeptides were used as starting material. Rüdiger *et al.* (1980) extracted chromopeptides with cold HBr gas and compared the product with authentic phytochromobilin prepared by total synthesis (Weller and Gossauer, 1980), providing final proof of the structure of phytochromobilin.

Chromopeptide analysis has provided important evidence on the position within the polypeptide chain at which the chromophore is attached, and on the nature of the linkage (Fry and Mumford, 1971; Lagarias and Rapoport, 1980). An undecapeptide was obtained which carried the chromophore, and its sequence was determined (see p. 261). The blue colour of the phytochromobilin was lost at the sequencing degradation step that cleaved cysteine; it was therefore concluded that the linkage to the peptide was via the thiol group (Lagarias and Rapoport, 1980), as was already known for phycocyanobilin. Using synthetic thioether compounds as models, Schoch *et al.* (1974) had shown that a sulphur substituent on the C-3 of ring A would yield a maleimide after the CrO_3-NH_3 oxidation/elimination reaction, whilst substitution at C-3[1] would yield the ethylidene succinamide. Only the latter product has been

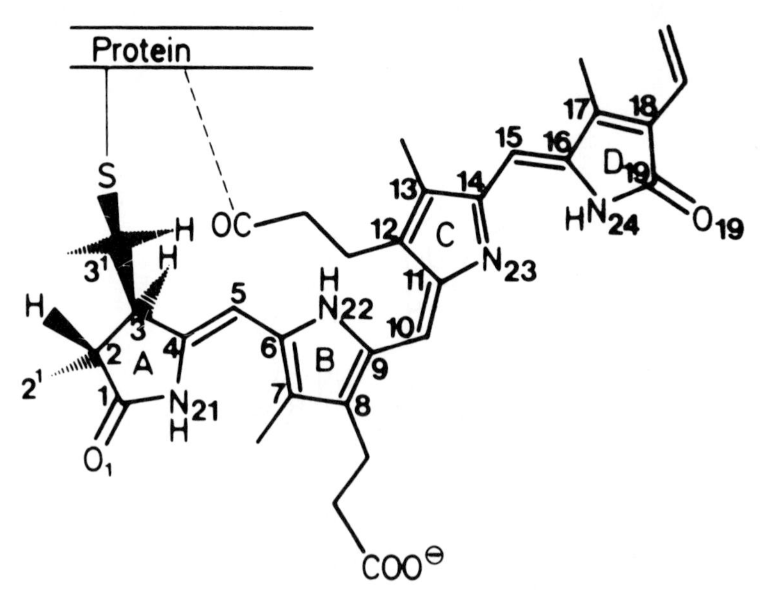

Fig. 6.5 Proposed extended conformation for Pr-phytochromobilin (from Rüdiger and Scheer, 1983 with permission).

obtained from phytochrome (Klein *et al.*, 1977), confirming the localization of the thioether linkage to the protein at C-3[1] (Fig. 6.5). High resolution H[1]-NMR (nuclear magnetic resonance) spectroscopy of the phytochromobilin undecapeptide (Lagarias and Rapoport, 1980) confirmed the hydrogenated ring A and the substitution at C-3[1], and showed that the structure of ring A and the thioether connection to the peptide chain are identical in phytochrome and phycocyanin. It is still not certain whether or not a further covalent linkage of phytochromobilin to the apoprotein exists via one of the propionic acid side chains.

In the intact, native molecule it is likely that the Pr chromophore has an extended conformation, rather than the quasi-cyclic conformation shown here (Burke *et al.*, 1972; Chae and Song, 1975; Song *et al.*, 1979; Rüdiger and Scheer, 1983). Comparison of the spectral properties of native phytochrome with those of proteolytically derived chromopeptides indicates that conformational changes of the chromophore occur on partial degradation. For example, Thümmler and Rüdiger (1983) analysed a Pr chromopeptide in which the absorption band in the red was both broadened and flattened compared to native Pr, with a pronounced reduction of the oscillator strength ratio of the red to blue absorption bands (Fig. 6.3C). The wavelength of maximum absorption differs only slightly, even though the chromopeptide was analysed as the cation at $pH<3$, whereas the native Pr was at pH 7.8. The coincidence of the peak position of absorption by the chromopeptide cation and native Pr was considered by Lagarias and Rapoport (1980) to indicate that the native Pr chromophore exists in a protonated state but this, by itself, does not account for the high chromophore absorption in native Pr, which must be due principally to non-covalent interactions of the chromophore with the native protein. A range of theoretical studies of bile pigments has shown that the oscillator strength ratio of the long-wavelength to the short-wavelength band is low in quasi-cyclic conformations (Scheer *et al.*, 1982) but relatively much higher in chemically fixed extended conformations (Bois-Choussy and Barbier, 1978). Thus, the sharp peak of absorbance in the red, together with the high extinction relative to that in the blue/UV, leads to the conclusion that the Pr chromophore has an extended conformation. Recently, the exact conformation of phycocyanobilin in native phycocyanin has been determined by X-ray crystallography (Schirmer *et al.*, 1985); taking this as a model supports the concept of an extended conformation for phytochromobilin (6, Fig. 6.5). On this view, the native protein is seen as being responsible for the remarkable spectral characteristics of Pr through firstly, stretching the chromophore (i.e. stabilizing an extended configuration) and secondly, protonating the chromophore (i.e. stabilizing the chromophore cation; Rüdiger *et al.*, 1985).

B The Pfr chromophore

Upon photoconversion of native Pr the wavelength of maximum absorption is red-shifted by *ca.* 70 nm (Fig. 6.3A). This was originally thought to indicate that the Pfr chromophore has an extended conjugated double bond system compared to Pr, but currently the most favoured proposal is that the Pr→Pfr photoconversion is a photoisomerization, with the spectral changes being due to non-covalent Pfr-chromophore/protein interactions (Rüdiger *et al.*, 1985). The suggestion that the geometric isomerization of a double bond between either rings A and B or rings C and D is at least partially responsible for the difference between Pr and Pfr was first made by Kroes (1970) and Mumford and Jenner (1971). Z,E (*cis-trans*) isomerization about the bonds between C-4 and C-5 (i.e. rings A and B) or between C-15 and C-16 (rings C and D) is thermodynamically unfavoured, but can be effected photochemically; the photoproducts are then stable. The methine bridge between rings B and C, on the other hand, although also capable of photochemical Z,E isomerization, readily undergoes thermal reversion (Braslavsky *et al.*, 1983) and this, together with other possibilities for internal conversion, competes effectively with the photochemistry at C-4,5 and C-15,16. In the native molecule, therefore, the protein must prevent the processes of internal conversion in the chromophore, or there would be no photoproduct (Rüdiger, 1986). Even so, the photoconversion quantum yield (15–20%) and the low fluorescence yield (<1%; Schaffner *et al.*, 1985) indicate a substantial likelihood (>80%) of apparently unavoidable radiation-less deactivation processes (Braslavsky, 1984).

Investigation of chromopeptides that have different spectral properties depending on whether proteolysis is carried out from Pr or Pfr provides supporting evidence for the Z,E isomerization hypothesis (Rüdiger, 1983; Rüdiger *et al.*, 1983; Thümmler *et al.*, 1983). Pr and Pfr chromopeptides produced in this way by Mumford and Jenner (1971) were found not to differ, but Thümmler *et al.* (1981) showed that the pepsin hydrolysis needed to be carried out in the dark and between pH 2 to 4 in order to prevent chemical conversion of the chromophore from the Pfr to the Pr form. The UV-visible spectra of Pr and Pfr chromopeptides prepared in this manner (Fig. 6.3C; Thümmler and Rüdiger, 1983) were shown to resemble those of Z,E isomeric biliverdin chromophores (Thümmler *et al.*, 1981). Comparison of high-resolution ^{1}H-NMR analyses of chromopeptides derived from phytochrome and from phycocyanin, which could be chemically converted from the all-Z configuration to specific E-isomers, proved that the Pfr-chromopeptide was 15E (Fig. 6.6), whilst the Pr-chromopeptide was 15Z (Fig. 6.6; Rüdiger, 1983; Rüdiger *et al.*, 1983; Thümmler *et al.*, 1983).

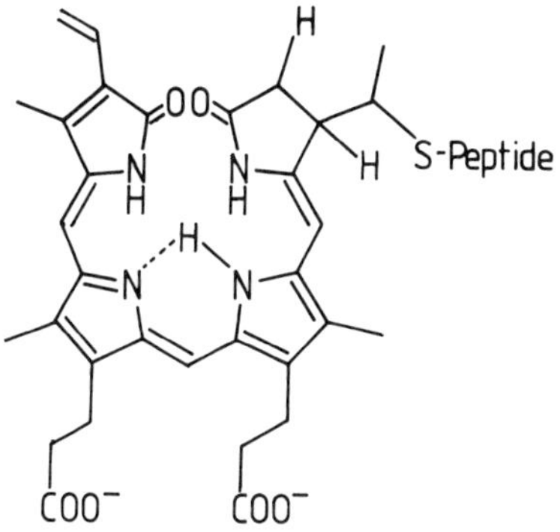

15Z Pr-Peptide

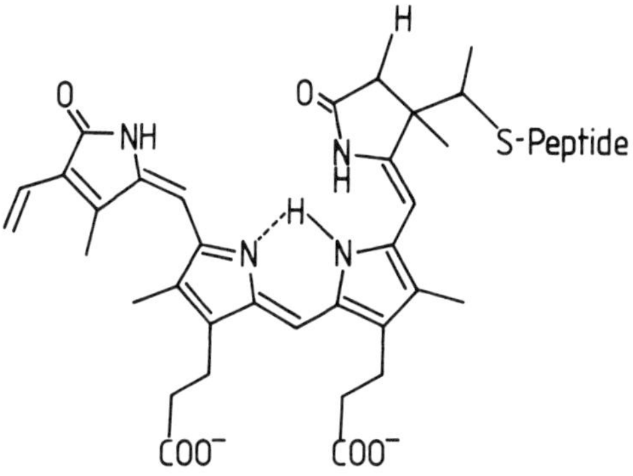

15E Pfr-Peptide

Fig. 6.6 Phytochromobilin structure in a Pr-chromopeptide and a Pfr-chromopeptide (redrawn with permission from Rüdiger, 1986).

C The photoconversion processes

The properties of the phytochrome molecule described so far provide no information on the biological action of the photoreceptor. Since the biological action appears to result from the photoconversion of Pr to Pfr it seems likely that photoresponse will involve recognition of some molecular difference between the two forms of the molecule. For this reason a number of studies aimed at identifying differences in the structure of Pr and Pfr have been initiated.

1 CHROMOPHORE PHOTOCONVERSION INTERMEDIATES

From the above it is clear that in the phototransformation of Pr to Pfr and vice versa a number of changes must be brought about, both to the chromophore, and to the apoprotein. This is reflected in the number of short-lived intermediates that has been recognized, by a variety of techniques, between the two stable forms. Recognition of intermediates has been based principally on spectral properties, thus reflecting either chemical changes of the chromophore, or perturbations of the chromophore brought about by conformational changes of the apoprotein. Other potential alterations of the protein moiety which would not result in changed spectral properties would not be detected by the methods used; some of these are discussed below. Early investigations of partially degraded phytochrome revealed different sequential pathways of intermediates for the Pr→Pfr and the Pfr→Pr pathways. As drawn together by Kendrick and Spruit (1977), the Pr→Pfr pathway was: Pr → lumi-R → meta-Ra → meta-Rb → Pfr, whilst the Pfr→Pr pathway was: Pfr → lumi-F → meta-Fa → meta-Fb → Pr. On this view, the photoreactions (formation of lumi-R and lumi-F) and the relaxations to meta-Ra and meta-Fa are considered to be very rapid events restricted to the chromophore, whilst the subsequent conversions to Pfr and Pr, respectively, are relatively slow conformational modifications involving both chromophore and apoprotein. Particularly slow is the relaxation of meta-Rb, a bleached form, to Pfr, such that under high fluence rates meta-Rb accumulates (Kendrick and Spruit, 1972). Recent work has concentrated on comparing intact and partially degraded phytochromes, principally from *Avena*.

Low temperature spectroscopy of intact 124 kD *Avena* phytochrome led Eilfeld and Rüdiger (1985) to modify the proposed pathway of the Pr→Pfr photoconversion. Irradiation at any temperature below −100°C stabilized a first intermediate, lumi-R, whose absorption spectrum over the 350–370 nm range was recorded. The peak of absorption was shifted to 693 nm, and the extinction coefficient was raised, compared to that of Pr. Raising the temperature to between −100 and −65°C (or irradiating Pr within this temperature range) yielded meta-Ra, with an absorption maximum at 663 nm and a reduced extinction coefficient. Between −65 and −25°C a further intermediate was stabilized, but this was not the bleached form (i.e. meta-Rb) previously reported for partially degraded phytochrome (see Kendrick and Spruit, 1977). Denoted meta-Rc by Eilfeld and Rüdiger (1985), this final intermediate relaxes to Pfr above −25°C.

Apparently, meta-Rb is only formed in 124 kD phytochrome upon irradiation at high fluence rate, but it decays immediately to meta-Rc in the dark. In the reverse direction, only two intermediates, lumi-F and meta-F have been recognized. These proposals for the intermediates are shown in Fig. 6.7. The differences in intermediate pathways for intact and for partially degraded phytochrome (Eilfeld and Rüdiger, 1985) indicate that the 6–10 kD region removed by partial proteolysis may be directly involved in the later stages of the Pr→Pfr photoconversions. In contrast, photophysical studies appear to confirm that the first steps in the phototransformations, in both directions, proceed with the same efficiency, whether or not the major part of the protein is present. Picosecond fluorescence decay measurements on "small", "large" and intact *Avena* phytochromes (Holzwarth *et al.*, 1984; Wendler *et al.*, 1984) showed an identical lifetime of 45 q 3 ps for all three species. Similarly, Ruzsicska *et al.* (1985), using nanosecond flash photolysis showed that the early non-photochemical steps in the Pr→Pfr transformation were unaffected by removal of much of the phytochrome protein by proteolysis. These same investigations (Ruzsicska *et al.*, 1985) provided evidence that for intact *Avena* phytochrome, as previously reported for "large" *Avena* and pea phytochrome (Shimazaki *et al.*, 1980; Furuya, 1983; Inoue *et al.*, 1982; Pratt *et al.*, 1984), the primary photoproduct formed from Pr decays with biexponential kinetics. In partially degraded pea phytochrome, the dark relaxation processes of the primary photoproducts in both transformation directions were shown to consist of three simultaneous components (Shimazaki *et al.*, 1980; Inoue and Furuya, 1985); recent

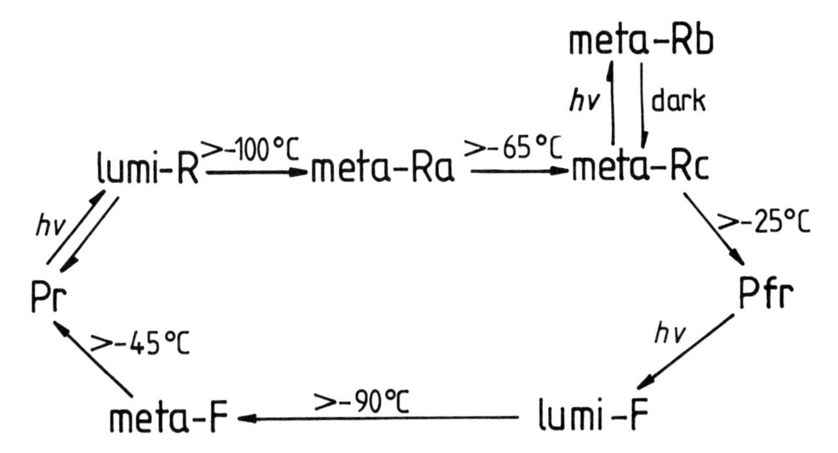

Fig. 6.7 Scheme of phytochrome intermediates as derived from low-temperature studies by Eilfeld and Rüdiger (1985, with permission).

data have confirmed the operation of parallel pathways in intact pea phytochrome (Inoue, 1986).

2 PROTEIN CONFORMATIONAL DIFFERENCES BETWEEN PR AND PFR

Apart from the changes in the configuration of the chromophore, a range of techniques has been used to provide evidence that photoconversion of Pr to Pfr is accompanied by light-induced conformational changes in the protein moiety. One of the most obvious manifestations of differences in the structures of Pr and Pfr is differential sensitivity of the two forms to proteolysis in crude extracts (Vierstra and Quail, 1982a,b, 1983a; Daniels and Quail, 1984). The proteolytic cleavage of the structurally important 6–10 kD NH_2-terminal fragments occurs more readily when phytochrome is in the Pr rather than in the Pfr form. This same region of the molecule is also susceptible to proteolytic cleavage by exogenous endoproteases with phytochrome in the Pr form (Lagarias and Mercurio, 1985). This suggests that several cleavage sites in the 6–10 kD NH_2-terminal subdomain are far more exposed in the Pr than in the Pfr form. This suggestion is supported by studies on the binding of monoclonal antibodies. A number of type 1 monoclonal antibodies, which bind to epitopes located in the NH_2-terminal 6 kD segment, exhibit considerably higher affinities for Pr than for Pfr in quantitative enzyme-linked immunosorbent assay (ELISA), indicative of a reduced accessibility of the epitope in the Pfr form (Cordonnier *et al.*, 1985; Vierstra and Quail, 1986; Holdsworth and Whitelam, 1987). Furthermore, monoclonal antibodies recognizing epitopes on the adjacent 4 kD sub-NH_2-terminal domain, designated type 1′, also show substantially higher affinities for Pr than for Pfr (Holdsworth and Whitelam, 1987). Significantly, one such antibody has been found only to bind to Pr, with the binding being fully reversible by red and far-red light (Holdsworth and Whitelam, 1987). Clearly, there are several sites within the NH_2-terminal 6–10 kD of the phytochrome polypeptide which undergo light-induced conformational changes. Furthermore, these sites are conserved in phytochrome from a range of plant species (Vierstra *et al.*, 1984; Vierstra and Quail, 1985).

In addition to the light-induced conformational changes observed in the NH_2-terminal region, the photoconversion of Pr to Pfr enhances the rate of proteolytic cleavage at sites located near the middle of the polypeptide and in the COOH-terminal half (Vierstra and Quail, 1983a; Daniels and Quail, 1984; Lagarias and Mercurio, 1985). Monoclonal antibodies which exhibit higher affinities for Pfr than for Pr have been mapped to the chromophore-bearing domain, designated type 2, confirming the existence of segments which are preferentially exposed in Pfr in this region of the molecule (Partis and Thomas, 1985; Shimazaki *et al.*, 1986).

Various techniques have been employed in an attempt to determine whether Pr to Pfr photoconversion leads to gross protein conformational changes, but the information obtained is rather contradictory. The migration rates of Pr and Pfr differ on size exclusion high-performance liquid chromatography (HPLC) columns, under a variety of elution conditions (Lagarias and Mercurio, 1985). This has been interpreted to indicate that photoconversion leads to a change in molecular dimensions of the phytochrome molecule. Similar differences in the retention times of Pr and Pfr on size exclusion HPLC columns have been reported by Jones and Quail (1986), although these differences were only observed using Tris buffers. The possibility that the differential retention of Pr and Pfr merely reflects differences in the interaction of the two forms with the chromatographic matrix (Vierstra and Quail, 1986) has not been rigorously tested, but Jones and Quail (1986) did observe the different retentions on other supports. These data seem to be consistent with an increase in the hydrated volume of phytochrome upon Pr to Pfr photoconversion. However, Sarkar *et al.* (1984) found no detectable difference in Stokes' radii of Pr and Pfr for intact, "large" or "small" oat phytochromes using quasi-elastic light-scattering measurements. In the same study, rotational relaxation time measurements for Pr and Pfr also showed little change in hydrated volume (Sarkar *et al.*, 1984). Similarly, evidence from ^{1}H-^{3}H exchange kinetics suggests that the gross conformations of intact and "large" Pr and Pfr forms are virtually identical (Song, 1985). The contradictory evidence from size exclusion HPLC and quasi-elastic light-scattering measurements for a photoconversion-induced change in the shape of the phytochrome molecule has yet to be resolved. It is also rather disturbing that the two methods give such widely different estimates of the Stokes' radius of phytochrome (see above); whether this reflects differences inherent in the methodologies or differences in the phytochrome preparations needs to be determined.

The reactivity of Pr and Pfr toward exogenously applied chemical probes has also provided information on photoconversion-induced changes in phytochrome structure, as well as on the accessibility of the chromophore. Tetranitromethane bleaches the Pfr form of intact oat phytochrome eight times faster than the Pr form (Hahn *et al.*, 1984). This differential oxidation of Pr and Pfr was also seen with degraded phytochrome, with "large" phytochrome showing a 40-fold difference for Pr and Pfr (Hahn *et al.*, 1984). Overall rates of bleaching of Pr and Pfr increase with increasing proteolysis. The differential bleaching of Pr and

Pfr could mean either that the chromophore is physically more accessible in the Pfr form, or that the Pfr chromophore is intrinsically more chemically reactive than the Pr chromophore. To distinguish between these possibilities, Thümmler *et al.* (1985) investigated the reactivity of the Pr and Pfr forms of intact oat phytochrome to ozone, a small molecule which should penetrate freely to the chromophore site. Since ozone was found to bleach the Pr and Pfr forms at virtually the same rate, it was concluded that the differential oxidation of intact Pr and Pfr by tetranitromethane was not the consequence of intrinsically different reactivities of Pr and Pr chromophores. This conclusion is supported by the observation that very small (*ca.* 2.5 kD) chromopeptides, derived by proteolysis of phytochrome in either the Pr or Pfr form, are bleached at the same rate by tetranitromethane (Thümmler *et al.*, 1985). The interpretation of these studies, that the Pfr chromophore is more exposed on the protein surface than is the Pr chromophore (Song, 1985) is not supported by the data of Baron and Epel (1983) for the permanganate oxidation of the Pr and Pfr forms of intact and degraded oat phytochrome. Although the Pfr form of "large" phytochrome was five times more sensitive to permanganate oxidation than the Pr form, such differences were not observed for the intact molecule. Baron and Epel (1983) suggest that in the intact molecule both the Pr and Pfr chromophores are in a protected environment and that only after proteolytic removal of the 6–10 kD NH_2-terminal fragment does the Pfr chromophore become relatively more exposed.

Tetranitromethane and permanganate are assumed to bring about bleaching of phytochrome through direct oxidation of the chromophore. However, since modification of the protein moiety of phytochrome can also induce bleaching, it is possible that at least some of the observed bleaching with these reagents is due to protein modification. The hydrophobic fluorescence probe, 8-anilino-naphthalene-1-sulphonate (ANS) bleaches phytochrome through modification of the protein moiety (Hahn and Song, 1981; Eilfeld and Rüdiger, 1984). With partially degraded "large" oat phytochrome, ANS causes rapid and almost complete bleaching of the Pfr form but has a much reduced effect on Pr (Hahn and Song, 1981; Thümmler and Rüdiger, 1984). This observation was incorporated into a model for phytochrome transformation in which photoconversion of Pr to Pfr was proposed to lead to the exposure of a hydrophobic region on the surface of the molecule, to which a hypothetical receptor could bind (Song, 1985). In contrast, with native phytochrome, the presence of ANS leads to only limited bleaching of the Pfr form and causes a slight blue shift in the Pr absorption spectrum (Eilfeld and Rüdiger, 1984). However, immediate bleaching occurs with ANS-treated

Pr upon photoconversion with red light. The product of photoconversion has spectral properties similar to that of a bleached intermediate which can be formed along the normal Pr to Pfr photoconversion pathway, and has been termed P^{640} (Eilfeld and Rüdiger, 1984). Based upon these observations, Eilfeld and Rüdiger (1984) proposed that the hydrophobic site to which ANS binds is also involved in Pfr chromophore–protein interaction, but is largely inaccessible *during* phototransformation such that, once bound, ANS prevents normal Pfr chromophore–protein interaction and no 730 nm absorption appears. The model of Eilfeld and Rüdiger (1984) also accounts for the contrasting observations on the effects of ANS on "large" phytochrome by proposing that the Pfr chromophore normally interacts with the 6–10 kD NH_2-terminal segment, and that loss of the segment leads to increased exposure of the hydrophobic site in Pfr, relative to Pr. This is consistent with several previous reports that for partially degraded phytochromes the Pfr form is more hydrophobic than the Pr form (e.g. Tokutomi *et al.*, 1981; Yamamoto and Smith, 1981).

Possible structural differences between Pr and Pfr have been investigated by spectroscopic techniques. CD spectra measurements on intact oat phytochrome show differences between Pr and Pfr in both the far-UV and UV-visible regions (Litts *et al.*, 1983; Vierstra *et al.*, 1987). In a recent study CD spectra of Pr and Pfr for intact and "large" oat phytochrome were compared. The data obtained show that the presence of the 6–10 kD NH_2-terminal segment was required in order to observe a photoreversible difference in the UV region of the spectrum (Vierstra *et al.*, 1987). Secondary structure predictions from the spectra suggest that Pr to Pfr photoconversion leads to a 3% increase in α helical content of phytochrome. These differences were not observed in earlier studies with intact oat phytochrome (Litts *et al.*, 1983).

Several differences between Pr and Pfr reported for partially degraded "large" phytochrome (for reviews see Pratt, 1982; W. O. Smith, 1983) have yet to be reinvestigated for the intact molecule. Since the 6–10 kD NH_2-terminal domain is known to have such a dramatic effect on many of the properties of the phytochrome molecule it would be premature to attach too much significance to these observations.

4 BIOSYNTHESIS AND DEGRADATION

Phytochrome is present at low levels in dry seeds and its content, as assayed spectrophotometrically, increases rapidly during imbibition owing to rehydration of existing phytochrome molecules, rather than synthesis of new (Spruit and Mancinelli, 1969; Tobin *et al.*, 1973). Further increases in phytochrome content occur during germination and seedling development as a result of *de novo* synthesis of phytochrome in the Pr form (Quail *et al.*, 1973b). Phytochrome accumulates to relatively high levels in dark-grown tissues (up to 0.5% of the total soluble protein) until a plateau is reached, representing a steady-state balance between synthesis and degradation of Pr (Quail *et al.*, 1973b). The transfer of dark-grown tissues to the light leads to a rapid decline in phytochrome content because of a much accelerated rate of degradation of Pfr compared to Pr (Quail *et al.*, 1973b; Hunt and Pratt, 1979). This rapid degradation of Pfr occurs with a half-time in the range of 20 min–4 h, depending upon species (Frankland, 1972; Kidd and Pratt, 1973; Schäfer *et al.*, 1973). Under continuous white light irradiation a new constant level of phytochrome, *ca.* 1–3% of the original dark content, is rapidly reached, presumably representing a balance between Pr synthesis and Pfr degradation (Hunt and Pratt, 1979). The half-time for phytochrome degradation in continuous white light is dependent upon photon fluence rate. At relatively high photon fluence rates phytochrome degradation is substantially reduced, presumably as a consequence of the accumulation of photoconversion intermediates which are not susceptible to degradation (Kendrick and Spruit, 1972; Smith *et al.*, 1987). Phytochrome reaccumulates, owing to *de novo* Pr synthesis, when light-treated tissues are returned to darkness (Quail *et al.*, 1973b; Shimazaki *et al.*, 1983).

From these and other data a model for the regulation of phytochrome levels has emerged which incorporates a constantly steady synthesis of Pr, slow degradation of Pr and rapid degradation of Pfr (e.g. Schäfer, 1975; Wall and Johnson, 1983). This view, in which phytochrome concentration is modulated strictly at the protein level, implies a more or less constant level of phytochrome mRNA supporting the constant synthesis of Pr. However, recent studies using *in vitro* translation of poly A^+RNA followed by immunoprecipitation of translation products, have shown that the levels of translatable phytochrome mRNA are regulated by phytochrome itself (Gottmann and Schäfer, 1982; Colbert *et al.*, 1983; Otto *et al.*, 1983, 1984). In etiolated seedlings of oat, sorghum and pea, irradiation with red light leads to a rapid decrease in translatable phytochrome mRNA levels whilst subsequent far-red irradiation wholly or partially reverses this effect.

More recently, the availability of cDNA hybridization probes has enabled this autoregulatory control to be studied in more detail. Indeed, phytochrome cDNA clones were initially selected for by virtue of the fact that certain clones prepared from mRNA isolated from etiolated oat seedlings showed reduced hybridization to cDNA prepared from red-irradiated tissue compared with cDNA from etiolated tissue (Hershey *et al.*, 1985). Colbert *et al.* (1985) investigated light-induced changes in the physical abundance of phytochrome mRNA in oat seedlings and used an SP-6-derived phytochrome-mRNA transcript to perform quantitative analyses. The phytochrome-mRNA represents about 0.1% of the total poly A^+RNA in etiolated oats and this level declines rapidly following a red light pulse such that, after 2 h, phytochrome-mRNA abundance is reduced to less than 10% of the dark level (Colbert *et al.*, 1985). The decline in phytochrome-mRNA abundance can be detected within 15–30 min. Significantly, a pulse of far-red light itself leads to a reduction in the level of phytochrome-mRNA, to about 50% of the dark level. This suggests that the phytochrome regulation of phytochrome gene expression is controlled by a very low fluence response (VLFR, see below), as observed for certain other phytochrome-controlled genes (Kaufman *et al.*, 1984, 1985). In order to establish whether or not the reduced abundance of the phytochrome-mRNA is a consequence of reduced transcription of the gene, run-off transcription assays have been performed with isolated nuclei. These experiments indicate that regulation of transcription of the phytochrome gene is only partially responsible for the reduction in phytochrome-mRNA levels, and that there is presumably some post-transcriptional regulation via either mRNA processing or mRNA stability (Quail, 1986; Vierstra and Quail, 1986).

The existence of autoregulatory control of phytochrome-mRNA abundance has necessitated the revision of schemes which account for control of phytochrome abundance. A more complete scheme would be:

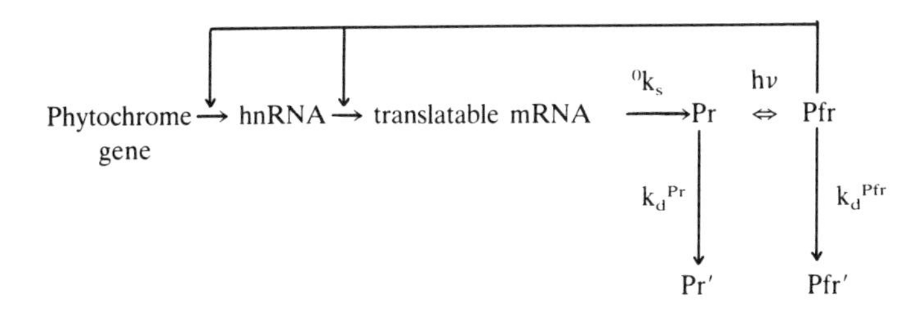

where: ^0k_s is the zero-order rate constant of phytochrome synthesis and k_d^{Pr} and k_d^{Pfr} are rate constants of Pr and Pfr degradation, respectively.

Although autoregulation has been observed for a number of dark-grown seedlings there are substantial differences in the extent to which it occurs. From *in vitro* translation studies it is apparent that far-red light alone decreases levels of translatable phytochrome-mRNA in etiolated oat seedlings (Colbert *et al.*, 1983) whilst it has no effect in etiolated pea seedlings (Otto *et al.*, 1984). Similarly, Northern blot analyses have revealed that in both maize and tomato seedlings red light leads to only a limited and comparatively short-lived decrease in phytochrome-mRNA abundance, in contrast to the dramatic reductions observed with oat seedlings (Quail, 1986). Clearly, the extent to which autoregulation may control the levels of phytochrome present in tissues is species-dependent.

The synthesis of phytochrome apoprotein could potentially be regulated by the synthesis of the tetrapyrrole chromophore. Coordination in the synthesis of prosthetic group and apoprotein has been demonstrated for several other porphyroproteins, including haemoglobin (Kruh and Borsook, 1956) and cytochrome C (Colleran and Jones, 1973). Similarly, synthesis of apoprotein and chlorophyll(ide) seem to be coordinated in the formation of the light-harvesting chlorophyll a/b protein (Apel and Kloppstech, 1980). The question of a similar coordination in the case of phytochrome has been addressed by use of the transaminase inhibitor, gabaculine (5-amino-1,3-cyclohexadienyl carboxylic acid). Gabaculine, which inhibits chlorophyll biosynthesis by means of irreversibly blocking the transamination of glutamate-1-semialdehyde to 5-aminolaevulinic acid, the porphyrin precursor (Flint, 1984), inhibits the accumulation of spectrophotometrically assayable phytochrome in a number of dark-grown seedlings (Gardner and Gorton, 1985). However, gabaculine does not affect synthesis of the phytochrome apoprotein, detected immunochemically, in pea seedlings (Jones *et al.*, 1986; Konomi and Furuya, 1986). In the presence of gabaculine, pea seedlings accumulate normal levels of the phytochrome apoprotein, implying that chromophore-less phytochrome is as stable as Pr and is not recognized as an aberrant protein by the cellular protein-degradation machinery. So, although the normal accumulation of Pr, in dark-grown seedlings, involves *de novo* synthesis of both the apoprotein and the chromophore, the synthesis of the protein moiety is not obligatorily coupled to the synthesis of the tetrapyrrole chromophore. Since the synthesis of phytochrome apoprotein is regulatable by light, it seems possible that chromophore synthesis may be dependent upon apoprotein synthesis, although this has yet to be tested.

The photoconversion-induced degradation of phytochrome, frequently referred to as “destruction”, was originally considered to be specific for the Pfr form (Mohr, 1972; Smith, 1975). It is now apparent that Pr, which has been cycled through Pfr, also undergoes rapid degradation (Dooskin and Mancinelli, 1968; Stone and Pratt, 1979; Schäfer, 1981). The rate of light-induced Pr degradation is the same as that for Pfr, but the extent of degradation is somewhat reduced (Stone and Pratt, 1979). This suggests that the photoconversion of Pr to Pfr in the cell may potentiate the proteolytic degradation of phytochrome irrespective of form, rather than there being a Pfr-specific protease. Such a potentiation mechanism could be the well characterized photoconversion-induced change in the cellular location of phytochrome (see below).

Much of our current understanding of the nature and kinetics of *in vivo* phytochrome degradation has come from spectrophotometric measurements of phytochrome *in vivo*, although, more recently, immunochemical techniques have been employed, and have confirmed that loss of spectrophotometrically detectable phytochrome is accompanied by loss of phytochrome protein (Coleman and Pratt, 1974; Hunt and Pratt, 1979). Spectrophotometric studies have shown that for several dicotyledonous plant species phytochrome degradation in the dark, following a pulse of red light, is a first order reaction with a half-time in the range of 20 min–4 h (Frankland, 1972; Schäfer *et al.*, 1973). Under continuous irradiation, phytochrome degradation rate is strongly dependent upon the relative amount of Pfr established by the irradiation (Pfr/[Pr + Pfr]), such that phytochrome loss is most rapid under red light and least rapid under far-red light (Kendrick and Frankland, 1969). Under some circumstances, deviations from first order kinetics have been observed. The rate of phytochrome degradation is reduced under continuous high fluence rate irradiation (Kendrick and Frankland, 1969; Kendrick and Spruit, 1972; Smith *et al.*, 1987), presumably due to the accumulation of photoconversion intermediates. Recently, reinvestigations of phytochrome degradation in a number of dicotyledons have revealed biphasic kinetics (Heim *et al.*, 1981; Brockmann and Schäfer, 1982). These data have been interpreted to indicate the existence of two pools of phytochrome: a large, labile pool of more than 90% of the phytochrome and a small, stable pool of a relatively constant size during seedling development. These two pools could be equivalent to the major phytochrome species predominating in etiolated and light-grown plants.

In those monocotyledonous species which have been examined, different degradation kinetics have been observed. The major differences are that in maize and oat seedlings, phytochrome degradation does not follow simple first order kinetics, and the degradation rate under continuous irradiation is saturated at low Pfr/[Pr + Pfr] (Butler *et al.*, 1963; Pratt and Briggs, 1966; Kendrick, 1972; Schäfer *et al.*, 1975). The precise mechanism by which photoconversion of Pr to Pfr in the cell leads to the proteolytic degradation of phytochrome is not known. It has been known for many years that the process is temperature- and oxygen-dependent (Butler and Lane, 1965) and is inhibited by the respiratory inhibitors, cyanide and azide (Butler and Lane, 1965). However, the insensitivity of phytochrome degradation to the respiratory uncoupler 2,4-dinitrophenol (Bradley and Hillman, 1966) suggests that the oxygen requirement is not related to the respiratory chain or oxidative phosphorylation. An attractive hypothesis to account for the observation that degradation is not specific for Pfr is that photoconversion of phytochrome in the cells of dark-grown tissues potentiates the degradation of phytochrome as a consequence of a change in its cellular location (Verbelen *et al.*, 1982; Whitelam *et al.*, 1984; McCurdy and Pratt, 1986b). It is well established that photoconversion of Pr to Pfr results in the "sequestering" of phytochrome from a diffuse cytoplasmic distribution to become associated with discrete structures within the cytoplasm (MacKenzie *et al.*, 1975; Epel *et al.*, 1980; Saunders *et al.*, 1983; McCurdy and Pratt, 1986a, 1986b; Speth *et al.*, 1986). This sequestering reaction, which occurs very rapidly upon irradiation of dark-grown tissues, could represent a mechanism by which phytochrome molecules are targeted for intracellular degradation. Circumstantial evidence in support of this hypothesis includes the observations that the kinetics for photoreversibility of the two processes are similar and, following red irradiation, sequestered phytochrome disappears without reappearing in the cytoplasm (Whitelam *et al.*, 1984). Although the structures with which sequestered phytochrome associates have yet to be identified are unknown, it is tempting to speculate that they are cytoplasmic sites of protein degradation and that targeting of proteins for degradation involves transfer to these sites (the "garbage can" hypothesis) as a crucial intermediate step.

More recent evidence, from studies of the light-induced degradation of phytochrome in oat seedlings, suggests that ubiquitination of phytochrome may also be an intermediate step in degradation (Vierstra *et al.*, 1986). Ubiquitin, as the name suggests, is a ubiquitous polypeptide (8500 D), known to be involved in targeting proteins for cellular degradation in animal systems (Hershko *et al.*, 1980; Wilkinson *et al.*, 1980). Ubiquitin-dependent protein degradation involves an ATP-dependent coupling of ubiquitin to the substrate protein, involving a number of intermediary proteins, followed by recognition of ubiquitin-protein conjugates by the cell's soluble proteases which degrade the attached protein (Hershko *et al.*, 1983). Vierstra *et al.* (1986) have shown that, following red light irradiation of oat seedlings, phytochrome appears as a series of bands, larger than its native size, on immunoblots. These larger bands are also immunostained by antibodies raised to oat- and human-ubiquitin, suggesting that they are phytochrome-ubiquitin conjugates. The half-life of these conjugates is shorter than that of phytochrome, consistent with the hypothesis that they may be degradation intermediates. It may be significant that antibodies against ubiquitin also immunostain the cytoplasmic structures with which sequestered phytochrome associates following irradiation *in vivo* (E. Schäfer, unpublished observations). One could speculate that these cytoplasmic structures are sites of ubiquitination, where the complex process of conjugation occurs, involving (in animals) at least three separate proteins and the consumption of ATP.

5 PHYTOCHROME FROM LIGHT-GROWN PLANTS

Phytochrome regulates development throughout the life cycle of the plant; indeed the regulation of vegetative growth in fully de-etiolated, green seedlings is amongst the most intensively studied phytochrome responses (H. Smith, 1982). However, very little work has been done on the characterization of phytochrome from light-grown plants. There are two principal reasons for this. First, light-grown plants provide a far less abundant source of phytochrome since they typically contain 50–100 times less phytochrome than etiolated plants of the same species (Hunt and Pratt, 1979; Pratt, 1982). Secondly, the chlorophylls present in extracts of light-grown, green seedlings prevent conventional spectrophotometric assays of phytochrome through screening and fluorescence. These problems can now be overcome with the advent of sensitive immunochemical assays for phytochrome (Pratt, 1982, 1984) and with the development of procedures for the removal of chlorophyll from extracts containing phytochrome (Bolton and Quail, 1981; Tokuhisa and Quail, 1983; Tokuhisa *et al.*, 1985). Using these techniques, evidence is accumulating that phytochromes extracted from etiolated and light-grown seedlings are distinct molecular species. Tokuhisa and Quail (1983) first reported that phytochrome isolated from green oat seedlings differed, both spectrally and immunochemically, from phytochrome from etiolated oats. Altered spectral characteristics of phytochrome from light-grown oat seedlings, with the Pr absorption maximum being blue-shifted by 10–15 nm, had previously been reported by Jabben and Deitzer (1978) using herbicide-bleached seedlings. Evidence for immunochemically distinct phytochrome species in dark- and light-grown seedlings has also been obtained by Shimazaki *et al.* (1983) and Thomas *et al.* (1984). Both groups observed that ELISA tests with antibodies directed against phytochrome isolated from etiolated oats failed to quantify accurately the spectrophotometrically assayable phytochrome isolated from green oat shoots.

These earlier observations have been extended by a number of laboratories, and definitive evidence for the existence of discrete molecular species of phytochrome in etiolated and light-grown seedlings has been achieved. Phytochrome isolated from green oat shoots has a Pr absorption maximum which is blue-shifted to 652 nm (Shimazaki and Pratt, 1985; Tokuhisa *et al.*, 1985) whilst the Pfr absorption maximum is unaltered at about 730 nm (Tokuhisa *et al.*, 1985; Cordonnier *et al.*, 1986a). As for phytochrome from etiolated tissue, that from green shoots is susceptible to rapid partial proteolysis in crude extracts, leading to a blue-shift in the Pfr absorption maximum (Shimazaki and Pratt, 1985; Cordonnier *et al.*, 1986a). However, in the case of green tissue extracts, this proteolysis does not depend upon whether phytochrome is in the Pr or Pfr form (Cordonnier *et al.*, 1986a). Immunochemical analyses reveal that little of the phytochrome extracted from green oat shoots is recognizable by antibodies, either monoclonal or polyclonal, raised to etiolated tissue phytochrome. A maximum of about 30% of the phytochrome in extracts of green oat shoots was immunoprecipitated by saturating levels of such antibodies (Shimazaki and Pratt, 1985; Tokuhisa *et al.*, 1985). This suggests that phytochrome isolated from green tissue is itself composed of at least two distinct molecular species of differing abundance and differing molecular masses. The most abundant species (*ca.* 70% of total) has an M_r of 118 kD, about 6 kD smaller than etiolated-tissue phytochrome, whilst the lower abundance species (*ca.* 30% of total) has an M_r of 124 kD (Tokuhisa *et al.*, 1985). The lower abundance 124 kD species may represent "contaminating" etiolated-tissue phytochrome or, alternatively, it may be yet another separate molecular species. Cordonnier *et al.* (1986a) and Shimazaki and Pratt (1985) have provided convincing immunochemical evidence that both the high and low abundance phytochrome species in green tissue extracts are distinct from etiolated tissue phytochrome. Significantly, in these studies the phytochrome species from green tissue have an M_r of 124 kD, the same as etiolated-tissue phytochrome. The reason for the discrepancy in the monomeric molecular weights reported by the two groups is not known; however Cordonnier *et al.* (1986a) do point out that 124 kD green-tissue phytochrome is readily degraded *in vitro* to yield approximately 118 kD fragments. Despite these differences both groups have shown that etiolated- and green-tissue phytochromes have quite different proteolytic peptide maps (Cleveland maps), indicative of possible differences in the primary structure of the two (or more) proteins (Tokuhisa *et al.*, 1985; Cordonnier *et al.*, 1986a).

These differences between phytochromes isolated from etiolated and green tissues are not confined to oat seedlings. Immunochemically distinct pools of phytochrome from etiolated and green pea seedlings were reported by Shimazaki and Pratt (1985). Similarly, phytochrome isolated from light-grown pea seedlings by Abe *et al.* (1985) was found to be composed of two spectrally identical, but immunochemically discrete species of phytochrome, which produced different digestion products by Cleveland mapping. One of these phytochromes from green seedlings appears to be identical to phytochrome from etiolated peas.

Although the evidence for the existence of structurally discrete forms of phytochrome is fairly convincing, the basis for these structural differences is not known. It is possible that the different molecular species are the products of separate and distinct genes or that the differences arise by post-transcriptional or post-translational modification of the product of a common gene. That the differences may have arisen by trivial post-homogenization modification of a single protein in green-tissue extracts has been discounted by a number of control experiments (Tokuhisa *et al.*, 1985; Cordonnier *et al.*, 1986a). No definitive evidence has yet been presented on the question of the origin of the differences between etiolated- and green-tissue phytochromes. However, Colbert *et al.* (1985) have shown by RNA blot analysis that poly A^+mRNA from green oat seedlings contains a single hybridizable mRNA species the same length (4.2 kb) as that in etiolated plants, but about 50 times less abundant. This quantitative difference in mRNA abundance correlates well with the difference in phytochrome abundance in etiolated and light-grown plants. The similarity in size of the hybridizable mRNA from etiolated- and green-tissue suggests a similar gene product, although the extent of sequence homology is not known.

6 INTERCELLULAR AND INTRACELLULAR LOCATION OF PHYTOCHROME

A variety of techniques, including spectrophotometry following whole plant fractionation, microbeam irradiation and immunocytochemistry, have been employed in order to determine the intercellular localization of phytochrome (Pratt, 1986). A general finding from studies using these techniques has been that the highest levels of phytochrome are found in relatively young expanding cells recently derived from meristematic zones. In particular, phytochrome is most abundant near coleoptile tips, at the coleoptilar node and in root caps and adventitious root tips of grass seedlings. In dicotyledonous seedlings phytochrome is most abundant in epicotyl and hypocotyl hooks (Pratt, 1986). Within tissues, phytochrome is distributed heterogeneously, being more prominent in some cell types than in others. This heterogeneity of distribution at the tissue level appears to be species-dependent.

Information on the intracellular localization of phytochrome has also been derived from techniques of microbeam irradiation, subcellular fractionation/spectrophotometry, and immunocytochemistry. Microbeam irradiation experiments have been limited mainly to the green alga *Mougeotia*, protonemata of the fern *Adiantum* and *Dryopteris*. In *Mougeotia*, the phytochrome that controls chloroplast orientation appears to be located at the cell periphery either in or on the plasmalemma. Furthermore, this phytochrome exhibits dichroic orientation, the absorption dipole changing orientation upon photoconversion, suggesting that the molecules are "fixed" to some cellular component which does not move as the chloroplast rotates (Haupt, 1970). The phytochrome controlling polarotropism in the protonemata of the ferns *Dryopteris* (Etzold, 1965) and *Adiantum* (Wada *et al.*, 1983) also exhibits dichroic orientation. Dichroic orientation of phytochrome is not observable in higher plants, although it would technically be very difficult to detect. In view of the currently prevailing view of a cytosolic location for phytochrome in higher plant cells (see below), the incontrovertible evidence for membrane association in some filamentous organisms is difficult to reconcile. One possibility is that phytochrome in such filamentous plants has taken on through evolution the rather specialized function of directional perception, and therefore has acquired a unique location at the cell membrane.

Subcellular fractionation followed by spectrophotometry has indicated that phytochrome is associated with most subcellular organelles (Pratt, 1986). The problem with these types of studies is in establishing whether a particular organelle fraction is enriched for phytochrome, over the levels found in crude extracts, and then of establishing whether the association is artifactual or not. In most cases the data are insufficient for any significance to be attached to them. There have been a few reports which indicate that not only is phytochrome associated with a particular organelle, it also modulates the activity of that organelle. For example, two independent groups have reported that phytochrome associated with isolated etioplasts regulates the efflux of gibberellins from the etioplasts (Cooke *et al.*, 1975; Cooke and Kendrick, 1976; Evans and Smith, 1976; Hilton and Smith, 1980). Although real progress in understanding the mechanism of phytochrome action must ultimately be dependent upon the elaboration of a reliable cell-free system in which phytochrome exerts its controlling effect, the levels of both phytochrome and gibberellin-like substances in these reports were so close to the theoretical limits of detection that further detailed analysis has been impossible. Attempts are currently being made to reinvestigate this intriguing phenomenon using immunochemical techniques (M. Malone and H. Smith, unpublished work). Phytochrome has also been reported to be associated with mitochondrial activities *in vitro* (Georgevich *et al.*, 1977; Cedel and Roux, 1980; Slocum and Roux, 1980). Phytochrome apparently only becomes associated with mitochondria following red irradiation *in vivo*. The mitochondria-associated phytochrome is reported to be resistant to trypsin and chymotrypsin proteolysis, and so it has been proposed that red irradiation leads to the import of phytochrome into mitochondria (Serlin and Roux, 1986). These authors have speculated that photoconversion of Pr to Pfr leads to the exposure of an internal insertion sequence which facilitates transport of the phytochrome molecule across the mitochondrial membrane. Primary sequence data for phytochrome show that there are no regions of the polypeptide with the features of an insertion sequence.

The idea that phytochrome may become associated with cellular organelles only following photoconversion of Pr to Pfr has been in circulation for many years. The phenomenon of light-induced pelletability, i.e. the association of previously soluble phytochrome with particulate material, was studied extensively during the 1970s. The basic observation is that conversion of Pr to Pfr *in vivo*, followed by extraction in buffers containing relatively high levels of Mg^{2+}, leads to up to 70% of the phytochrome becoming associated with pelletable material (Quail *et al.*, 1973a; Pratt and Marme, 1976). Since the presence of Mg^{2+} in the buffers, necessary to observe the effect, also leads to aggregation of cellular membranes and organelles it was not possible further to fractionate the particulate material. However, it was always assumed that phytochrome was associating with membrane material. The known primary structure of phytochrome makes a direct association of the molecule with the lipid component of membranes very unlikely. Indeed, recent immunocytochemical evidence suggests that the bulk of the pelletable phytochrome is not associated with membrane (McCurdy and Pratt, 1986b). Carrying out similar pelletability experiments with oat seedlings in the absence of Mg^{2+} reveals a much smaller amount of membrane-associated phytochrome, the level of which can be increased up to threefold by prior *in vivo* red irradiation, freely reversible by far-red (Watson and Smith, 1982a,b; Napier and Smith, 1987a,b). The phytochrome induced to associate by red-irradiation (*ca.* 8% of the total) could only be removed by detergent treatment, suggesting that a small population of cytosolic phytochrome becomes capable of associating with particulate/membrane material when converted from Pr to Pfr. A model based on the aggregation of phytochrome dimers was proposed to account for the precise stoichiometry of the associations (Napier and Smith, 1987b). Since the nature of the phytochrome-binding component in the crude particulate preparations is quite unknown, and since there was little, if any, enrichment of phytochrome compared to the crude extract, these data may represent a complex artifact; if so, the peculiar behaviour of phytochrome under the experimental conditions used requires explanation.

Phytochrome has been localized in the cell by both light and electron microscopy following immunolabelling (Coleman and Pratt, 1974;

McCurdy and Pratt, 1986a,b; Speth *et al.*, 1986). In unirradiated plant cells, phytochrome is distributed diffusely throughout the cytoplasm, not apparently associated with organelles and membranes (McCurdy and Pratt, 1986b; Speth *et al.*, 1986). After photoconversion of Pr to Pfr, the majority of the phytochrome becomes sequestered into electron-dense areas within the cytoplasm. This sequestering appears to occur in two distinct stages: a rapid sequestering into small loci within a few seconds of irradiation, followed by a gradual aggregation of these loci into larger areas (McCurdy and Pratt, 1986a,b). The electron-dense structures, to which sequestering phytochrome associates, are composed of amorphous, granular material and are apparently not found in unirradiated plant cells (McCurdy and Pratt, 1986a,b; Speth *et al.*, 1986). In darkness, after a red light pulse, the phytochrome-associated structures, as well as the phytochrome itself, gradually disappear, perhaps indicating the involvement of sequestering in the degradation of phytochrome (see p. 284). Significantly, similar structures with sequestered phytochrome have also been observed in the pelletable material obtained by homogenization of red light-irradiated tissue in Mg^{2+}-containing buffers (McCurdy and Pratt, 1986b). Since these structures were the only component of the particulate material to be immunostained for phytochrome it seems probable that the two phenomena of sequestering and pelletability are two manifestations of the same cellular event. The components of the amorphous, electron-dense structures have yet to be identified, as has the biological significance of sequestering, although the circumstantial evidence for its involvement in phytochrome degradation is fairly compelling.

7 THE MECHANISM OF ACTION OF PHYTOCHROME

The molecular mechanisms by which the light-induced changes in the properties of phytochrome are transduced within the cell to bring about photomorphogenesis are unknown. Elucidation of the primary events involved in signal transduction will require detailed knowledge of both the molecular properties of the photoreceptor and of the early events of phytochrome-controlled responses. Efforts towards establishing the link between phytochrome and its transduction chain would be substantially aided by a reliable *in vitro* system in which phytochrome could be shown to exert some biologically meaningful action. At present, a convenient and reproducible system does not exist and so the question of the molecular mechanism of action of phytochrome must be a matter of speculation rather than experimental evidence. Speculative notions on the mechanism of action of phytochrome are inevitably complicated by the diverse array of response types in both etiolated and light-grown plants and by variations within each response type. For example, phytochrome-mediated responses of etiolated tissues can be broadly divided into three categories, depending upon the fluence requirements (Kendrick and Kronenberg, 1986). Classical red/far-red reversible responses require red light fluences in the range 1–10^3 μmol m^{-2}, and are termed low fluence responses (LFR). Within this response type, even for an individual organism, there is great variability of kinetics and fluence requirements. Many phytochrome responses require continuous irradiation, and fluence rate rather than fluence becomes the determining factor. Such responses are termed high irradiance responses (HIR), but again variability is high, particularly with respect to the wavelengths which give maximum effect. In addition, a third response type is frequently observable in which the red light fluence required is in the range 10^{-4}–10^{-1} μmol m^{-2}. The extent of Pr→Pfr photoconversion under such low fluences of red light is barely detectable by spectrophotometry and is also achieved by low fluences of far red light. These very low fluence responses (VLFR) are clearly not red/far-red reversible, since far-red light itself induces sufficient photoconversion to saturate many of the responses. In fully de-etiolated, green plants a number of developmental responses appear to be correlated with the Pfr/[Pr + Pfr] ratio established by the incident light (for review see H. Smith, 1982). Any speculation of a unified mechanism to account for phytochrome action must account for all of these types of responses and the variations within each response type.

Much of the current research towards elucidating the components of the transduction chain between photoconversion and developmental change is being focused upon the phytochrome regulation of gene expression in etiolated plant tissues. It has been known for many years that the activities of a large number of enzymes are modulated by phytochrome (Schopfer, 1977; Lamb & Lawton, 1983) and it is now emerging that in many cases these changes in enzyme activities are associated with changes in mRNA activity and/or abundance (for reviews see Lamb and Lawton, 1983; Schopfer and Apel, 1983; Tobin and Silverthorne, 1985; Schäfer *et al.*, 1986). In a smaller number of cases the rate of transcription of specific mRNA species in isolated nuclei have been shown to be regulated by phytochrome action prior to nucleic isolation (Tobin and Silverthorne, 1985; Schäfer *et al.*, 1986). These *in vitro* transcription assays ("run-off" transcription) are thought to detect the elongation of transcripts previously initiated *in vivo* (Darnell, 1982); thus, phytochrome apparently regulates the rate of initiation of RNA transcripts. Molecular biological techniques have been used to locate 5′-upstream non-coding sequences from a number of phytochrome-regulated genes which contain the elements necessary to confer photoregulation (Broglie *et al.*, 1984; Herrera-Estrella *et al.*, 1984; Timko *et al.*, 1985). It seems reasonable to suppose that phytochrome regulation of transcription involves some form of interaction between Pfr and these regulatory sequences, but at present we have no information concerning the intervening steps. Kinetic analyses of phytochrome-regulated transcription suggest that in some cases the phytochrome effects are very rapid, even though expression of the response may be slower (Schäfer *et al.*, 1986). This could be interpreted to mean it is unlikely that a series of second messengers are involved between Pfr and the regulatory sequences, so suggesting a direct effect of phytochrome on nuclei. Interestingly, Ernst and Oesterhelt (1984) have reported that the overall transcription rate of isolated rye nuclei can be increased by about 70% by the addition of native rye phytochrome, in predominantly the Pfr form. Several other proteins, including partially degraded rye phytochrome and native Pr, could not reproduce this effect. Similar effects have been reported for oat nuclei, although the Pfr-enhancement of *in vitro* transcription is only seen for nuclei isolated from red light-irradiated tissues (Schäfer *et al.*, 1986). These data do seem to suggest a more or less direct interaction between Pfr and the transcription process. However, until the transcription rates of known positively and negatively regulated genes, as well as genes which are not phytochrome-regulated *in vivo*, are assayed, the potential significance of these observed effects on overall transcription rates cannot be fully assessed. Furthermore, the question of what processes in isolated nuclei are being influenced by added Pfr needs to be addressed. It is generally thought that transcription in isolated nuclei reflects "run-off" of already-initiated transcripts, although there is some evidence for initiation of new RNA chains in isolated nuclei (Thompson *et al.*, 1985). Also, it should be emphasized that although in some cases phytochrome-regulated changes in mRNA abundance can be correlated with changes in transcription rate, there is equally compelling evidence that phytochrome regulates the process of gene expression at other levels, such as mRNA stability (Schäfer *et al.*, 1986).

Acceptance of the speculative hypothesis that phytochrome interacts directly with the genome to bring about photomorphogenesis presents additional problems. For example, in *Mougeotia*, phytochrome-mediated chloroplast rotation appears to involve regulation of ion fluxes at the plasmalemma which initiate a series of reactions, not involving altered gene expression, to elicit chloroplast movement. Similar effects on the electrical properties of membranes in higher plants have also been observed and these are amongst the fastest known phytochrome responses (Roux, 1986). In principle, these effects on membrane properties, though not obviously related to photomorphogenesis, could represent intermediate steps in the control of gene expression, but this would not be consistent with direct effects of phytochrome on the nucleus, unless multiple primary mechanisms are proposed. Clearly, any hypothesis for phytochrome action must account for these effects at the membrane level, as well as those which may be controlled via the regulation of gene expression. Additionally, models for phytochrome action should also accommodate multiple molecular species of phytochrome, which vary in relative abundance during development and may therefore be involved in different facets of photomorphogenesis.

REFERENCES

Abe, H., Yamamoto, K. T., Nagatani, A. and Furuya, M. (1985). *Plant Cell Physiol.* **26**, 1387–99.
Apel, K., and Kloppstech, K. (1980). *Planta* **150**, 426–30.
Baron, O. and Epel, B. (1983). *Plant Physiol.* **73**, 471–4.
Boeshore, M. L., and Pratt, L. H. (1980). *Plant Physiol.* **66**, 500–4.
Bois-Choussy, M., and Barbier, M. (1978). *Heterocycles* **9**, 677–99.
Bolton, G. W., and Quail, P. H. (1981). *Plant Physiol.* **67**, S104.
Bolton, G. W., and Quail, P. H. (1982). *Planta* **155**, 212–17.
Bradley, M. O. and Hillman, W. S. (1966). *Nature* **210**, 838.
Braslavsky, S. E. (1984). *Pure Appl. Chem.* **56**, 1153–65.
Braslavsky, S. E., Holzwarth, A. R., and Schaffner, K. (1983). *Angew. Chemie.* **95**, 670–89.
Briggs, W. R., and Rice, H. V. (1972). *Annu. Rev. Plant Physiol.* **23**, 293–334.
Brockmann, J. and Schäfer, E. (1982). *Photochem. Photobiol.* **35**, 555–8.
Broglie, R., Coruzzi, G., Fraley, R. T., Rogers, S. G.,and Horsch, R. B. (1984). *Science* **224**, 838–43.
Butler, W. L. and Lane, H. C. (1965). *Plant Physiol.* **40**, 13–17.
Butler, W. L., Norris, K. H., Siegelman, H. W., and Hendricks, S. B. (1959). *Proc. Natl. Acad. Sci. USA* **45**, 1703–8.
Butler, W. L., Lane, H. C. and Siegelman, H. W. (1963). *Plant Physiol.* **38**, 514–19.
Burke, M. J., Pratt, D. C., and Moscovitz, A. (1972). *Biochemistry* **11**, 4025–31.
Capaldi, R. A., and Vanderkooi, G. (1972). *Proc. Natl. Acad. Sci. USA* **69**, 930–2.
Cedel, T. E. and Roux, S. J. (1980). *Plant Physiol.* **66**, 704–9.
Chae, Qu., and Song, P-S. (1975). *J. Am. Chem. Soc.* **97**, 4176–9.
Colbert, J. T., Hershey, H. P., and Quail, P. H. (1983). *Proc. Natl. Acad. Sci. USA* **80**, 2248–52.
Colbert, J. T., Hershey, H. P. and Quail, P. H. (1985). *Plant Mol. Biol.* **5**, 91–101.
Coleman, R. A. and Pratt, L. H. (1974). *J. Histochem. Cytochem.* **11**, 1039–47.

Colleran, E. M., and Jones, O. T. G. (1973). *Biochem. J.* **134**, 89–96.
Cooke, R. J., and Kendrick, R. E. (1976). *Planta* **131**, 303–7.
Cooke, R. J., Saunders, P. F., and Kendrick, R. E. (1975). *Planta* **124**, 319–28.
Cordonnier, M-M., and Pratt, L. H. (1982). *Plant Physiol.* **69**, 360–5.
Cordonnier, M-M., Greppin, H. and Pratt, L. H. (1985). *Biochemistry* **24**, 3246–53.
Cordonnier, M-M., Greppin, H. and Pratt, L. H. (1986a). *Biochemistry* **25**, 7657–66.
Cordonnier, M-M., Greppin, H. and Pratt, L. H. (1986b). *Plant Physiol.* **80**, 982–7.
Daniels, S. M. and Quail, P. H. (1984). *Plant Physiol.* **76**, 622–6.
Darnell, J. E. Jr (1982). *Nature* **297**, 365–71.
Dooskin, R. H. and Mancinelli, A. L. (1968). *Bull. Torrey Bot. Club* **95**, 474–87.
Eilfeld, P. and Rüdiger, W. (1984). *Z. Naturforsch.* **39c**, 742–5.
Eilfeld, P. and Rüdiger, W. (1985). *Z. Naturforsch.* **40c**, 109–14.
Epel, B. L., Butler, W. L., Pratt, L. H. and Tokuyasu, K. T. (1980). In: *Photoreceptors and Plant Development* (Ed. J. A. DeGreef), pp. 121–133. Antwerpen University Press, Antwerpen.
Ernst, D. and Oesterhelt, D. (1984). *EMBO J.* **3**, 3075–8.
Etzold, H. (1965). *Planta* **64**, 254–80.
Evans, A. and Smith, H. (1976). *Proc. Natl. Acad. Sci. USA* **73**, 138–42.
Flint, D. H. (1984). *Plant Physiol.* **75**, S170.
Frankland, B. (1972). In: *Phytochrome* (Eds. E. Mitrakos and W. Shropshire, Jnr.), pp. 197–225. Academic Press, New York.
Fry, K. T. and Mumford, F. E. (1971). *Biochem. Biophys. Res. Commun.* **45**, 1466–73.
Furuya, M. (1983). *Phil. Trans. R. Soc. Lond.* **B303**, 361–75.
Gardner, G. and Gorton, H. L. (1985). *Plant Physiol.* **77**, 540–3.
Gardner, G., Pike, C. S., Rice, H. V. and Briggs, W. R. (1971). *Plant Physiol.* **48**, 686–93.
Georgevich, G., Cedel, T. E. and Roux, S. J. (1977). *Proc. Natl. Acad. Sci. USA* **74**, 4439–43.
Gottmann, K. and Schäfer, E. (1982). *Photochem. Photobiol.* **35**, 521–5.
Grimm, R., Lottspeich, F., Schneider, H. A. W. and Rüdiger, W. (1987). *Z. Naturforsch.* **41C**, 993–1000.
Grombein, S. and Rüdiger, W. (1976). *Hoppe-Seyler's Z. Physiol. Chem.* **357**, 1015–18.
Grombein, S., Rüdiger, W. and Zimmermann, H. (1975). *Hoppe-Seyler's Z. Physiol. Chemie.* **356**, 1709–14.
Hahn, T-R. and Song, P-S. (1981). *Biochemistry* **20**, 2602–9.
Hahn, T-R., Song, P-S., Quail, P. H. and Vierstra, R. D. (1984). *Plant Physiol.* **74**, 755–8.
Haupt, W. (1970). *Z. Pflanzenphysiol.* **62**, 287–98.
Heim, B., Jabben, M. and Schäfer, E. (1981). *Photochem. Photobiol.* **34**, 89–93.
Herrera-Estrella, L., Van den Broeck, G., Maenhaut, R. *et al.* (1984). *Nature* **310**, 115–20.
Hershey, H. P., Barker, R. F., Idler, K. B., Lissemore, J. L. and Quail, P. H. (1985). *Nucleic Acids Res.* **13**, 8543–60.
Hershko, A., Ciechanover, A., Heller, H., Haas, A. L. and Rose, I. A. (1980). *Proc. Natl. Acad. Sci. USA* **77**, 1783–6.

Hershko, A., Heller, H., Elias, S. and Ciechanover, A. (1983). *J. Biol. Chem.* **258**, 8206–14.
Hilton, J. R. and Smith, H. (1980). *Planta* **148**, 312–18.
Holdsworth, M. and Whitelam, G. C. (1987). *Planta* **172**, 539–547.
Holzwarth, A. R., Wendler, J., Ruzsicska, B. P., Braslavsky, S. E. and Schaffner, K. (1984). *Biochim. Biophys. Acta* **791**, 265–73.
Hunt, R. E. and Pratt, L. H. (1979). *Plant Physiol.* **64**, 327–31.
Hunt, R. E. and Pratt, L. H. (1980). *Biochemistry* **19**, 390–4.
Inoue, Y. (1986). *Proc. XVI Yamada Conf. Abstr.*, 40.
Inoue, Y. and Furuya, M. (1985). *Plant Cell Physiol.* **26**, 813–19.
Inoue, Y., Konomi, K. and Furuya, M. (1982). *Plant Cell Physiol.* **23**, 731–6.
Jabben, M. and Deitzer, G. F. (1978). *Planta* **143**, 309–13.
Jones, A. M. and Quail, P. H. (1986). *Biochemistry* **25**, 2987–95.
Jones, A. M., Vierstra, R. D. and Quail, P. H. (1985). *Planta* **164**, 501–6.
Jones, A. M., Allen, C. D., Gardner, G. and Quail, P. H. (1986). *Plant Physiol.* **81**, 1014–16.
Kaufman, L. S., Thompson, W. F. and Briggs, W. R. (1984). *Science* **226**, 1447–9.
Kaufman, L. S., Briggs, W. R. and Thompson, W. F. (1985). *Plant Physiol.* **78**, 388–93.
Kendrick, R. E. (1972). *Planta* **102**, 286–93.
Kendrick, R. E. and Frankland, B. (1969). *Planta* **86**, 21–32.
Kendrick, R. E. and Kronenberg, G. H. M. (eds.) (1986). *Photomorphogenesis in Plants*. Martinus Nijhoff, Dordrecht.
Kendrick, R. E. and Spruit, C. J. P. (1972). *Nature* **237**, 281–2.
Kendrick, R. E. and Spruit, C. J. P. (1977). *Photochem. Photobiol.* **26**, 201–4.
Kidd, G. H. and Pratt, L. H. (1973). *Plant Physiol.* **52**, 309–11.
Klein, G., Grombein, S. and Rüdiger, W. (1977). *Hoppe-Seyler's Z. Physiol. Chemie.* **358**, 1077–9.
Komeda, Y., Tomizawa, K., Sato, N., Furuya, M. and Iino, T. (1986). *Proc. XVI Yamada Conf. Abstr.*, 78.
Konomi, K. and Furuya, M. (1986). *Plant Cell Physiol.* **27**, 1507–12.
Kroes, H. H. (1970). *Meded. Landbouwhogeschool Wageningen* **70–18**, 1–112.
Kruh, J. and Borsook, H. (1956). *J. Biol. Chem.* **220**, 905–15.
Lagarias, J. C. and Mercurio, F. M. (1985). *J. Biol. Chem.* **260**, 2415–23.
Lagarias, J. C. and Rapoport, H. (1980). *J. Am. Chem. Soc.* **102**, 4821–8.
Lamb, C. J. and Lawton, M. A. (1983). In: *Encyclopedia of Plant Physiology*, vol. 16A (Eds. W. Shropshire, Jr., and H. Mohr), pp. 213–257. Springer-Verlag, Berlin.
Litts, J. C. (1980). *Ph.D. Dissertation*, University of Minnesota.
Litts, J. C., Kelly, J. M. and Lagarias, J. C. (1983). *J. Biol. Chem.* **258**, 11025–31.
McCurdy, D. W. and Pratt, L. H. (1986a). *Planta* **167**, 330–6.
McCurdy, D. W. and Pratt, L. H. (1986b). *J. Cell Biol.* **103**, 2541–50.
MacKenzie, J. M., Coleman, R. A., Briggs, W. R. and Pratt, L. H. (1975). *Proc. Natl. Acad. Sci. USA* **72**, 799–803.
Mohr, H. (1972). *Lectures in Photomorphogenesis*. Springer-Verlag, Berlin.
Mumford, F. E. and Jenner, E. L. (1971). *Biochemistry* **10**, 98–101.
Napier, R. M. and Smith, H. (1987a). *Plant Cell Environ.* **10**, 383–9.
Napier, R. M. and Smith, H. (1987b). *Plant Cell Environ.* **10**, 391–6.
Otto, V., Mösinger, E., Sauter, M. and Schäfer, E. (1983). *Photochem. Photobiol.* **38**, 693–700.

Otto, V., Schäfer, E., Nagatani, A., Yamamoto, K. T. and Furuya, M. (1984). *Plant Cell Physiol.* **25**, 1579–84.
Parker, M. W., Hendricks, S. B. and Borthwick, H. A. (1950). *Bot. Gaz.* **111**, 242–52.
Partis, M. D. and Thomas, B. (1985). *Biochem. Soc. Trans.* **13**, 110.
Pratt, L. H. (1982). *Ann. Rev. Plant Physiol.* **33**, 557–82.
Pratt, L. H. (1984). In: *Techniques in Photomorphogenesis* (Eds. H. Smith and M. G. Holmes), pp. 175–200. Academic Press, London.
Pratt, L. H. (1986). In: *Photomorphogenesis in Plants* (Eds. R. E. Kendrick and G. H. M. Kronenberg), pp. 61–81. Martinus Nijhoff, Dordrecht.
Pratt, L. H. and Briggs, W. R. (1966). *Plant Physiol.* **41**, 467–74.
Pratt, L. H. and Marme, D. (1976). *Plant Physiol.* **58**, 686–92.
Pratt, L. H., Inoue, Y. and Furuya, M. (1984). *Photochem. Photobiol.* **39**, 241–6.
Quail, P. H. (1986). *Proc. XVI Yamada Conf. Abstr.*, 28.
Quail, P. H., Marme, D. and Schäfer, E. (1973a). *Nature* **245**, 189–90.
Quail, P. H., Schäfer, E. and Marme, D. (1973b). *Plant Physiol.* **52**, 128–34.
Roux, S. J. (1986). In: *Photomorphogenesis in Plants* (Eds. R. E. Kendrick and G. H. M. Kronenberg), pp. 115–134. Martinus Nijhoff, Dordrecht.
Roux, S. J., Lisansky, S. G. and Stoker, B. M. (1975). *Physiol. Plant.* **35**, 85–90.
Rüdiger, W. (1969). *Hoppe-Seyler's Z. Physiol. Chemie.* **350**, 1291–1300.
Rüdiger, W. (1983). *Phil. Trans. Roy. Soc. London B* **303**, 377–85.
Rüdiger, W. (1986). In: *Photomorphogenesis in Plants* (Eds. R. E. Kendrick and G. H. M. Kronenberg), pp. 17–33. Martinus Nijhoff, Dordrecht.
Rüdiger, W. and Correll, D. L. (1969). *Liebigs Ann. Chem.* **723**, 208–12.
Rüdiger, W. and Scheer, H. (1983). In: *Encyclopedia of Plant Physiology*, vol. 16A (Eds. W. Shropshire, Jr. and H. Mohr), pp. 119–151. Springer-Verlag, Berlin.
Rüdiger, W., Brandlmeier, T., Blos, I., Gossauer, A. and Weller, J. P. (1980). *Z. Naturforsch.* **35c**, 763–9.
Rüdiger, W., Thümmler, F., Cmiel, E. and Schneider, S. (1983). *Proc. Natl. Acad. Sci. USA* **80**, 6244–8.
Rüdiger, W., Eilfeld, P. and Thümmler, F. (1985). In: *Optical Properties of Tetrapyrroles* (Eds. G. Blauer and H. Sund), 349–66. Walter de Gruyter, Berlin.
Ruzsicska, B. P., Braslavsky, S. E. and Schaffner, K. (1985). *Photochem. Photobiol.* **41**, 681–8.
Sarkar, H. K., Moon, D-K., Song, P-S., Chang, T. and Yu, H. (1984). *Biochemistry* **23**, 1882–8.
Saunders, M. J., Cordonnier, M-M., Palevitz, B. A. and Pratt, L. H. (1983). *Planta* **159**, 545–53.
Schäfer, E. (1975). *J. Math. Biol.* **2**, 41–56.
Schäfer, E. (1981). In: *Plants and the Daylight Spectrum* (Ed. H. Smith), pp. 461–80. Academic Press, London.
Schäfer, E., Schmidt, W. and Mohr, H. (1973). *Photochem. Photobiol.* **18**, 331–4.
Schäfer, E., Lassig, T-U. and Schopfer, P. (1975). *Photochem. Photobiol.* **22**, 193–202.
Schäfer, E., Apel, K., Batschauer, A. and Mösinger, E. (1986). In: *Photomorphogenesis in Plants* (Eds. R. E. Kendrick and G. H. M. Kronenberg), pp. 83–98. Martinus Nijhoff, Dordrecht.
Schaffner, K., Braslavsky, S. E. and Holzwarth, A. R. (1985). In: *Optical*

Properties and Structure of Tetrapyrroles (Eds. Blauer, G. and Sund, H.), pp. 367–382. W. de Gruyter, Berlin.
Scheer, H., Formanek, H. and Schneider, S. (1982). *Photochem. Photobiol.* **36**, 259–72.
Schirmer, T., Bode, W. and Huber, R. (1985). In: *Optical Properties and Structure of Tetrapyrroles* (Eds. Blauer, G. and Sund, H.), pp. 445–9. W. de Gruyter, Berlin.
Schoch, S., Klein, G., Linsenmeier, U. and Rüdiger, W. (1974). *Tetrahedron Lett.* 2465–8.
Schopfer, P. (1977). *Annu. Rev. Plant Physiol.* **28**, 223–52.
Schopfer, P. and Apel, K. (1983). In: *Encyclopedia of Plant Physiology*, vol. 16A (Eds. W. Shropshire Jr. and H. Mohr), pp. 258–88. Springer-Verlag, Berlin.
Schram, B. L. and Kroes, H. H. (1971). *Eur. J. Biochem.* **19**, 581–94.
Serlin, B. S. and Roux, S. J. (1986). *Biochim. Biophys. Acta* **848**, 372–7.
Sharrock, R. A., Lissemore, J. L. and Quail, P. H. (1986). *Gene* **47**, 287–95.
Shimazaki, Y. and Pratt, L. H. (1985). *Planta* **164**, 333–44.
Shimazaki, Y., Inoue, Y., Yamamoto, K. T. and Furuya, M. (1980). *Plant Cell Physiol.* **21**, 1619–25.
Shimazaki, Y., Cordonnier, M-M. and Pratt, L. H. (1983). *Planta* **159**, 534–44.
Shimazaki, Y., Cordonnier, M-M. and Pratt, L. H. (1986). *Plant Physiol.* **82**, 109–13.
Shropshire, W. Jr. and Mohr, H. (eds.) (1983). *Encyclopedia of Plant Physiology*, vol. 16. Springer-Verlag, Berlin.
Siegelman, H. W., Turner, B. C. and Hendricks, S. B. (1966). *Plant. Physiol.* **41**, 1289–92.
Slocum, R. D. and Roux, S. J. (1980). *Plant Physiol.* **65**, S101.
Smith, H. (1982). *Annu. Rev. Plant Physiol.* **33**, 481–518.
Smith, H., Whitelam, G. C. and Jackson, M. J. (1987). *Planta* (in press).
Smith, W. O. (1975). Ph.D. Dissertation, Univ. Kentucky.
Smith, W. O. Jr. (1983). In: *Encyclopedia of Plant Physiology*, vol. 16A (Eds. W. Shropshire, Jr. and H. Mohr), pp. 96–118. Springer-Verlag, Berlin.
Song, P-S. (1985). In: *Optical Properties and Structure of Tetrapyrroles* (Eds. Blauer, G. and Sund, H.), pp. 331–48. W. de Gruyter, Berlin.
Song. P-S., Chae, A. and Gardner, J. D. (1979). *Biochim. Biophys. Acta* **576**, 479–95.
Speth, V., Otto, V. and Schäfer, E. (1986). *Planta* **168**, 299–304.
Spruit, C. J. P. and Mancinelli, A. L. (1969). *Planta* **88**, 303–10.
Stone, H. J. and Pratt, L. H. (1979). *Plant Physiol.* **63**, 680–2.
Thomas, B., Cook, N. E. and Penn, S. E. (1984). *Physiol. Plant.* **60**, 409–15.
Thompson, W. F., Kaufman, L. S. and Watson, J. C. (1985). *Bioassays* **3**, 153–9.
Thümmler, F. and Rüdiger, W. (1983). *Tetrahedron* **39**, 1943–51.
Thümmler, F. and Rüdiger, W. (1984). *Physiol. Plant.* **60**, 378–82.
Thümmler, F., Brandlmeier, T. and Rüdiger, W. (1981). *Z. Naturforsch.* **36c**, 440–9.
Thümmler, F., Rüdiger, W., Cmiel, E. and Schneider, S. (1983). *Z. Naturforsch.* **38c**, 359–68.
Thümmler, F., Eilfield, P., Rüdiger, W., Moon, D-K. and Song, P-S. (1985). *Z. Naturforsch.* **40c**, 215–18.
Timko, M. P., Kausch, A. P., Castresana, C. *et al.* (1985). *Nature* **318**, 579–93.

Tobin, E. M. and Briggs, W. R. (1973). *Photochem. Photobiol.* **18**, 487–95.
Tobin, E. M. and Silverthorne, J. (1985). *Annu. Rev. Plant Physiol.* **36**, 569–93.
Tobin, E. M. and Briggs, W. R. and Brown, P. K. (1973). *Photochem. Photobiol.* **18**, 497–503.
Tokuhisa, J. G. and Quail, P. H. (1983). *Plant Physiol.* **75**, S73.
Tokuhisa, J. G., Daniels, S. M. and Quail, P. H. (1985). *Planta* **164**, 321–32.
Tokutomi, S., Yamamoto, K. and Furuya, M. (1981). *FEBS Lett.* **134**, 159–62.
Verbelen, J-P., Pratt, L. H., Butler, W. L. and Tokuyasu, K. (1982). *Plant Physiol.* **70**, 867–71.
Vierstra, R. D. and Quail, P. H. (1982a). *Planta* **156**, 158–65.
Vierstra, R. D. and Quail, P. H. (1982b). *Proc. Natl. Acad. Sci. USA* **79**, 5272–6.
Vierstra, R. D. and Quail, P. H. (1983a). *Biochemistry* **22**, 2498–505.
Vierstra, R. D. and Quail, P. H. (1983b). *Plant Physiol.* **72**, 264–7.
Vierstra, R. D. and Quail, P. H. (1985). *Plant Physiol.* **77**, 990–8.
Vierstra, R. D. and Quail, P. H. (1986). In: *Photomorphogenesis in Plants* (Eds. R. E. Kendrick and G. H. M. Kronenberg), pp. 35–60. Martinus Nijhoff, Dordrecht.
Vierstra, R. D., Cordonnier, M-M., Pratt, L. H. and Quail, P. H. (1984). *Planta* **160**, 521–8.
Vierstra, R. D., Shanklin, J. and Jabben, M. (1986). *Proc. VI Yamada Conf. Abstr.*, 76.
Vierstra, R. D., Quail, P. H., Hahn, T-R. and Song, P-S. (1987). *Photochem. Photobiol.* **45**, 429–32.
Wada, M., Kadota, A. and Furuya, M. (1983). *Plant Cell Physiol.* **24**, 1441–7.
Wall, J. K. and Johnson, C. B. (1983). *Planta* **159**, 387–97.
Watson, P. J. and Smith, H. (1982a). *Planta* **154**, 128–34.
Watson, P. J. and Smith, H. (1982b). *Planta* **154**, 121–7.
Weller, J. P. and Gossauer, A. (1980). *Chem. Ber.* **113**, 1603–11.
Wendler, H., Holzwarth, A. R., Braslavsky, S. E. and Schaffner, K. (1984). *Biochim. Biophys. Acta* **786**, 213–21.
Whitelam, G. C., Napier, R. M. and Smith, H. (1984). In: *Membranes and Compartmentation in the Regulation of Plant Functions* (Eds. A. M. Boudet, G. Alibert, G. Marigo and P. J. Lea), pp. 127–46. Clarendon Press, Oxford.
Wilkinson, K. D., Urban, M. K. and Haas, A. L. (1980). *J. Biol. Chem.* **225**, 7529–32.
Yamamoto, K. and Smith, W. O. (1981). *Biochim. Biophys. Acta* **668**, 27–34.

7
The Flavonoids: Recent Advances

JEFFREY B. HARBORNE

Plant Science Laboratories, University of Reading, Reading, Berkshire RG6 2AH, U.K.

1 INTRODUCTION

The water-soluble flavonoid pigments are universally distributed in vacuolar plants; one class of flavonoids, the anthocyanins, is the source

PLANT PIGMENTS
ISBN 0-12-289847-8

of orange to blue colours in petals, fruits, leaves and roots. Flavonoids also contribute to yellow flower colour, either by co-occurring with yellow carotenoids or by replacing them in about 15% of plant species. The so-called colourless flavones and flavonols make an important contribution to plant colour either by acting as copigments to the anthocyanins or by providing 'body' to cream and ivory flowers. Both yellow and colourless flavonoids can also provide ultraviolet(UV)-absorbing honey guides in yellow petals; these guides are perceived by bees, which are then able to collect nectar and pollinate the flower more effectively. In green leaves, UV-absorbing flavonoids are universally present either at the surface or in the epidermal cells and here they appear to provide protection from the potentially damaging effect of atmospheric UV radiation.

The 3000 known flavonoids share a common biosynthetic origin (see section 4) and are conveniently divided into some 12 classes according to the oxidation level of the central pyran ring. Typical representatives of these 12 classes are illustrated in Fig. 7.1, while the numbers of known structures in each class are given in Table 7.1. Most flavonoids occur *in vivo* conjugated with sugar and a bewildering array of glycosides is known. Other flavonoid conjugates may contain aliphatic or aromatic acyl substituents or sulphate groups. In the case of cyanidin, the commonest anthocyanidin pigment (see Fig. 7.1), at least 77 different glycosidic (or acylated glycosidic) forms are known.

O-methylation of hydroxyl groups is a common structural feature and highly methylated flavonoids occur, without sugar substitution, in the leaf waxes or on other surfaces of many plants. *C*-methylation and isopentenyl substitution have the same effects as *O*-methylation in increasing the lipophilic character of what are otherwise polar molecules and such derivatives are also present at leaf surfaces or in exudates.

Flavonoid molecules have the ability to link together through, for example, the reactive carbon centres at the 6- and 8-positions of the A-ring (see Fig. 7.1) to form dimers, trimers and eventually polymers. In the flavone series, dimers are frequently produced; these compounds are known as biflavonyls and they occur very widely in the leaves of conifers. The most important and widespread polymers are the proanthocyanidins (or condensed tannins) which are formed from leucoanthocyanidin and catechin precursors and which may contain between 20 and 30 flavan units. Although colourless when pure, proanthocyanidins are so called because on acid treatment they yield the anthocyanidins, cyanidin and delphinidin, through cleavage of carbon–carbon bonds. The coloured anthocyanins do not usually occur otherwise than as monomers, but polymerization can take place in solution; thus, during the ageing of red wine, the grape anthocyanins slowly form polymeric pigments.

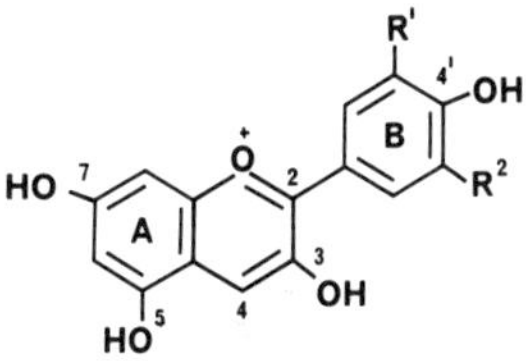

Anthocyanidins
I Pelargonidin, $R^1=R^2=H$
II Cyanidin, $R^1=OH$, $R^2=H$
III Delphinidin, $R^1=R^2=OH$

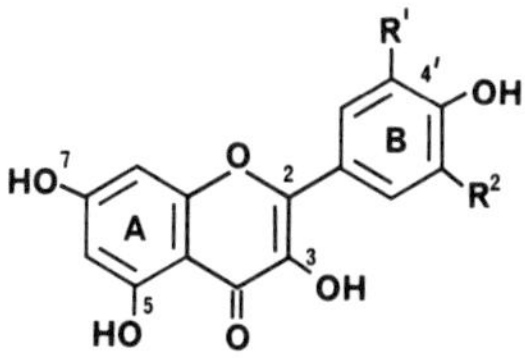

Flavonols
IV Kaempferol $R^1=R^2=H$
V Quercetin, $R^1=OH$, $R^2=H$
VI Myricetin, $R^1=R^2=OH$

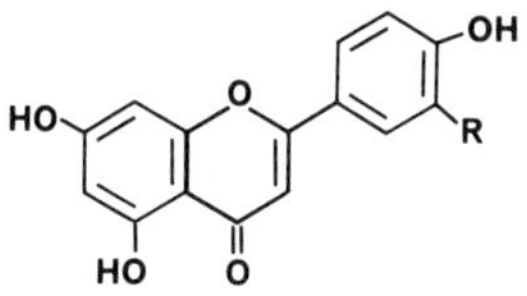

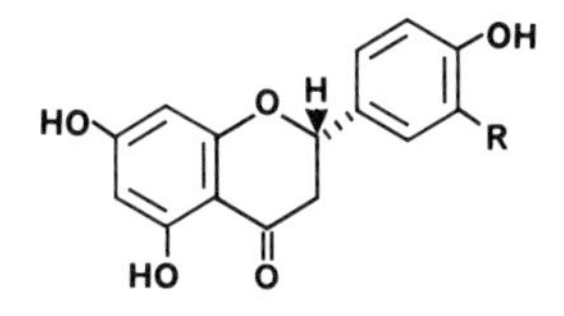

Flavones
VII Apigenin, R=H
VIII Luteolin, R=OH

Flavanones
IX Naringenin, R=H
X Eriodictyol, R=OH
stereochemistry at C−2 (2R)

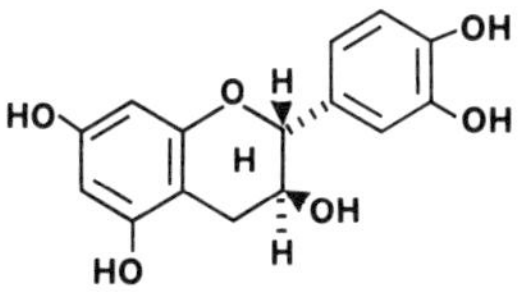

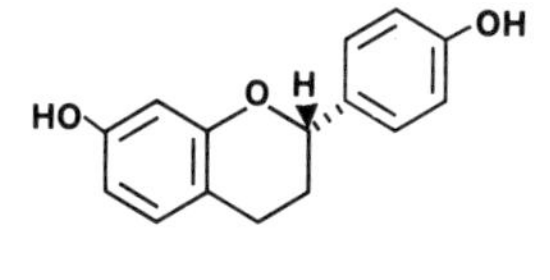

Catechins
XI (+)-catechin
(2R, 3S) as shown
XII (−)-epicatechin
(2R, 3R) stereoisomer

Flavan
XIII 7,4′-Dihydroxyflavan

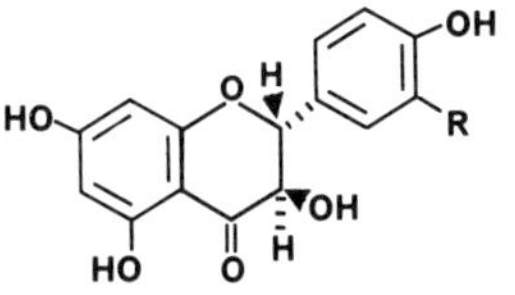

Dihydroflavonols
XIV Dihydrokaempferol, R=H
XV Dihydroquercetin, R=OH

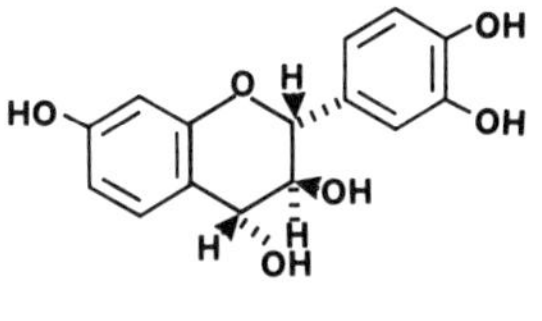

Flavan-3, 4-diols
XVI Mollisacacidin
(2R, 3S, 4R) as shown
XVII (+)-Leucofisetinidin
(2S, 3R, 4S) stereoisomer

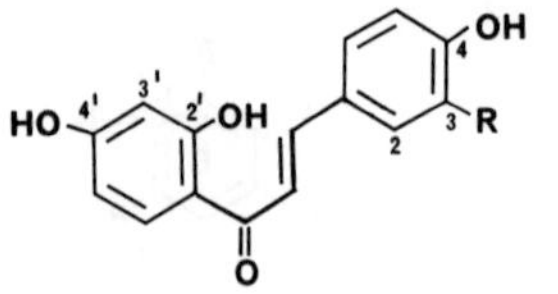

Chalcones
XVIII Isoliquiritigenin, R=H
XIX Butein, R=OH

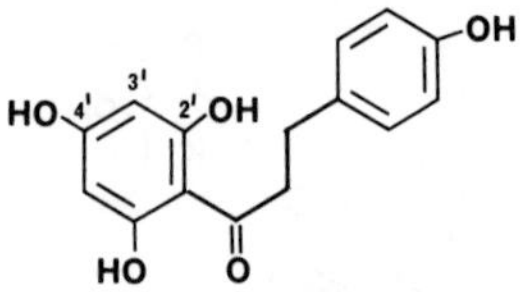

Dihydrochalcone
XX Phloretin

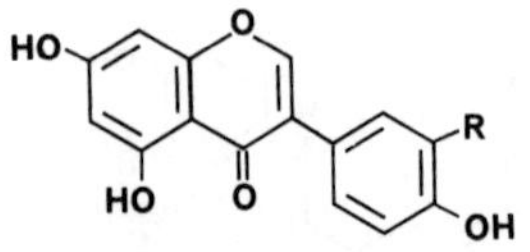

Isoflavones
XXI Genistein, R=H
XXII Orobol, R=OH

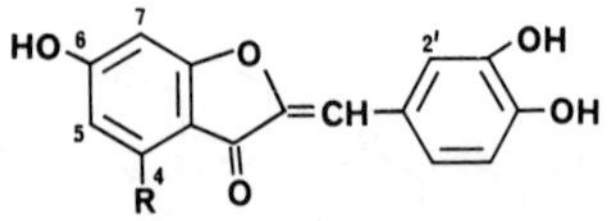

Aurones
XXIII Sulphuretin, R=H
XXIV Aureusidin, R=OH

Fig. 7.1 Structures of common flavonoid aglycones.

Table 7.1 The major classes of flavonoid

Class	Number of known structures[1]	Role of pigmentation
Anthocyanins	260	Provides cyanic colour from scarlet, red, mauve to blue
Aurones	20	Visible yellow colour and/or UV patterning in yellow flowers
Chalcones	90	
Yellow flavonols	40	
Colourless flavones and flavonols	1600	Copigments to anthocyanins Body to cream/ivory flowers UV patterning in some yellow flowers
Flavanones	200	Mostly colourless, but may interact with the other flavonoids; may act as UV screen in leaves; some are anti-feedant or anti-microbial in activity
Dihydroflavonols	70	
Dihydrochalcones	35	
Leucoanthocyanidins	40	
Catechins	70	
Flavans	30	
Isoflavonoids	650	

[1]Including aglycones and glycosides, but excluding oligomers and polymers.

In order to cope with the enormous numbers of flavonoid structures, it is convenient to divide them arbitrarily into those compounds which are common and those which are rare. When this is done, it is found that only a handful of structures (e.g. compounds I–XXIV, Fig. 7.1) are widely present in plants. In the case of the anthocyanins, there are only three basic chromophores: the scarlet pelargonidin, the crimson cyanidin and the purple delphinidin. Almost all known pigments are derived from these three molecules. Three other common anthocyanidins are cyanidin 3′-methyl ether (peonidin), delphinidin 3′-methyl ether (petunidin) and delphinidin 3′,5′-dimethyl ether (malvidin). The flavonols, which are closely related in structure to the anthocyanidins, are represented mainly by kaempferol, quercetin and myricetin (see Fig. 7.1); again simple *O*-methylated derivatives such as isorhamnetin (quercetin 3′-methyl ether) are also common. Flavones only differ from flavonols in lacking the 3-hydroxyl group. Two flavones, apigenin and luteolin are common; tricetin, the flavone corresponding to myricetin or delphinidin in hydroxylation pattern, is quite rare. Flavones are unusual in occurring not only with sugars linked through the phenolic groups by ether linkage but also with carbon-linked sugars attached at the 6- and/or 8-positions.

One structural feature which distinguishes common flavonoids from the rarer substances is a change in the standard hydroxylation pattern through the insertion or removal of one or more hydroxyl functions. This may affect the colour properties. For example, introduction of a further hydroxyl at the 6- or 8-position of colourless flavones or flavonols causes a significant shift in colour to yellow and pigments such as 6-hydroxyquercetin (quercetagetin) and 8-hydroxyquercetin (gossypetin) are important petal constituents in a number of yellow-flowered species. Removal of a hydroxyl will tend to shift the visible spectrum of a flavonoid towards shorter wavelengths. For example, removal of the 3-hydroxyl in cyanidin leads to the rare pigment, luteolinidin, which is orange rather than red in colour.

Flavonoids do not occur in isolation from other plant constituents and their contribution to *in vivo* colour is often modified by other cellular constituents. They have the ability to chelate with certain metals, and such chelates have been isolated intact from several blue-flowered species (see section 6). The cellular concentrations of flavonoids in flowers affects the expression of flower colour and there are now very accurate methods available for determining the amounts of individual components (section 2). Anthocyanins are natural indicators and their *in vivo* colours vary with the pH of the cell sap. Knowledge of the reactions that anthocyanins undergo in aqueous solution is directly relevant to determining the way in which anthocyanins exist in the cell vacuole. The synthesis of anthocyanin is under tight genetic control and the study of flavonoids in flower petal mutants has contributed both to our knowledge of the relationship between structure and colour and to our understanding of the biosynthetic pathway (section 4).

The biochemistry of the flavonoids and their role in plant colour has been reviewed in earlier editions (Goodwin, 1965, 1976) of this volume. Since then, research has revealed the very real complexity in structure of blue anthocyanins and has shown how these pigments are stabilized in the petal, in an otherwise hostile aqueous environment. The purpose of this chapter, then, is to review briefly the progress that has been made since 1976 in our understanding of the biochemistry of these fascinating, intensely coloured pigments.

For more detailed listings of the occurrence of flavonoids in plants, the reader is referred to the three-volume series '*The Flavonoids*' (Harborne *et al*., 1975; Harborne and Mabry, 1982; Harborne, 1988). Other books on plant pigments with up-to-date accounts of the flavonoids are those of Czygan (1980) and of Britton (1983). There is also a recent compilation on the biological and medicinal properties of the flavonoids (Cody *et al*., 1986).

2 ANALYTICAL METHODS

The general procedures for detecting and identifying plant flavonoids have been succinctly described by Swain (1976). In his review, this author considers methods for extraction, isolation and analysis and he points out the value of UV, proton nuclear magnetic resonance (NMR) and mass spectral techniques for the structural elucidation of flavonoids. Another excellent guide to flavonoid analysis, written especially for a biological audience, is the handbook of Markham (1981).

Since 1976, the most important developments in techniques have been the increasing application of high-performance liquid chromatography (HPLC) to flavonoid separation and quantitation, coupled with the use of carbon-13 NMR spectroscopy and of fast atom bombardment mass spectrometry (FAB–MS) for flavonoid identification. The present section will be devoted to these and related advances in analytical procedures.

A High-performance liquid chromatography (HPLC)

HPLC has added a new dimension to the investigation of flavonoids in plant extracts, as it has for most other classes of plant constituent (Daigle and Conkerton, 1983; Harborne, 1983, 1985). Particular advantages are the improved resolution of pigment mixtures compared to other chromatographic techniques, the ability to obtain both qualitative and accurate quantitative data in one operation and the great speed of analysis. The latter point is important for pigments such as the anthocyanins, which tend to break down slowly in solution in the presence of light.

For the HPLC of anthocyanins, reversed-phase C_8 or C_{18} columns are most suitable, used with solvent mixtures such as water–acetic acid–methanol and tetrahydrofuran–0.05 M phosphoric acid (pH 1.8) for gradient elution. The low sensitivity of detector units in the visible range (490–540 nm) can be overcome by monitoring anthocyanin separations both in the visible and in the UV at around 280 nm.

A practical application of HPLC to anthocyanin separations is the quantification of pigment concentrations in petals or bracts of ornamental plants, as a means of biochemical identification of different cultivars. For example, it is possible to distinguish 28 garden forms of *Euphorbia pulcherrima* from the varying amounts of five anthocyanins present in the bracts (Stewart *et al.*, 1979). Similar results have been obtained with flowers of azaleas (van Sumere and Van de Casteele, 1985) and of gladioli (Akavia *et al.*, 1981). Because of its sensitivity, HPLC is particularly useful for identifying anthocyanins when only small amounts of plant tissue are available, as in the case of rare flower mutants (Schram *et al.*, 1983) or when working with these pigments in tissue culture (Colijn *et al.*, 1981). It has also proved valuable for monitoring the acidic hydrolysis of complex anthocyanins for the detection of fugitive intermediates which may not be apparent on paper chromatograms or thin-layer chromatography (TLC) plates.

HPLC has been applied extensively to the separation and analysis of other classes of flavonoid as well as the anthocyanins. An additional refinement here is the on-line identification by UV spectroscopy of the various peaks as they separate from the column. This can be achieved with a photodiode array detector, which measures the complete spectrum in a few microseconds. Since most flavone and flavonol glycosides have similar spectral characteristics, the UV spectrum alone may be insufficient for identification. However, it is possible to add suitable reagents (e.g. aluminium chloride or alkali) to the solvent flow so that the glycosides undergo derivatization and characteristic shifts in the UV spectra can also be recorded. This procedure has been used to identify the flavone *C*-glycosides present in *Gentiana* extracts (Hostettmann *et al.*, 1984).

HPLC can also be used to determine accurately the concentrations of individual flavonoids in the mixtures present in crude plant extracts (Harborne *et al.*, 1984). Coupled with the use of an ion-pairing reagent such as 0.01 M tetrabutylammonium phosphate, HPLC nicely separates flavonoid sulphates from other flavonoid conjugates (Harborne and Boardley, 1985b). Finally, HPLC provides through the measurement of retention times an additional criterion in the chromatographic identification of flavonoids; co-HPLC can be added to co-paper chromatography (PC) and co-TLC for matching up newly isolated compounds with authentic markers.

Naturally, other chromatographic techniques continue to be developed for flavonoid analysis. For large-scale separations, there is the choice of preparative HPLC, medium-pressure liquid chromatography or counter current chromatography (Hostettmann *et al.*, 1986). TLC is still important; TLC on polyamide plates, followed by spraying with Naturstuff A reagent, has proved invaluable for separating the complex mixtures of flavonoid aglycones found on leaf surfaces or in the crystalline deposits of certain ferns (Wollenweber, 1982).

B Carbon-13 NMR spectroscopy

Just as proton NMR measurements on flavonoids became routine by about 1970 (Markham and Mabry, 1975), so the measurement of carbon-13 NMR spectra became common about 1980. An extensive compilation of such data, illustrated with many spectral diagrams, has been published (Markham *et al.*, 1982). Measurements are usually made in either DMSO-d_6 or in D_2O. The chief value of carbon-13 NMR is in the structural elucidation of the more complex flavonoid pigments, particularly those with several sugar and/or acylated substituents. For example, it has been applied successfully to the elucidation of the acylated anthocyanins cinerarin (Goto *et al.*, 1984) and monardaein (Kondo *et al.*, 1985).

Carbon-13 NMR spectroscopy has been used with success to define the sequence, position and configuration of sugar residues in flavonoid glycosides. It can also pick out acyl substituents (e.g. acetic acid) which may not otherwise be apparent. One example is the flavone, 6″-acetyl-8-galactosylapigenin, from *Briza media*, where the acetyl signals are clearly separated from those of the flavone and the sugar (Fig. 7.2; Chari *et al.*, 1980). Furthermore, spectral comparison of the acetate with that of the deacylated compound shows that the acetyl is specifically located at the 6-position of the galactose, since there is a downfield shift in the C-6″ resonance and an upfield shift in the C-5″ resonance as a result of this attachment.

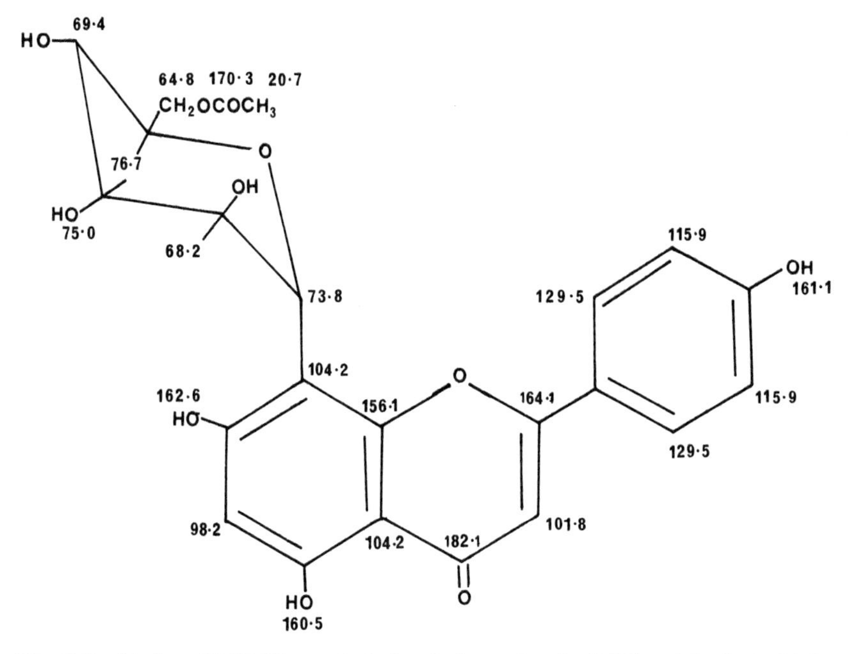

Fig. 7.2 Carbon-13 NMR spectral signals for apigenin 8-(6″-acetylgalactoside).

C Fast atom bombardment–mass spectrometry (FAB–MS)

Mass spectrometry has long been used in flavonoid analysis; it yields molecular ions and characteristic fragmentations for most flavonoid aglycones (Mabry and Markham, 1975). It has turned out to be particularly useful for differentiating flavone *C*-glycosides, as their permethyl ethers (Chopin and Bouillant, 1975), but has had limited success with *O*-glycosides. Even after derivatization of *O*-glycosides, it proved difficult to detect molecular ions. With the advent of FAB–MS, the situation has completely changed and it is now possible to observe intense molecular ions for flavonoids which were either too unstable or too involatile for analysis by standard procedures.

For FAB–MS, glycerol is commonly used to solubilize the flavonoid sample and the sample is bombarded with energized xenon or argon atoms. The technique was first shown to be producing molecular ions for a number of common flavonol glycosides, but proved to be particularly successful with glucuronides and sulphates (Domon and Hostettmann, 1985).

In the anthocyanin series, FAB–MS is indispensable for characterizing malonated pigments where the acyl group is otherwise very labile. Intense molecular ions are obtainable and characteristic fragments are detectable (Table 7.2). Dimalonated and monomalonated pigments are clearly distinguishable in their behaviour by this technique (Takeda *et al.*, 1986).

3 CHEMISTRY

A Anthocyanins

The most important advance in our understanding of the chemistry of anthocyanins has been the clarification of the various reactions that these pigments undergo in weak acidic solutions (Brouillard, 1983). The various species of anthocyanin molecule that can exist in solution are illustrated in Fig. 7.3, with respect to malvin (malvidin 3,5-diglucoside) as an example. Proton transfer reactions occur very rapidly at pH values of between 4 and 5, with the flavylium cation (AH^+) giving rise to two quinonoid bases (A), through loss of a proton at C-7 and C-4′ respectively. Further protonation can take place at pH between 6 and 7 with the production of a resonance-stabilized quinonoid anion (A^-).

An even more significant reaction in aqueous solution is the covalent hydration of the flavylium cation (AH^+) with the formation of a carbinol pseudobase (AOH) at pH values between 3 and 6. Addition of water takes place essentially at the 2-position, with the transient production of both epimers (Cheminat and Brouillard, 1986). The kinetic and thermodynamic conditions are usually unfavourable for the production of the 4-adduct, although this can be detected in the case of a flavylium cation which lacks free phenolic groups.

Since the carbinol pseudobase is colourless, this reaction explains the extensive loss of colour intensity that takes place when the pH of pure anthocyanins in solution is changed from 2.0 to 4.5, the pH of most plant cell saps. Since anthocyanins are thus so vulnerable to hydration, this means that they rarely exist *in vivo* as such. They must be protected in some way, e.g. by copigmentation (see section 6), so that they can be maintained in the cell sap as the flavylium (AH^+) cation.

Anthocyanins undergo one further reaction in aqueous solution: the opening of the pyranol ring with the formation of a chalcone pseudobase (Fig. 7.3). At room temperature, the equilibrium of this reaction does not favour chalcone formation, but by increasing the temperature it is possible to demonstrate chalcone production by HPLC. Preston and Timberlake (1981) have prepared chalcones from anthocyanins using this technique.

Table 7.2 FAB–MS of some malonated anthocyanins

Pigment	$[M]^+$	Loss of malonic $[M-86]^+$	Loss of glucose $[M-162]^+$	Loss of malonyl-glucose $[M-248]^+$	Loss of 2 malonyl $[M-172]^+$	Aglycone $[M]^+$
Pg 3-(6″-malonylglucoside)	519	433	—	—	—	271
Pg 3-(6″-malonylglucoside) -5-glucoside	681	595	519	433	—	271
Pg 3-(6″-malonylglucoside-5-malonyl-glucoside	767	681	—	519	595	271
Cy 3-dimalonylglucoside	621	535	—	—	—	287
Cy 3-malonylglucuronosylglucoside	711	625	535′	—	—	287
Cy 3-(6″-malonylglucoside)-5-glucoside	697	611	535	449	—	287
Cy 3-(6″-malonylglucoside) -5-malonylglucoside	783	697	—	535	611	287
Dp 3-(6″-malonylglucoside) -5-malonylglucoside	799	713	—	551	—	303

Pg = pelargonidin; Cy = cyanidin; Dp = delphinidin
′Loss of glucuronic acid $[M-176]^+$.

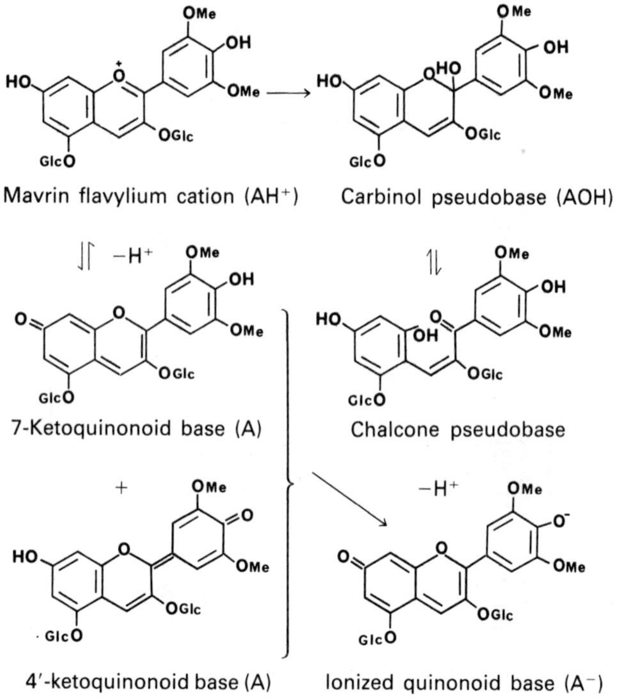

Mavrin flavylium cation (AH^+) → Carbinol pseudobase (AOH)

$-H^+$

7-Ketoquinonoid base (A)

Chalcone pseudobase

+

$-H^+$

4′-ketoquinonoid base (A)

Ionized quinonoid base (A^-)

Fig. 7.3 Anthocyanin reactions in aqueous media.

Since 1976, the number of known anthocyanidins has increased by two from 15 to 17. 5-Methylcyanidin (XXV) has been obtained as the 3-glucoside from the water plant *Egeria densa* (Momose *et al.*, 1977) and 6-hydroxycyanidin (XXVI) has been reported as the 3-glucoside and 3-rutinoside in red flowers of *Alstroemeria* cultivars (Saito *et al.*, 1985). It has long been recognized that anthocyanin-like pigments occur in red *Sphagnum* mosses, attached to the cell walls; three compounds have now been fully characterized, sphagnorubins A-C (XXVII–XXIX; Mentlein and Vowinkel, 1984).

A wide range of new glycosides of known anthocyanidins have been described recently. Most are structurally related to pigments described earlier, but three new features are apparent. These are the presence of sugars attached to B-ring hydroxyl groups, the repeated substitution at the same position of an acylated sugar, such as caffeoylglucose, and the presence of aliphatic acids, particularly malonic, as an acyl substituent.

One of the first anthocyanins carrying B-ring hydroxyls to be isolated was the acylated derivative of delphinidin 3-rutinoside-5,3′,5′-triglucoside

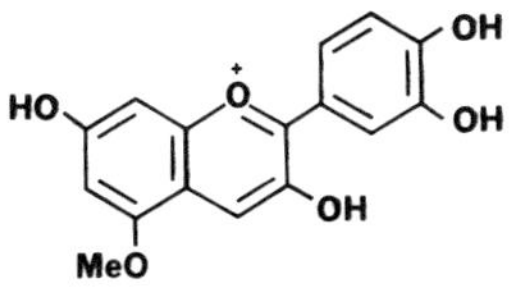

XXV 5-Methylcyanidin

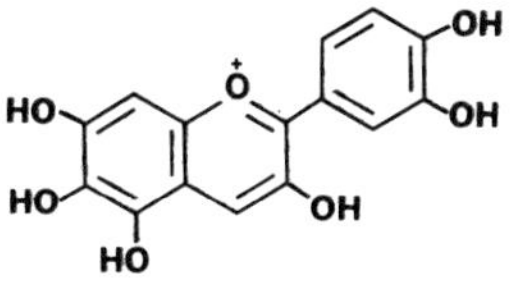

XXVI 6-Hydroxycyanidin

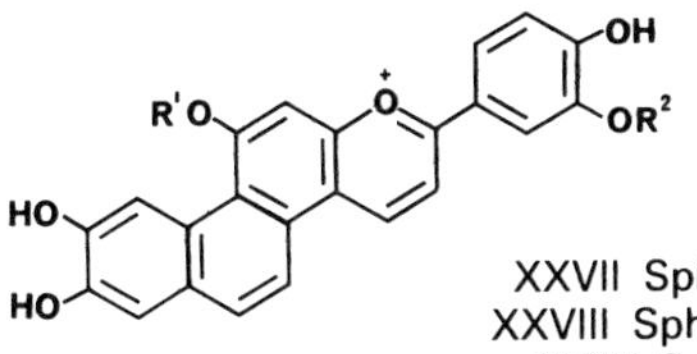

XXVII Sphagnorubin A, $R^1=R^2=H$
XXVIII Sphagnorubin B, $R^1=Me_1$, $R^2=H$
XXIX Sphagnorubin C, $R^1=R^2=Me$

(XXX), which occurs in blue *Lobelia* flowers (Yoshitama, 1977). About 10 others have been reported since then; one example is cyanidin 3,5,3′-triglucoside (XXXI) which occurs quite widely in plants of the Bromeliaceae (Saito and Harborne, 1983). Addition of glucose to the B-ring causes a hypsochromic shift in colour, and this is matched by the effect of 6-hydroxylation, as in the case of 6-hydroxycyanidin, mentioned above. Table 7.3 indicates the effects that different substitutions have on the spectral maxima of cyanidin derivatives. How far these different

Table 7.3 Effect of sugar and hydroxyl substitution on the spectral maxima of cyanidin derivatives

Pigment	Visible maximum (in nm) in MeOH-HCl
Cyanidin	535
3-glucoside	528
3,5-diglucoside	525
3,7-diglucoside	524
3,3′-diglucoside	519
3,5,3′-triglucoside	518
3,7,3′-triglucoside	515
6-Hydroxycyanidin	518
3-glucoside	514

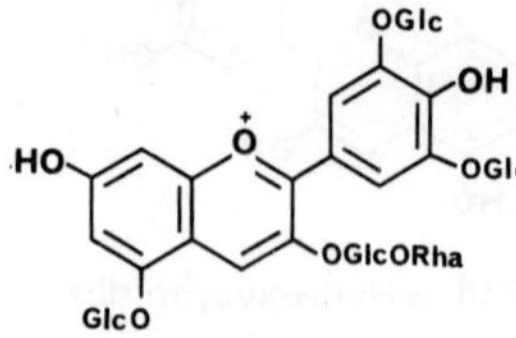

XXX Delphinidin 3-rutinoside 5, 3′, 5′-triglucoside

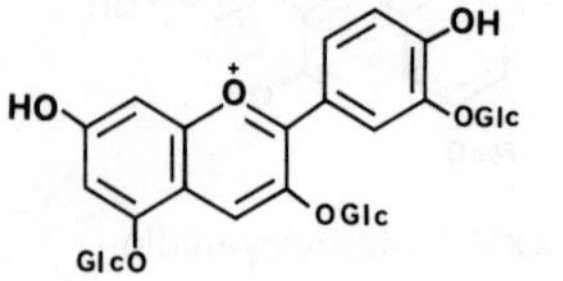

XXXI Cyanidin 3′,5,3′-triglucoside

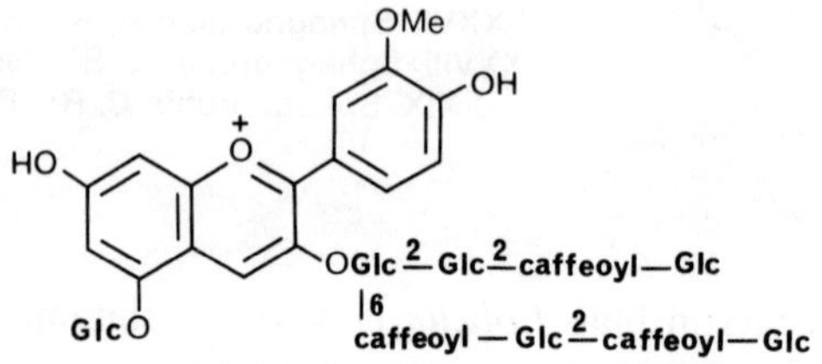

XXXII Heavenly blue anthocyanin

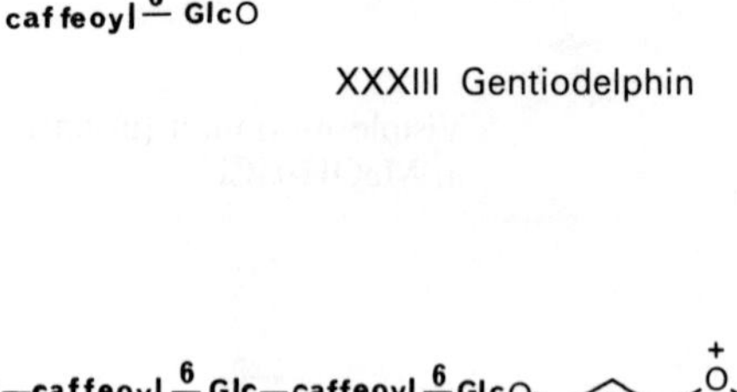

XXXIII Gentiodelphin

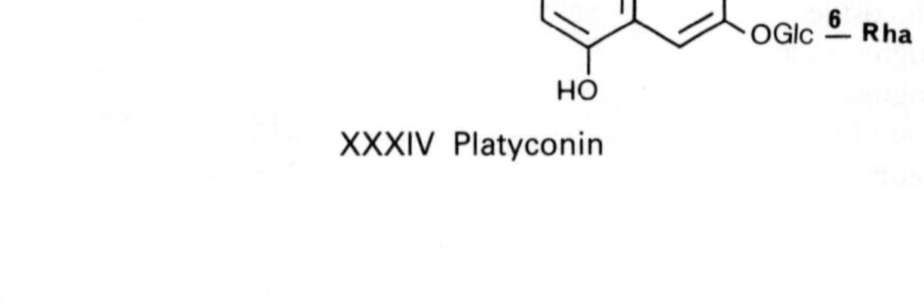

XXXIV Platyconin

substitution patterns affect *in vivo* flower colours is problematic, because of the many other factors involved (see section 6).

Perhaps the most notable series of new anthocyanins are those with repeated caffeoylglucose substitutions. All these pigments are remarkably stable in aqueous solution due to intramolecular copigmentation (see section 6). One is 'heavenly blue anthocyanin' (XXXII) from flowers of *Ipomoea tricolor*, which is a peonidin 3-sophoroside-5-glucoside substituted through sugar with three caffeoylglucose residues. This pigment is probably the largest anthocyanin yet reported, since it has a molecular weight of 1759. It is also remarkable that the basic chromophore is peonidin, since most other blue flowers are based on delphinidin. Two related delphinidin pigments are, indeed, gentiodelphin (XXXIII), from the deep blue flowers of *Gentiana makinoi*, and platyconin (XXXIV), from the sky blue flowers of *Platycodon grandiflorum* (Goto, 1984).

The third type of new anthocyanin is that containing malonyl substitution. Anthocyanins substituted through sugar (usually through the 6-position of glucose) by malonic acid are now recognized to be common and they may occur in some 30% of all higher plant species (Harborne, 1986). The recognition of such substitution has been delayed by the fact that if anthocyanins are extracted with the standard acidic solvents of the classical period of anthocyanin chemistry, these labile groups are lost. By using milder extraction procedures with solvents based on acetic rather than hydrochloric acid, it is possible to isolate malonated pigments in good yield. Because of the remaining negative charge on the malonyl residue, malonates behave like zwitterions and can be distinguished from other anthocyanins by paper electrophoresis (Harborne and Boardley, 1985). They are then readily identified by hydrolytic procedures, by H_2O_2 oxidation to give a malonated sugar and by FAB–MS and NMR techniques.

Three examples of malonated anthocyanins are: monardaein (XXXV), the newly revised structure of the pigment of *Monarda didyma* (Kondo *et al.*, 1985); malonylawobanin (XXXVI), the newly revised structure for the *Commelina communis* blue flower pigment (Goto *et al.*, 1983); and cyanidin 3-malonylglucuronosylglucoside (XXXVII) from red flowers of *Helenium autumnale*, a new pigment and the first anthocyanin with glucuronic acid as one of its sugars (Takeda *et al.*, 1986). While malonic is the most common aliphatic acid found in this association, succinic, malic and oxalic acids have also been found. Succinic is present, for example, in the pigment of the blue cornflower, *Centaurea cyanus*; originally thought to be cyanin, this is now known to be the 6″-succinate (XXXVIII) (Takeda and Tominaga, 1983).

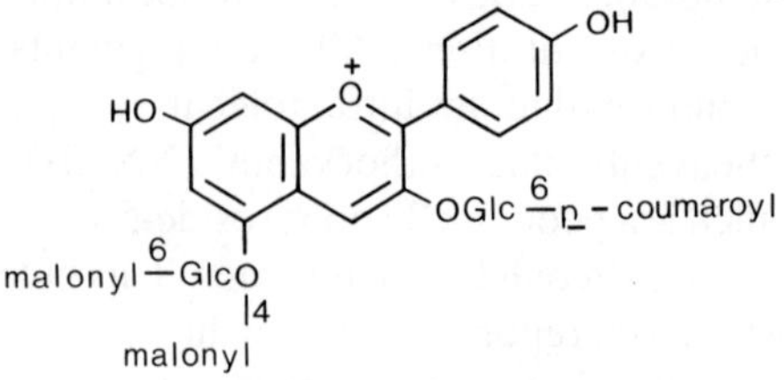

XXXV Monardaein

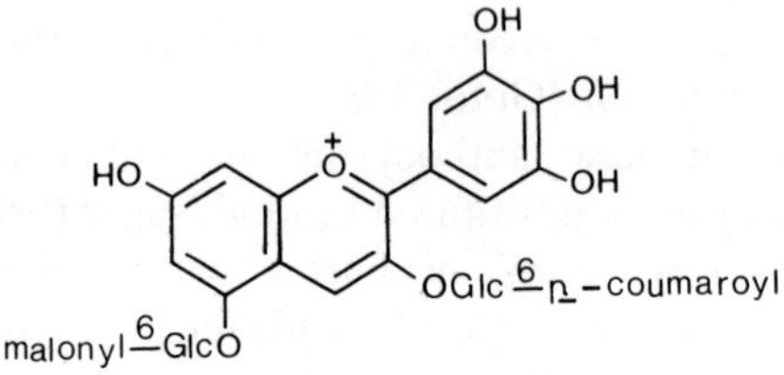

XXXVI Malonylawobanin

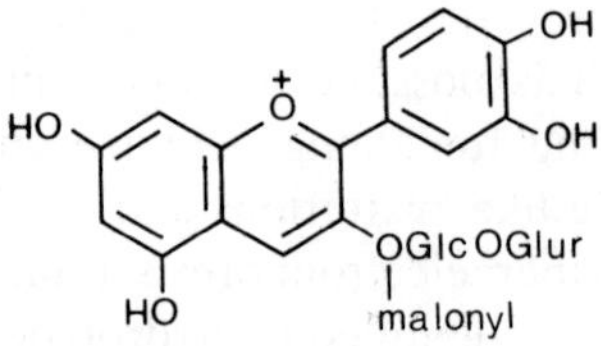

XXXVII Cyanidin 3-malonyl glucuronosyl glucoside

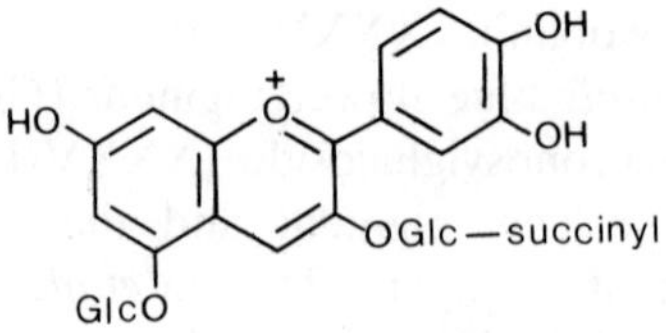

XXXVIII Cyanidin3-(6″-succinylglucoside)-5-glucoside

B Yellow flavonoids

The three main classes of flavonoid which contribute to yellow flower colour (and sometimes to UV patterning as well) are chalcones, aurones and yellow flavonols (Harborne, 1976; Swain, 1976). To these three classes can now be added a fourth, yellow flavones. The flavone isoetin (XXXIX) has been found as the 7-glucoside and a mixture of 2′-glycosides as the major yellow pigment of *Heywoodiella oligocephala* and of two other composite flowers (Harborne, 1978). The yellow colour results from the introduction of a hydroxyl group at the 2′-position of luteolin, with a shift in spectrum from 350 to 379 nm. In view of the very large bathochromic shift caused by 2′-substitution, it is conceivable that the *p*-quinone form of isoetin contributes to the yellow colour *in vivo*.

Among the many new chalcone structures reported recently (Bohm, 1982, 1988) are the *C*-methyl derivative (XL) from *Comptonia peregrina* (Wollenweber *et al.*, 1985) and the prenylated chalcone (XLI) from *Helichrysum rugulosa* (Bohlmann and Misra, 1984). Retrochalcones are a new group of yellow pigments, so called because of their biosynthetically anomalous structures. One is echinatin (XLII) from *Glycyrrhiza* plants, which is known to be formed from isoliquiritigenin (XVIII), the functionality of the C_3 chain being reversed during subsequent formation of (XLII) (Dewick, 1984). Another biosynthetically odd pigment is 2′,3,4-trihydroxychalcone (XLIII) from *Semecarpus vitiensis* (Pramono *et al.*, 1981).

Carthamin, the red pigment of safflower *Carthamus tinctorius*, long used as a dye, has been the subject of numerous investigations and was thought to be a monomeric chalcone (Swain, 1976). Recent structural studies indicate that it is a dimer (XLIV) (Takahashi *et al.*, 1982). *Morus alba* seems to be another source of dimeric chalcones, kuwanon J (XLV) being one of a number of such pigments present in this plant (Ikuta *et al.*, 1986).

C Flavone and flavonol glycosides

An increasing number of new flavone and flavonol aglycones have been described since 1976, the great majority being related fairly closely to previously described molecules. The number of known glycosides of these aglycones has also expanded and there are now about 1000 of them (Harborne and Williams, 1988). The only novel sugar is allose, but 30 disaccharides and 20 trisaccharides have been characterized in association with these flavonoids.

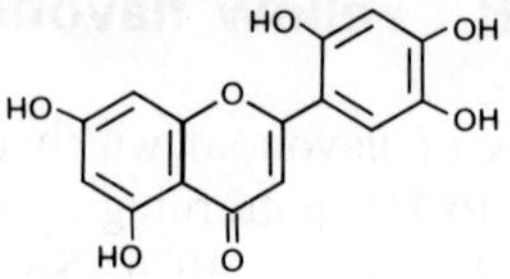

XXXIX Isoetin

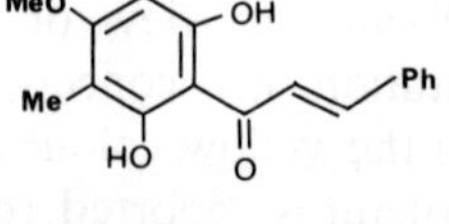

XL

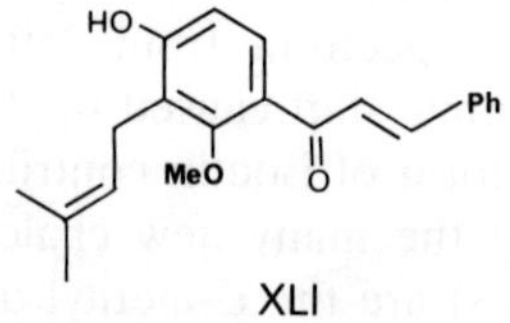

XLI

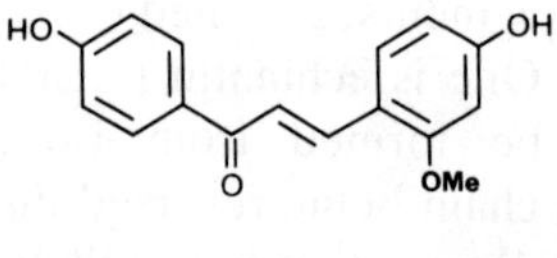

XLII Echinatin

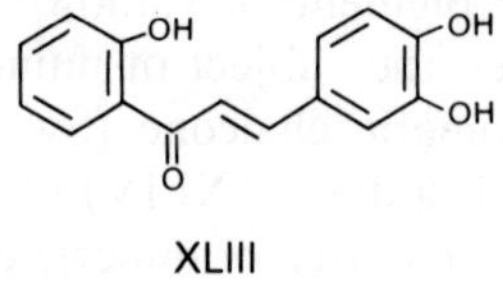

XLIII

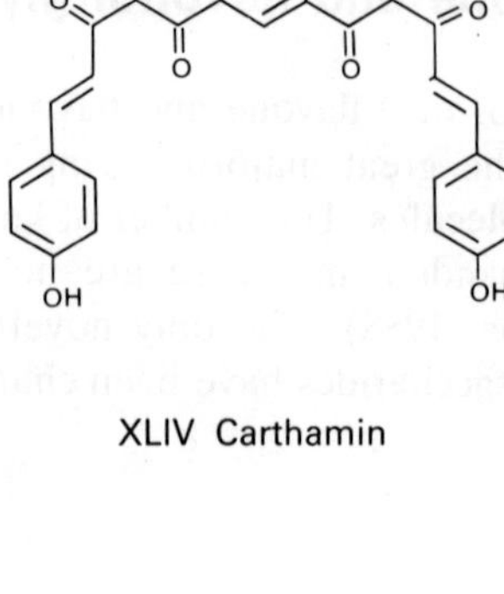

XLIV Carthamin

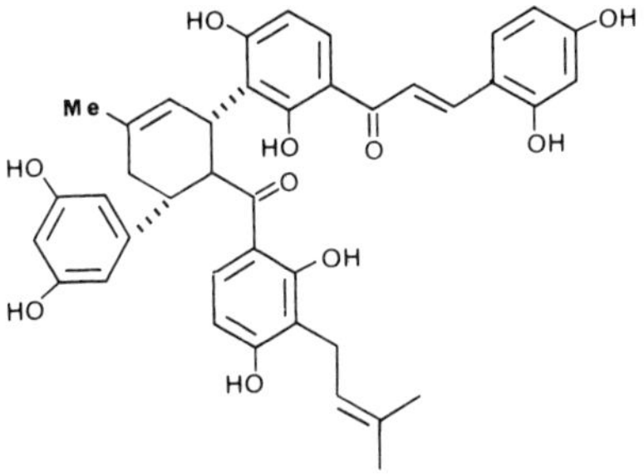

XLV Kuwanon J

Typical of the many new structures reported as floral constituents are apigenin 7-glucuronide-4′-(6″-malonylglucoside) (XLVI) from *Centaurea cyanus* (Tamura *et al.*, 1983), scutellarein 7-(6″-feruloylglucuronide) (XLVII) from *Holmskioldia sanguinea* (Nair and Mohandoss, 1982), kaempferol 3-rhamnoside-7,4′-diglucoside (XLVIII) from *Coronilla emerus* (Harborne and Boardley, 1983) and larycitrin 3,7,5′-triglucoside (XLIX) from *Medicago arborea* (Torck *et al.*, 1983).

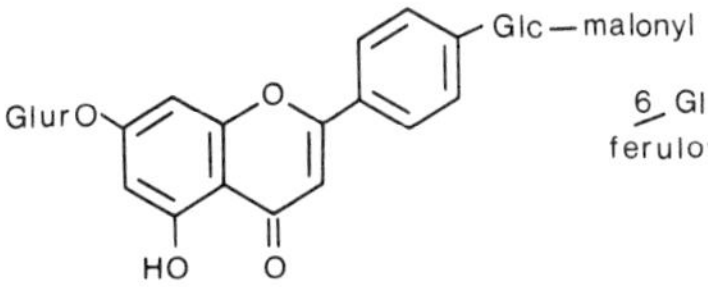

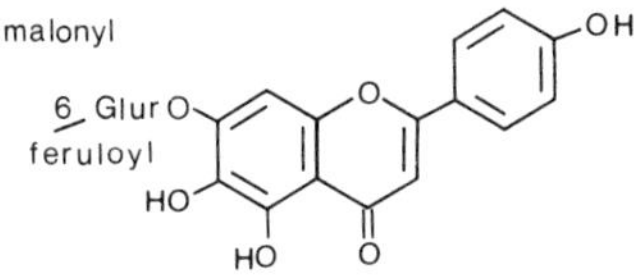

XLVI Apigenin 7-glucuronide 4′-(malonyl glucoside)

XLVII Scutellarein7 -(6″-feruloylglucuronide)

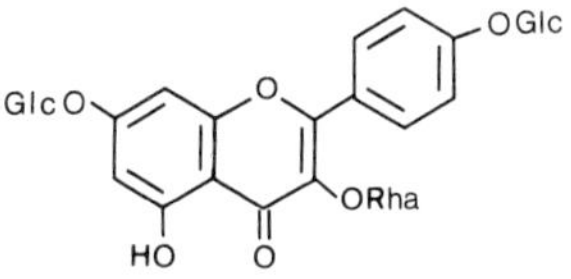

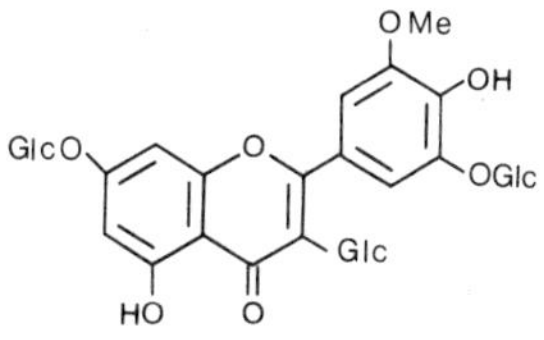

XLVIII Kaempferol 3-rhamnoside 7, 4′-diglucoside

XLIX Larycitrin3,7,5′-triglucoside

4 BIOSYNTHESIS

A Introduction

The main pathway of flavonoid biosynthesis was established some 20 years ago, based on data from chemicogenetical studies (Harborne, 1962) and on the results of feeding radioactively labelled precursors (Grisebach, 1965). The main emphasis since then has been to elucidate the fine details of the pathway from enzymological studies (Fig. 7.4). The early work on the pathway exemplified the importance of tissue culture preparations, since the first enzymes were obtained from cell cultures of parsley (Hahlbrock and Grisebach, 1975). During the period under review, attention has returned to the intact plant and, in particular, the petals of known genotypes. The extensive genetic information available has thus been turned to advantage in linking the presence/absence of particular enzymes with known genetic blocks along the pathway.

The number of enzymes that have been described since 1976 has increased and almost all the key steps in biosynthesis have now been accounted for (Table 7.4). The recognition that flavan-3,4-diols (leucoanthocyanidins) are key intermediates to anthocyanidins, catcchins and proanthocyanidins has been followed by the characterization of a flavan-3,4-cisdiol-4-reductase, which catalyses catechin synthesis. The enzyme concerned in isoflavonoid synthesis, namely 2-hydroxyisoflavanone synthase, has also just been found.

This account of flavonoid biosynthesis describes advances since the review of Wong (1976). Other recent reviews include those of Ebel and Hahlbrock (1982), Grisebach (1985) and Heller (1986).

B Enzymes of the main pathway

The central step of flavonoid biosynthesis is the condensation of three molecules of malonyl-CoA with 4-coumaroyl-CoA to form the C_{15} intermediate 4,2′,4′,6′-tetrahydroxychalcone (Fig. 7.4). The enzyme catalysing this step is chalcone synthase and much recent work has concentrated on establishing its properties. For all chalcone synthases tested so far, 4-coumaroyl-CoA is the best substrate and in general it appears that the second B-ring hydroxyl is inserted at a later stage to give flavonoids with 3′,4′-diOH substitution. Indeed, the enzyme flavonoid 3′-hydroxylase has now been characterized from a range of plants.

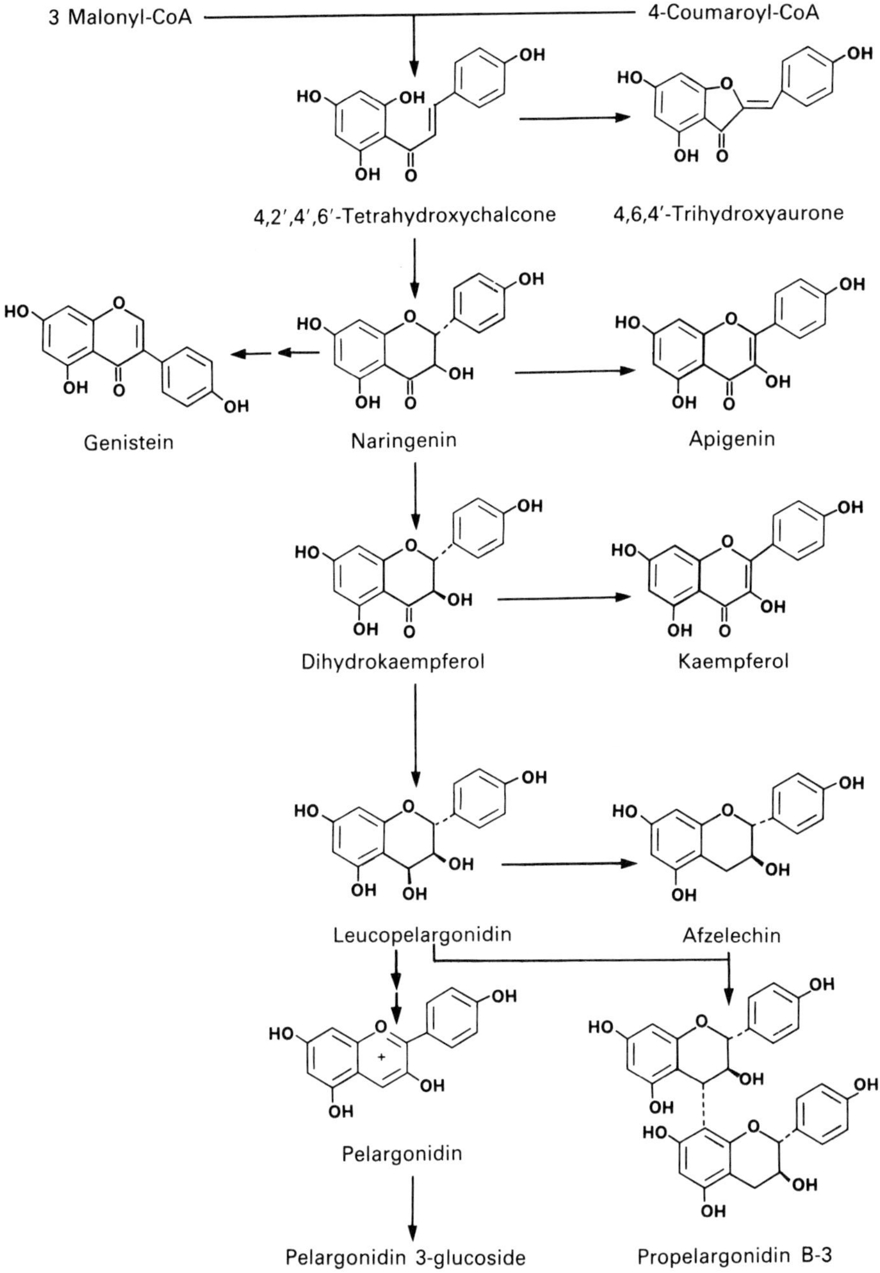

3 Malonyl-CoA

4-Coumaroyl-CoA

4,2′,4′,6′-Tetrahydroxychalcone

4,6,4′-Trihydroxyaurone

Genistein

Naringenin

Apigenin

Dihydrokaempferol

Kaempferol

Leucopelargonidin

Afzelechin

Pelargonidin

Propelargonidin B-3

Pelargonidin 3-glucoside

Fig. 7.4 Biosynthetic pathway of flavonoids.

Table 7.4 Enzymes of flavonoid biosynthesis

Enzymes	
Precursor biosynthesis	
I	Acetyl-CoA carboxylase
II	Phenylalanine ammonia-lyase
III	Cinnamate 4-hydroxylase
IV	4-Coumarate:CoA ligase
Synthesis of flavonoid classes	
1	Chalcone synthase
2	Chalcone isomerase
3	2-Hydroxyisoflavanone synthase
4	Flavone synthase
5	Flavanone 3-hydroxylase
6	Flavonol synthase
7	Dihydroflavonol 4-reductase
8	Flavan-3,4-*cis*-diol 4-reductase
9	Anthocyanidin/flavonol 3-O-glucosyl-transferase
Flavonoid modification	
1	B-ring hydroxylases
2	Glycosyl transferases
3	Methyl transferases
4	Acyl transferases
5	Sulphate transferases

Some chalcone synthases show *in vitro* activity towards caffeoyl-CoA and feruloyl-CoA, with the formation of eriodictyol and homoeriodictyol chalcones respectively, but it is doubtful whether these reactions normally occur *in vivo*. For the chalcone synthase in *Dianthus* flowers, caffeoyl-CoA can be ruled out as a physiological substrate, since in competition experiments, 4-coumaroyl-CoA was used exclusively. Indeed, these flowers also contain a flavonoid-specific 3′-hydroxylase (Spribille and Forkmann, 1982). Exceptionally, the chalcone synthase of scarlet *Verbena hybrida* flowers may act *in vivo* with both 4-coumaroyl- and caffeoyl-CoA, but these tissues are unusual in lacking 3′-hydroxylase activity (Stotz *et al.*, 1984).

Although chalcone synthase has no cofactor requirements, it is usual to add 2-mercaptoethanol to the buffer in order to reduce the formation of byproducts of the condensation reaction. It has now been found possible to eliminate byproducts by the use of oxygen-free buffers and the addition of ascorbate (Britsch and Grisebach, 1986).

The molecular weights of those chalcone synthases that have been purified range from 77 000 to 90 000. That the enzyme is dimeric has been confirmed through sequence analysis of the genes coding for the enzyme. In *Antirrhinum majus*, the chalcone synthase gene codes for a protein subunit of 42 665 (Sommer and Saedler, 1986) and in *Petroselinum crispum*, sequence analysis of the cDNA gives an enzyme subunit with M_r 43 682 (Reimold *et al.*, 1983). More than one gene for chalcone synthase has been recorded in several plants (e.g. *Petunia*; Reif *et al.*, 1985) but the only plant in which isozymes have been characterized is *Spinacia oleracea* (Beerhues and Wiermann, 1985). The two isozymes of spinach have similar M_r but differ in their Pi values (4.65 and 4.30).

The next step in biosynthesis after chalcone formation is its stereospecific isomerization to a (2S)-flavanone, naringenin, which is catalysed by chalcone isomerase. This isomerism can occur spontaneously to give a racemic product, but this reaction is considerably repressed if high protein concentrations are present (Mol *et al.*, 1985). The ability of some plants to accumulate yellow chalcones in their floral tissues must depend not only on the absence of chalcone isomerase (Chmiel *et al.*, 1983) but also on the repression of spontaneous ring closure.

Two types of chalcone isomerase are known. In plants containing both 5-deoxy- and 5-hydroxyflavonoids (e.g. *Phaseolus vulgaris*), the enzyme isomerizes both resorcinol- and phloroglucinol-based chalcones, whereas in plants lacking 5-deoxyflavonoids (e.g. *Tulipa*) only phloroglucinol-based chalcones act as substrates. Two isozymes of chalcone isomerase have been detected in *Petunia*, but they are selectively expressed in flower and anther respectively (van Weeley *et al.*, 1983). Molecular weights of between 27 000 and 62 500 have been reported for this enzyme, but it has not yet been sequenced.

Flavanones represent a branch point in the biosynthetic sequence since they may be converted to either flavones (e.g. apigenin) or to isoflavones (e.g. genistein; see Fig. 7.4). The search for an 'isoflavone synthase', which catalyses the oxidative rearrangement of flavanones and in the process produces a 2,3-aryl shift, has at last borne fruit. Hagman and Grisebach (1984) found enzymic activity in elicitor-challenged cell cultures and etiolated seedlings of *Glycine max* capable of transforming (2S)-naringenin to genistein and (2S)-liquiritigenin to daidzein.

In fact, two enzymes are required; the first step, an oxidation and rearrangement involving a transient intermediate such as (L), is catalysed by a cytochrome P-450-dependent monooxygenase and produces 2-hydroxy-2,3-dihydrogenistein (LI) (Kochs and Grisebach, 1986). The second enzyme, which catalyses the removal of water from (LI) to give genistein, has yet to be fully characterized.

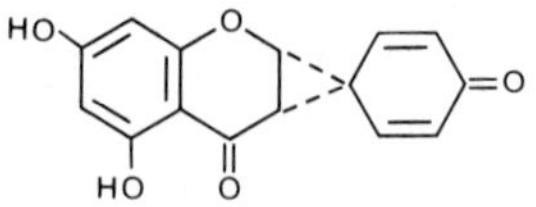

L

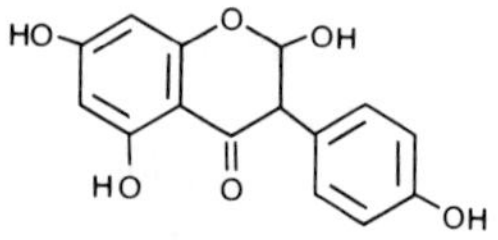

LI

The next step in the main pathway, from flavanone to flavone, can be catalysed by two different enzymes. In parsley cell suspension cultures, the enzyme requires 2-oxoglutarate, ferrous ion and possibly ascorbate as cofactors and is classified as a dioxygenase (Britsch *et al.*, 1981). By contrast, the enzyme when isolated from various floral tissues is a microsomal monooxygenase requiring NADPH as cofactor (Stotz and Forkmann, 1981). The reaction may proceed via a 2-hydroxyflavanone, but all attempts so far to detect such an intermediate have failed.

The next enzyme along the pathway, flavanone-3-hydroxylase, is similar to the flavone synthase from parsley in being a 2-oxoglutarate-dependent dioxygenase and it has been detected as such from a variety of plants. The purified enzyme from *Petunia* flowers has an M_r of 74 000 and catalyses the conversion of (2S)-naringenin to (2R, 3R)-dihydrokaempferol and also of (2S)-eriodictyol to (2R,3R)-dihydroquercetin (Britsch and Grisebach, 1986). The enzyme flavonol synthase, which converts dihydrokaempferol to kaempferol, is yet another 2-oxoglutarate-dependent dioxygenase, but it has not yet been studied in any detail (Spribille and Forkmann, 1984).

The first evidence that leucoanthocyanidins (or flavan-3,4-diols) are intermediates in proanthocyanidin synthesis was obtained when dihydroquercetin was converted into *cis*-leucocyanidin by a preparation from *Pseudotsuga menziesii* cell suspension culture (Stafford and Lester, 1982). Subsequent experiments with white flowering mutants of *Matthiola incana* showed that leucoanthocyanidins are also intermediates in anthocyanin biosynthesis (Heller *et al.*, 1985). The dihydroflavonol 4-reductase which catalyses this conversion can be either NADPH- or NAD-dependent, according to plant source. In barley, the enzyme has two pH optima (6.0 and 7.0) and may exist as two tissue-specific isozymes, one specific for anthocyanin and the other for proanthocyanidin biosynthesis (Kristiansen, 1986).

The further reduction of leucoanthocyanidins to catechins has been shown to be catalysed by a NADPH-dependent reductase (Stafford and Lester, 1985). No evidence, however, has yet been obtained for the existence of a condensing enzyme for producing the polymeric proanthocyanidins. This final step, in which catechins are linked to leucoanthocyanidins, may in fact occur spontaneously via a carbocation or quinone-methide intermediate.

The involvement of leucoanthocyanidins in anthocyanin synthesis has been nicely demonstrated by a series of supplementation experiments, in which these precursors were fed to acyanic flowers with different gene blocks; pigment production was the outcome in genotypes lacking, for example, chalcone synthase. However, the enzymic conversion of leucoanthocyanidin to anthocyanidin has not yet been achieved. Since anthocyanidins do not normally occur in plant tissues, it is assumed that the final step in the pathway is 3-glucoside formation. Indeed, anthocyanidin 3-glucosyltransferase activity has been demonstrated in many anthocyanidin-producing tissues and the genetic control of this last step has been actively investigated in several ornamental plant species, including *Petunia* (Gerats *et al.*, 1985).

B Enzymes of flavonoid modification

1 HYDROXYLATION

While the first B-ring hydroxyl group in flavonoids is introduced at the C_9 level, through the incorporation of 4-coumaroyl-CoA as the normal substrate for biosynthesis, the other two B-ring hydroxyl groups are usually introduced at the C_{15} level, with a flavanone or flavanonol as substrate. A flavonoid 3′-hydroxylase has been characterized from a variety of plants (Table 7.5); it behaves as a monooxygenase in its cofactor requirements. It is found in plants with a dominant allele for cyanidin production and likewise is absent from recessive genotypes which make pelargonidin. The alternative route to cyanidin via a chalcone synthase capable of utilizing caffeoyl-CoA as substrate cannot be ruled out completely, since it may operate in a minority of plant species. Thus, there is both genetic and enzymic evidence in *Verbena hybrida* (Stotz *et al.*, 1984) and in *Silene pratensis* (Kamsteeg *et al.*, 1981) that this minor route operates in these two plants.

Table 7.5 Enzymes of flavonoid hydroxylation

Enzyme	Reactions catalysed	Plants and Ref.
Flavonoid 3′-hydroxylase (Cytochrome P-450 mono-oxygenase)	Naringenin → eriodictyol Dihydrokaempferol → dihydroquercetin Kaempferol → quercetin Apigenin → luteolin	Many spp., e.g. *Petunia* (Stotz *et al.*, 1985)
Flavonoid 3′,5′-dihydroxylase	Naringenin → 5,7,3′,4′,5′-penta OH flavanone Dihydrokaempferol → dihydromyricetin Dihydroquercetin → dihydromyricetin	Several spp., e.g. *Verbena* (Stotz and Forkmann, 1982)

An enzyme capable of oxidizing flavonoids with a catechol B-ring (e.g. dihydroquercetin) to the pyrogallol derivative (e.g. dihydromyricetin) has yet to be described. In fact, it appears that delphinidin (and myricetin) are formed from a 4′-hydroxylated C_{15} precursor through the activity of a flavonoid 3′,5′-dihydroxylase. Such an enzyme was first isolated from delphinidin-producing *Verbena* flowers (Stotz and Forkmann, 1982) and it has subsequently been reported from several other plants. Like the 3′-hydroxylase, its presence in flowers is strictly correlated with the production of 3′,4′,5′-trihydroxylated pigments and of genes responsible for delphinidin synthesis.

2 METHYLATION

O-methylation of flavonoids occurs at a late stage in biosynthesis and may take place either before or after *O*-glycosylation (see below). *O*-methyl transferases have been characterized from a variety of plants (Table 7.6). Most are relatively substrate-specific, operate only at the C_{15} level (and *not* the C_9 level) and introduce a single methyl substitution. All *O*-methyltransferases have a pH optimum at 7.0 or above and a molecular weight around 50 000; about half the described enzymes (Table 7.6) require magnesium for full activity.

One of the best studied of these enzymes is *S*-adenosyl-L-methionine:-flavonoid 3′-hydroxylase from parsley cell cultures, which methylates luteolin to chrysoeriol and quercetin to give isorhamnetin (Ebel and Hahlbrock, 1982). Most *O*-methyltransferases described since 1982 are

Table 7.6 Flavonoid *O*-methyltransferases

Source	Substrate	Position(s) of methylation	pH optimum	M_r	Reference
Avena sativa leaf	Vitexin 2″-rhamnoside	7	7.5	52 000	Knogge and Weissenbock (1984)
Chrysosplenium americanum shoot	Quercetin Isorhamnetin	3	8.0–8.5	*ca.* 61 000	de Luca and Ibrahim (1985)
	Quercetin 3ME	7	8.0–9.0	*ca.* 61 000	
	Quercetin 37diME Quercetagetin 37 diME	4′	8.0–8.5	*ca.* 61 000	
	Quercetagetin 37 diME Quercetagetin	6	8.0–9.0	*ca.* 61 000	
Lotus corniculatus flowers	Herbacetin Gossypetin	8	7.5–8.4	55 000	Jay *et al.* (1985)
	Quercetin Gossypetin	3′	7.7	nd	Jay *et al.* (1983)
Petunia hybrida petals	Anthocyaninidin 3-(4-coumaroyl rutinoside)-5-glucoside	3′ and 5′	7.7	*ca.* 48 000 (4 enzymes)	Jonsson *et al.* (1984)
Silene pratensis	Iso-orientin 2″-rhamnoside	3′	8.0–8.5	nd	Brederode *et al.* (1982)

ME = methyl ether, nd = not determined.

similar in operating more effectively with flavonoid than with cinnamic acid substrates. Some are highly specific towards their flavone substrate. For example, the *O*-methyltransferase in *Avena sativa* leaf will only catalyse the 7-*O*-methylation of vitexin 2″-rhamnoside and is inactive towards vitexin or vitexin 2″-arabinoside (Knogge and Weissenbock, 1984). A similar enzyme from *Silene pratensis* is highly specific for iso-orientin 2″-rhamnoside (see Table 7.6).

In the case of polymethoxylated flavonoids, there appear to be individual enzymes for introducing the different methyl groups. This is true in *Chrysosplenium americanum*, from which separate enzymes have been obtained which sequentially methylate quercetin and quercetagetin (6-hydroxyquercetin) in the 3-,7-,4′- and 6-positions to yield finally quercetin 3,7,4′-trimethyl ether and quercetagetin 3,6,7,4′-tetramethyl ether respectively (de Luca and Ibrahim, 1985). Even when only one or two hydroxyl groups of the flavonoid undergo methylation, there may be a multiplicity of enzymes. For example, in *Petunia hybrida*, there are four *O*-methyltransferases, each controlled by a separate gene, which are all capable of catalysing the methylation of delphinidin in the 3′- and 5′-positions to give petunidin and malvidin (Table 7.6).

3 GLYCOSYLATION

The occurrence in nature of such an enormous range of flavonoid glycosides suggests the presence in plants of a similar range of *O*-glycosyltransferases for their synthesis and this has been borne out in practice. The many enzymes of this class that have been characterized so far share a pronounced specificity with regard to substrate and the position and nature of the sugar to be transferred (Wong, 1976). One exception to this is the anthocyanidin 3-*O*-glucosyltransferases, since they have been found in almost all plants to operate with flavonols as well as with anthocyanidins. Even here, greater specificity has been noted in individual cases. Thus, the flavonol 3-glucosyltransferase from tulip anthers will not catalyse the glucosylation of the corresponding anthocyanidins (Kleinenollenhorst *et al.*, 1982).

In the case of flavonoid di- and triglycosides, sugars are added in a logical sequence, each *O*-glycosyltransferase being specific for a particular intermediate. For example, three *O*-glucosyltransferases have been obtained from *Pisum sativum* flowers by Jourdan and Mansell (1982) controlling the sequential addition of one, two and three glucose units to the 3-position of kaempferol and quercetin. In the case of the biosynthesis of acylated glycosides, glycosyltransferases may sometimes exhibit specificity for an acylated substrate. Thus the 5-*O*-glucosylation of delphinidin 3-(4-coumaroylrutinoside) is catalysed by an enzyme from *Petunia* flowers which shows no activity whatsoever towards delphinidin 3-glucoside or even delphinidin 3-rutinoside. Thus acylation must precede 5-glycosylation in this biosynthetic pathway (Jonsson *et al.*, 1984).

Finally, it is worth recording that the first flavonol glucuronosyl transferase has been described from cell cultures of *Anethum graveolens* (Mohle *et al.*, 1985). Again, the first enzymic synthesis of *C*-glycosylflavones has been reported in buckwheat seedlings. The substrates here for *C*-glucosyltransferases are 2-hydroxyflavanones, and not the expected flavanones or flavones (Kerscher and Franz, 1986).

4 ACYLATION

Acylated flavonoids, in which one or more acyl groups are usually linked through sugars, have been found with increasing frequency in recent years. The enzymes controlling the addition of aliphatic or aromatic acids to flavonoids appear to be almost as specific in their substrate requirements as *O*-glycosyltransferases. Thus *O*-malonyltransferases specific respectively for flavone glycosides, isoflavone glucosides and anthocyanins have now been obtained from *Petroselinum crispum* (Matern *et al.*, 1983), from *Cicer arietinum* (Koster *et al.*, 1984) and from *Callistephus chinensis* (Teusch and Forkmann, 1987). The latter enzyme catalyses the transfer of malonate from malonyl CoA to pelargonidin, cyanidin and delphinidin 3-glucosides to give the corresponding 3-(6″-malonylglucosides) but has no activity with the corresponding 3-xylosylglucosides. While malonyl CoA is the most efficient acyl donor, the CoA esters of methylmalonate, succinate and glutarate were also used to some extent.

The first enzyme catalysing the aromatic acylation of anthocyanins was demonstrated in *Silene pratensis* extracts by Kamsteeg and coworkers in 1981. It catalyses the transfer of 4-coumarate or caffeate from the CoA esters to the 4-position of the rhamnose moiety in anthocyanidin 3-rutinosides or 3-rutinoside-5-glucosides. An aromatic acyltransferase has also been characterized from *Matthiola incana* flowers (Teusch *et al.*, 1986). This adds the same two cinnamic acids to anthocyanidin 3-glucosides and 3-xylosylglucosides, but it has no activity with the corresponding 5-glucosides (i.e. the 3,5-diglucoside or 3-xylosylglucoside-5-glucoside).

The enzymology of one other modification in flavonoid biosynthesis deserves mention, that of sulphation. Over 60 flavone and flavonol sulphates have been recorded variously in some 30 plant families but their biosynthesis has only just been examined. A sulphate transferase, using adenosine 3′-phosphate-5′-phosphosulphate (PAPS) as donor, which catalyses quercetin 3-sulphate synthesis, has been found in an extract of *Flaveria bidentis* by Varin *et al.* (1987).

5 NATURAL DISTRIBUTION

A Localization within the plant

While it is well established that anthocyanins are located in the vacuoles of petal tissues, little attention has been given to the question of whether all cells of the average petal are pigmented or not. In fact, an extensive survey of petals of 201 species from 60 families has shown that flavonoids are nearly always located specifically and exclusively in the epidermal cells (Kay *et al.*, 1981). Thus, they are located in those cells where they can most effectively contribute to flower colour. Anthocyanins were found exceptionally in mesophyll cells in most members of the Boraginaceae and a few species of Liliaceae; in these instances, localization of anthocyanin in mesophyll is correlated with morphological differences in the shape of the epidermal cells.

The same survey (Kay *et al.*, 1981) showed that co-occurring flavones were also located in the epidermal cells, which adds useful confirmation to the idea that flavones have an essential function in coloured petals as copigments to the anthocyanins (see section 6A). Even in white or cream flowered species, the flavones are similarly restricted to epidermal cells, so that such petals absorb rather than reflect light in the UV. The recognition of white flowered species by insect pollinators would seem to depend therefore on their ability to perceive in the UV as well as in the visible spectral range.

How far anthocyanins are restricted to particular cell layers in other plant parts is as yet uncertain. With anthocyanin pigmentation in leaves, it has been found that an *argenteum* mutant of the garden pea, *Pisum sativum*, contains the pigment just in the epidermal cells (Hrazdina *et al.*, 1982). On the other hand, a similar investigation of the leaves of a coloured cultivar of rye, *Secale cereale*, showed that the anthocyanin is not in the epidermis, but is restricted to mesophyll cells (McClure, 1986). In other species where pigmentation is intense, e.g. copper beech or red cabbage leaves, one suspects that anthocyanin is present in both epidermal and mesophyll cells.

The enzymes of anthocyanin synthesis are probably located in the cytoplasm attached to the vacuolar membrane and it is likely that in the final stage of biosynthesis, the attachment of a sugar, the pigment is transported into the vacuole (Grisebach, 1982). This hypothesis is at some variance with the observations of Pecket and Small (1980) and Yasuda and Shinoda (1985) of transient production of 'anthocyanoplasts' within the vacuoles of hypocotyls of red cabbage and radish, respectively. These red dots are only apparent when pigment synthesis is operating and they disappear as the whole vacuole fills with colour. Ultrastructural studies in red cabbage cells indicate that anthocyanoplasts are bound by a single tripartite membrane 10 nm thick.

Formation of anthocyanoplasts within the vacuole appears to be a general phenomenon in that they have been observed in 70 species representing 33 families (Small and Pecket, 1982). While it is possible that the later steps in anthocyanin biosynthesis actually occur at the membrane of anthocyanoplasts, Neumann (1983) has argued that the existence of functional organelles within the vacuole is unlikely.

To turn to the localization of flavone and flavonol glycosides in the leaves, it has usually been assumed that because of their water solubility they would be confined to vacuoles. In the early 1970s, however, there were reports from several laboratories of these flavonoids being found in plant chloroplasts (see Harborne, 1976). In one instance it was suggested that the flavonoids found in spinach chloroplasts might actually have a role in photosynthesis (Oettmeier and Heupel, 1972). An obvious criticism of such observations is the likelihood that the flavonoids so detected came from the vacuole as a result of irreversible adsorption on to the outer chloroplast membrane during organelle fractionation. This now seems to be the explanation for these reports, since recently Charriere-Ladreik and Tissut (1981) were able to prepare spinach chloroplasts essentially free of flavonoids.

How far flavonoids are distributed within the average plant leaf is another question which has yielded some results in investigations involving the stripping of epidermal cells from other tissues. In *Vicia faba*, flavonoids are exclusively found in the vacuoles of epidermal cells. No less than 59% of total leaf flavonoid is in the upper epidermis, with only 7% in the lower epidermis and the remainder in the guard cells (Weissenbock *et al.*, 1984). The situation in two other legumes, the pea and soya bean, is similar (McClure, 1986).

Cereal leaves seem to differ from legumes in having flavonoids throughout the leaf. There are, however, differences between epidermal and mesophyll constituents. In *Avena sativa*, glycosylflavones occur in both epidermis and mesophyll, while kaempferol glycosides are restricted to mesophyll (Proksch *et al.*, 1981). In barley, *Hordeum vulgare*, the main glycosylflavone of the epidermis, saponarin, differs from the glycosylflavone of the mesophyll, lutonarin (McClure, 1986). Finally, in *Secale cereale*, glycosylflavones are limited to the epidermis, while flavone *O*-glucuronides replace them in the mesophyll layer (Schulz *et al.*, 1985).

Flavonoids also occur on the leaf surface in waxes, in leaf resins and in bud exudates. One survey (Wollenweber and Dietz, 1981) showed that these flavonoids are almost always in the free state, the lipophilic character of the flavonoid aglycones being enhanced in most cases by *O*-methylation. Glycosides may not, however, be excluded completely from the leaf surface. In tomato plants, for example, the flavonol glycoside rutin has been found to accumulate in the leaf hairs (Isman and Duffey, 1982).

B Patterns of occurrence

Much new information on the distribution of anthocyanins in plants has accumulated in recent years. Surveys at the family level (Table 7.7) have indicated that the type of glycoside present is often characteristic, even if the anthocyanidin type is not. In some families (e.g. Commelinaceae, Compositae) more than one type of glycoside may be present. Detailed listings of the occurrence of these pigments at specific and generic level can be found in Timberlake and Bridle (1975), Hrazdina (1982) and Harborne and Grayer (1988).

Reports of the presence of chalcones, aurones and flavonols in yellow flowers continue to appear (Table 7.8). Most work centres on plants of the Compositae, where yellow flower colour is so prevalent, and where anthochlor pigments (chalcones and aurones) are particularly characteristic.

Colour in other plant organs may be due to yellow flavonoids. Thus the yellow flavonols herbacetin 8-methyl ether and gossypetin 8,3′-dimethyl ether have been identified in the pollen of *Paeonia* and *Rumex* species (Wiermann *et al.*, 1981). A survey of 140 species for pollen pigments confirmed that this is a rare phenomenon. Although chalcones have occasionally been reported to provide yellow colour in pollen, the usual source of pollen coloration is carotenoid.

Aurones have recently been identified for the first time in bryophytes. The deep yellow pigment bracteatin adds colour to the sporophyte tissues of the moss *Funaria hygrometrica* (Weitz and Ikan, 1977) and aureusidin 6-glucuronide occurs in the antheridiophores of four liverwort species (Markham and Porter, 1978). Aurones are widely present in the free state in plants of the Cyperaceae. They are more frequent in the inflorescences and seeds, where they add colour, than in the leaves, where their colour is largely masked by chlorophyll (Harborne *et al.*, 1985).

Table 7.7 Recent family surveys of anthocyanin pigmentation

Family	Pigment types	Reference
Aquifoliaceae (8 spp.)	Pg and Cy 3-glucosides and 3-xylosylglucosides	Ishikura and Sugahara (1979)
Araceae (59 spp.)	Pg and Cy 3-glucosides and 3-rutinosides	Williams *et al.* (1981)
Berberidaceae (14 spp.)	Cy-Mv 3-glucosides, Dp 3-(4 coumaroylsophoroside)-5-glucoside	Yoshitama (1984)
Bromeliaceae (34 spp.)	Cy 3,5,3′-triglucoside common	Saito and Harborne (1983)
Commelinaceae (28 spp.)	Acylated Dp 3,5-diglucoside and acylated Cy 3,7,3′-triglucoside	Stirton and Harborne (1980)
Compositae (50 spp.)	Malonyl anthocyanins in most tribes; succinyl anthocyanins in the Cynareae	Takeda *et al.* (1986); Sukyok and Raszlo-Benczik (1985)
Melastomataceae (24 spp.)	Acylated 3,5-diglucosides of Pg, Cy, Dp and Mv	Lowry (1976)
Myrtaceae (140 spp.)	3-Glucosides and 3,5-diglucosides of Cy and Mv	Lowry (1976)
Orchidaceae (20 spp.)	Cy 3-oxalylglucoside	Strack *et al.* (1986)
Polemoniaceae (34 spp.)	Pg, Cy and Dp 3-glucosides, 3,5-diglucosides and acylated derivatives	Harborne and Smith (1978)
Polygonaceae (33 spp.)	Pg, Cy and Dp 3-glucosides, Mv 3,5-diglucoside	Yoshitama *et al.* (1984)
Sterculiaceae (13 spp.)	Cy and Dp 3-glucosides	Scogin (1979)

Pg = pelargonidin; Cy = cyanidin; Pn = peonidin; Dp = delphinidin; Pt = petunidin; Mv = malvidin.

Table 7.8 Plant species with yellow flavonoids in the flowers

Family, genus and species	Pigment type	Pigment(s)	Reference
Compositae			
Eriophyllum spp.	Flavonol	Quercetagetin, patuletin	
Geraea spp.	Flavonol	Gossypetin 8-ME	Harborne and Smith (1978)
Helianthus spp.	Anthochlor	Coreopsin, sulphurein	
Viguiera laciniata	Anthochlor	Coreopsin, sulphurein	
Pyrrhopappus spp.	Anthochlor	Coreopsin	Harborne, 1977
Heywoodiella olicocephala	Flavone	Isoetin	Harborne, 1978
Zinnia linearis	Anthochlor	Marein, maritimein	Harborne *et al*. 1983
Leguminosae			
Acacia dealbata	Aurone	Cernuoside	Imperato (1982)
Coronilla valentina	Flavonol	Gossypetin 3′Me Gossypetin 8,3′-diMe	Harborne (1981)
Rosaceae			
Potentilla spp.	Chalcone	Isosalipurposide	Harborne and Nash (1984)
Rubiaceae			
Mussaenda hirsutissimum	Aurone	Aureusin, cernuoside	Harborne *et al*. (1983)

6 FUNCTIONS

A Cyanic colour

As mentioned earlier (section 3A), anthocyanins in aqueous solution at the pH range of cell sap (2.8–6.2, with most about 4.5) are subject to attack by hydroxyl ions, being then converted to colourless pseudobases. The intensity of colour is considerably reduced and hence, for anthocyanins to provide deep colours *in vivo*, there must be some way in which the flavylium cations are protected from hydrolytic attack. Research of the last decade has indicated that there are at least four ways in which anthocyanins are stabilized *in vivo*. These involve self-association, intramolecular copigmentation, intermolecular copigmentation and metal complexing. A fifth factor, malonylation, also appears to stabilize these pigments (Saito *et al.*, 1985) but it will not be mentioned further, since it is not yet known how it contributes to stability.

1 SELF-ASSOCIATION

The idea that anthocyanins are able to self-associate in solution was first suggested by Asen *et al.* (1972) to explain why the visible absorbance of a solution of cyanin (cyanidin 3,5-diglucoside) deviated from the Beer-Lambert law with increasing concentration. Timberlake (1980) observed the same effect with malvidin 3-glucoside but it was Hoshino and his coworkers (1982) who first investigated the phenomenon using both absorption spectrophotometry and circular dichroism. Cyanin and malvin exhibit opposite Cotton effects and association occurs much faster for malvin than for cyanin. A free 4′-hydroxyl is necessary, since 4′-*O*-methylmalvin failed to associate. Sugars at both the 3- and 5-positions increase the opportunities for self-association. Proton NMR measurements on malvin in solution at pH 7 are consistent with the idea that the molecules are vertically stacked one above another.

Self-association is thus a mechanism by which some anthocyanins, especially partly methylated 3,5-diglucosides, are protected *in vivo* from hydrolytic attack. Above a certain critical concentration (perhaps about 10^{-2} M), the flavylium cations which are known to be nearly planar (Saito and Ueno, 1985) form molecular aggregates. How far this is important *in vivo* is not yet known. Since anthocyanins are almost always accompanied in petals by colourless flavones, it is likely that intermolecular copigmentation will operate (see below). It is therefore difficult to separate the two factors which may both contribute to stabilization. An increasing concentration of pigment in the vacuole could lead to crystallization and pigment crystals have been observed *in vivo* (Pecket and Small, 1980). This would be yet another method of avoiding the attack of water on the flavylium cation.

2 INTERMOLECULAR COPIGMENTATION

The copigmentation of anthocyanins by colourless flavonoids has long been suspected to contribute to flower colour (Robinson and Robinson, 1931; Asen, 1976). Recent investigations based on HPLC have confirmed the fact that flavone copigments usually co-occur with anthocyanins in most angiosperm flowers (Asen, 1984). The protective value of copigmentation has recently been considered by Brouillard (1982). Calculations indicate that the stability constant of the complex between the flavylium cation of cyanin and the flavonol glycoside quercitrin is close to 2×10^{-3} M^{-1}, which leads to an apparent reduction in the hydration constant for the cation from 10^{-2} M to 7×10^{-4} M. Hence, the copigment–pigment complex will be much less subject to hydration by water.

The stability of copigment complexes will be affected by pH so that pH changes may alter the shade of colour. Yazaki (1976) was able to account for the colour changes that take place in *Fuchsia* flowers during ageing on the basis of the pH changes affecting a solution of malvin and quercetin glycoside. Other phenolic constituents besides flavonoids are capable of forming complexes with anthocyanin, and it is likely that in *Hydrangea* flowers the copigment is 3-caffeoylquinic acid (Takeda *et al.*, 1985).

The isolation of intact pigment–copigment complexes using very mild methods from a range of blue flowered plant species (e.g. *Commelina communis*, see below) confirms the view that such complexes are common *in vivo*. However, it would be more satisfactory if it was possible to demonstrate the presence of such complexes *in vivo* by a non-destructive technique. And this has been accomplished in *Malva sylvestris* by means of resonance Raman spectroscopy. These measurements on individual epidermal cells of mallow petals establish that malvin occurs exclusively as the flavylium cation. Likewise, the partial quenching *in vivo* of the fluorescence emission of malvin, as compared with pure malvin in solution, strongly suggests that the flavylium cation is complexed in the petal with flavone (Merlin *et al.*, 1985).

3 INTRAMOLECULAR COPIGMENTATION

While normal anthocyanins lose most of their colour when dissolved in aqueous solutions at pHs 4–6, a blue pigment platyconin from *Platycodon grandiflorum* flowers was discovered to be exceptionally stable, retaining its colour in solution even at pH 7 (Saito *et al.*, 1971). Similar stable pigments were then isolated from other blue flowers: heavenly blue anthocyanin from *Ipomoea tricolor*, cinerarin from *Senecio cruentus* and an acylated cyanidin 3,7,3′-triglucoside from *Zebrina pendula* (for structures, see section 3A). All these pigments were united in having more than one aromatic acyl substituent linked through sugar and Brouillard (1981) was able to show that these pigments in solution adopt a configuration with the acyl groups on either side of the pyrylium ring, thus protecting the flavylium cation from nucleophilic water addition (see Fig. 7.5).

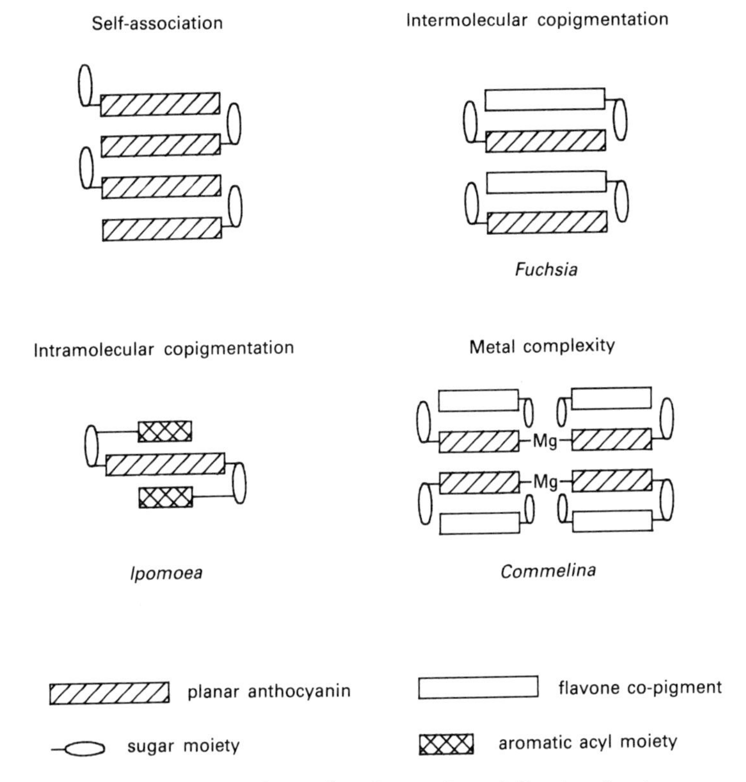

Fig. 7.5 Mechanisms of anthocyanin stabilization *in vivo*.

While anthocyanins with two or more acyl substituents are most effectively stabilised *in vivo* through this 'intramolecular copigmentation', monoacylated anthocyanins may also be similarly stabilized, but to a lesser extent.

There is no doubt that intramolecular copigmentation exists in the petal cell, since such anthocyanins have a very characteristic absorption in the visible region at pH 4 to 7, with three bands and a shoulder. The visible spectrum of a pure aqueous solution of heavenly blue anthocyanin matches exactly the visible spectrum of an intact petal of *Ipomoea tricolor* (Asen *et al.*, 1977). This characteristic absorption pattern can be used to survey petals directly for the presence of these intramolecularly copigmented quinonoid bases.

4 METAL COMPLEXES

The fact that anthocyanins with catechol B-rings readily form metal complexes in solution with the shifting of colour to the blue region has led to the idea that such complexing may well stabilize anthocyanins *in vivo*. The difficulty of isolating such complexes intact from petal extracts has seriously limited progress and the subject has become controversial with claims and counterclaims regarding the presence of metal ions in different pigment preparations. There are now two blue flowered plant species where the evidence of metal complexing is reasonably secure, namely *Commelina communis* and *Hydrangea macrophylla*. However, in both cases, other stabilizing features are present and it is not yet clear whether metal complexing is essential to the stability of the pigment *in vivo*.

The pigment–metal complex which has been most extensively studied is commelinin, from blue flowers of *Commelina communis*. The components are the acylated anthocyanin, malonylawobanin, and the flavone flavocommelinin (7-methylisovitexin 4′-glucoside) with magnesium as metal. The most recent experiments suggest that the complex is composed of four anthocyanin and four flavone units held together by the presence of two magnesium atoms, as illustrated in Fig. 7.5 (Goto *et al.*, 1986), but the position of attachment of the metal to the complex is still uncertain.

The second pigment complex that has been reinvestigated recently is that of the red and blue sepals of *Hydrangea macrophylla*. The blueing of the red sepals has long been considered to be due to the complexing of the anthocyanin with aluminium ion. This has been confirmed by Takeda *et al.* (1985) who found that 3-caffeoylquinic acid is also required as a copigment to the blueing process. Curiously, *Hydrangea* petals also contain 5-caffeoylquinic acid (chlorogenic acid) but this is quite ineffective in replacing the 3-caffeoyl isomer in the blue complex of delphinidin 3-glucoside and aluminium. Presumably the aluminium cation stabilizes the interaction between quinic ester and anthocyanin, but exactly how is not yet clear.

B Yellow flower colour and UV patterning

UV patterning is apparent in a range of yellow flowered species, where the principal pigments are usually carotenoids. Analyses of the UV regions of the petal or ray have shown that yellow flavonols or anthochlors are exclusively located here and they quench the light reflectance of the carotenoids present in chromoplasts (Harborne, 1976). Thus one type of yellow pigment reinforces the effectiveness of another in the visible region, at the same time providing honey guide patterning, which guides the pollinator to the centre of the flower and the nectaries.

There is, of course, no absolute necessity for the flavonoids in such petals to be visibly yellow, since all flavonoids, irrespective of their absorption characteristics in the visible region, absorb strongly in the UV. Recent experiments have now shown that UV patterning can be provided by colourless flavonol glycosides in certain species. This is true, for example, in *Coronilla* flowers. Yellow gossypetin derivatives provide UV patterning in the wings of the petal of *C. valentina*, while colourless kaempferol and quercetin glycosides take over this role in the wings of *C. emerus* (Harborne, 1981; Harborne and Boardley, 1983). This also happens in petals of *Potentilla*, where a range of different UV patterns can be observed. The chalcone isosalipurposide provides UV absorption in six species, whereas quercetin glycosides are responsible for UV patterning in eight other species (Harborne and Nash, 1984).

C Feeding deterrence

Earlier, it was pointed out (Harborne, 1976) that the accumulation of flavonoids in leaves could be related to the fact that these substances can have a significant effect on insect feeding responses and that the *raison d'être* might lie, in part, in providing feeding deterrency, although clearly not all insects will necessarily be deterred from feeding. Various examples were given in Harborne (1976) of flavonol, flavone and flavonol glycosides which acted on insect feeding behaviour. To these examples can now be added anthocyanin pigments.

Hedin *et al.* (1983) have observed that cyanidin 3-glucoside is an important factor in the resistance of cotton leaves to the feeding of the tobacco budworm in field experiments. Proanthocyanidins are also present in cotton leaves, but their distribution is not correlated with resistance, as it is with cyanidin 3-glucoside. Laboratory experiments (Table 7.9) confirm the effectiveness of anthocyanins in reducing larval growth in this insect *Heliothis viriscens*. The mechanism by which anthocyanins exert this effect is not yet clear.

This discovery does raise the question of whether leaf anthocyanins generally have a protective role against insect feeding. Can some insects perceive the colour of red leaves and avoid feeding, without even tasting them? Further experiments along these lines would be of interest, comparing for example herbivory on ordinary beech leaves with that on copper beech leaves.

Table 7.9 Inhibitory effects of flavonoids on larval growth in the tobacco budworm

Flavonoid	ED_{50}[1]
(+)-Catechin	0.052
Proanthocyanidin	0.063
Quercetin	0.042
Quercetin 3-glucoside	0.060
Cyanidin	0.166
Delphinidin	0.138
Cyanidin 3-glucoside	0.070

[1]ED_{50} = percentage concn. in the diet which reduces larval growth of *Heliothis viriscens* by 50%.

REFERENCES

Akavia, N., Strack, D. and Cohen, A. (1981). *Z. Naturforsch.* **36c**, 378.
Asen, S. (1976). *Acta Hortic.* **63**, 217.
Asen, S. (1984). *Phytochemistry* **23**, 2523.
Asen, S., Stewart, R. N. and Norris, K. H. (1972). *Phytochemistry* **11**, 1139.
Asen, S., Stewart, R. N. and Norris, K. H. (1977). *Phytochemistry* **16**, 1118.
Beerhues, J. and Wiermann, R. (1985). *Z. Naturforsch.* **40c**, 160.
Bohlmann, F. and Misra, L. N. (1984). *Planta Medica* **50**, 271.
Bohm, B. A. (1982). In: *Flavonoids: Advances in Research* (Eds. Harborne, J. B. and Mabry, T. J.), pp. 313–416. Chapman & Hall, London.
Bohm, B. A. (1988). In: *Flavonoids: Advances in Research since 1980* (Ed. Harborne, J. B.) pp. 329–388. Chapman & Hall, London.

Brederode, J. van, Kamps-Heinsbroek, R. and Mastenbroek, O. (1982). *Z. Pflanzenphysiol.* **106**, 43.
Britsch, L. and Grisebach, H. (1985). *Phytochemistry* **24**, 1975.
Britsch, L., Heller, W. and Grisebach, H. (1981). *Z. Naturforsch.* **36c**, 742.
Britton, G. (1983). *The Biochemistry of Natural Pigments*. Cambridge University Press, Cambridge.
Brouillard, R. (1981). *Phytochemistry* **20**, 143.
Brouillard, R. (1982). In: *Anthocyanins as Food Colors* (Ed. P. Markakis), pp. 1–40. Academic Press, New York.
Brouillard, R. (1983). *Phytochemistry* **22**, 1311.
Chari, V. M., Harborne, J. B. and Williams, C. A. (1980). *Phytochemistry* **19**, 983.
Charriere-Ladreix, Y. and Tissut, M. (1981). *Planta* **151**, 309.
Cheminat, A. and Brouillard, R. (1986). *Tetrahedron Lett.* 4457.
Chmiel, E., Sutfeld, R. and Wiermann, R. (1983). *Biochem. Physiol. Pflanzen.* **178**, 139.
Chopin, J. and Bouillant, M. L. (1975). In: *The Flavonoids* (Eds. Harborne, J. B., Mabry, T. J. and Mabry, H.) pp. 632–691. Chapman & Hall, London.
Cody, V., Middleton, E. and Harborne, J. B. (eds) (1986). *Plant Flavonoids in Biology and Medicine*. Alan R. Liss, New York.
Colijn, C. M., Jonsson, L. M. V., Schram, A. W. and Kool, A. J. (1981). *Protoplasma* **107**, 63.
Czygan, F. C. (ed.) (1980). *Pigments in Plants*. Gustav Fischer, Stuttgart.
Daigle, D. J. and Conkerton, E. J. (1983). *J. Liquid Chromatogr.* **6**, 105.
Dewick, P. M. (1984). *Natural Product Reports* **1**, 451.
Domon, B. and Hostettmann, K. (1985). *Phytochemistry* **24**, 575.
Ebel, J. and Hahlbrock, K. (1982). In: *Flavonoids: Advances in Research* (Eds. Harborne, J. B. and Mabry, T. J.) pp. 641–680. Chapman & Hall, London.
Gerats, A. G. M., Vrijlandt, E., Wallroth, M. and Schram, A. W. (1985). *Biochem. Genet.* **23**, 591.
Goodwin, T. W. (ed.) (1965). *Chemistry and Biochemistry of Plant Pigments*. Academic Press, London.
Goodwin, T. W. (ed.) (1976). *Chemistry and Biochemistry of Plant Pigments*, 2nd edn. Academic Press, London.
Goto, T. (1984). *Proc. Fifth Asian Symp. Medicinal Plant Spices*, 593.
Goto, T., Kondo, T., Tamura, H. and Takase, S. (1983). *Tetrahedron Lett.* 4863.
Goto, T., Kondo, T., Kawai, T. and Tamura, H. (1984). *Tetrahedron Lett.* 6021.
Goto, T., Tamura, H., Kawa, T., Hoshino, T., Harada, N. and Kondo, T. (1986). *Annals N.Y. Acad. Sci.* **471**, 155.
Grisebach, H. (1965). In: *Chemistry and Biochemistry of Plant Pigments* (Ed. Goodwin, T. W.) pp. 279–308. Academic Press, London.
Grisebach, H. (1982). In: *Anthocyanins as Food Colors* (Ed. Markakis, P.) pp. 69–96. Academic Press, New York.
Grisebach, H. (1985). *Ann. Proc. Phytochem. Soc. Eur.* **25**, 183.
Hagman, M. L. and Grisebach, H. (1984). *FEBS Lett.* **175**, 199.
Hahlbrock, K. and Grisebach, H. (1975). In: *The Flavonoids* (Eds. Harborne, J. B., Mabry, T. J. and Mabry, H.) pp. 866–915. Chapman & Hall, London.
Harborne, J. B. (1962). In: *The Chemistry of Flavonoid Compounds* (Ed. Geissman, T. A.) pp. 593–617. Pergamon Press, Oxford.
Harborne, J. B. (1976). In: *Chemistry and Biochemistry of Plant Pigments* (Ed. Goodwin, T. W.), vol. 1, pp. 736–779. Academic Press, London.

Harborne, J. B. (1977). *Phytochemistry* **16**, 927.
Harborne, J. B. (1978). *Phytochemistry* **17**, 915.
Harborne, J. B. (1981). *Phytochemistry* **20**, 1117.
Harborne, J. B. (1983). In: *Chromatography, Part B. Applications* (Ed. Heftmann, E.) pp. 407–434. Elsevier, Amsterdam.
Harborne, J. B. (1985). In: *Advances in Medicinal Plant Research* (Eds. Vlietinck, A. J. and Dommisse, R. A.) pp. 135–151. Wissenschaftliche Verlagsgesellschaft, Stuttgart.
Harborne, J. B. (1986). *Phytochemistry* **25**, 1887.
Harborne, J. B. (ed.) (1988). *Flavonoids: Advances in Research since 1980* Chapman & Hall, London.
Harborne, J. B. and Boardley, M. (1983). *Z. Naturforsch.* **38c**, 148.
Harborne, J. B. and Boardley, M. (1984). *J. Chromatogr.* **299**, 377.
Harborne, J. B. and Boardley, M. (1985a). *Z. Naturforsch.* **40c**, 305.
Harborne, J. B., Boardley, M. and Linder, H. P. (1985b). *Phytochemistry* **24**, 273.
Harborne, J. B. and Grayer, R. (1988). In: *Flavonoids: Advances in Research since 1980* pp. 1–20. Chapman & Hall, London.
Harborne, J. B. and Mabry, T. J. (eds.) (1982). *Flavonoids: Advances in Research*. Chapman & Hall, London.
Harborne, J. B. and Nash, R. J. (1984). *Biochem. System. Ecol.* **12**, 315.
Harborne, J. B. and Smith, D. M. (1978). *Biochem. System. Ecol.* **6**, 287.
Harborne, J. B. and Williams, C. A. (1988). In: *Flavonoids: Advances in Research since 1980* (Ed. Harborne, J. B.) pp. 304–328. Chapman & Hall, London.
Harborne, J. B., Mabry, T. J. and Mabry, H. (eds.) (1975). *The Flavonoids*. Chapman & Hall, London.
Harborne, J. B., Girija, A. R., Devi, H. M. and Lakshmi, N. K. M. (1983). *Phytochemistry* **22**, 2741.
Harborne, J. B., Williams, C. A. and Wilson, K. L. (1985). *Phytochemistry* **24**, 751.
Hedin, P. A., Jenkins, J. N., Collum, D. H., White, W. H. and Parrott, W. L. (1983). In: *Plant Resistance to Insects* (Ed. Hedin, P. A.) p. 347. American Chemical Society, Washington.
Heller, W. (1986). In: *Plant Flavonoids in Biology and Medicine* (Eds. Cody, V., Middleton, E. and Harborne, J. B.) pp. 25–42. Alan R. Liss, New York.
Heller, W., Forkmann, G., Britsch, L. and Grisebach, H. (1985). *Planta* **165**, 284.
Hoshino, T., Matsumoto, U., Goto, T. and Harada, N. (1982). *Tetrahedron Lett.* 433.
Hostettmann, K., Domon, B., Schaufelberger, D. and Hostettmann, M. (1984). *J. Chromatogr.* **283**, 137.
Hostettmann, K., Hostettmann, M. and Marston, A. (1986). *Preparative Chromatography Techniques*. Springer-Verlag, Berlin.
Hrazdina, G. (1982). In: *Flavonoids: Advances in Research* (Eds. Harborne, J. B. and Mabry, T. J.) pp. 135–188. Chapman & Hall, London.
Hrazdina, G., Marx, G. A. and Hoch, H. C. (1982). *Plant Physiol.* **70**, 745.
Ikuta, J., Fukai, T., Nomura, T. and Ueda, S. (1986). *Chem. Pharm. Bull.* **34**, 2471.
Imperato, F. (1982). *Phytochemistry* **21**, 480.

Ishikura, N. and Sugahara, K. (1979). *Bot. Mag. Tokyo* **92**, 157.
Isman, M. B. and Duffey, S. S. (1982). *Ent. Exp. and Appl.* **31**, 370.
Jay, M., De Luca, V. and Ibrahim, R. K. (1983). *Z. Naturforsch.* **38c**, 413.
Jay, M., De Luca, V. and Ibrahim, R. K. (1985). *Eur. J. Biochem.* **153**, 321.
Jonsson, L. M. V., Aarsman, M. E. G., de Vlaming, P. and Schram, A. W. (1984). *Theor. Appl. Genet.* **68**, 459.
Jourdan, P. S. and Mansell, R. L. (1982). *Arch. Biochem. Biophys.* **213**, 434.
Kamsteeg, J., Brederode, J., Verschuren, P. M. and Nigtevecht, G. van (1981). *Z. Pflanzenphysiol.* **102**, 435.
Kay, Q. O. N., Daoud, H. S. and Stirton, C. H. (1981). *Bot. J. Linn. Soc.* **83**, 57.
Kerscher, F. and Franz, G. (1986). *Abstr. 34th Ann. Congress Medicinal Plant Research* (Ed. Reinhard, E.) p. 29. Thieme, Stuttgart.
Kleinenollenhorst, G., Behrens, H., Pegels, G., Srunk, N. and Wiermann, R. (1982). *Z. Naturforsch.* **37c**, 587.
Knogge, W. and Weissenbock, G. (1984). *Eur. J. Biochem.* **140**, 113.
Kochs, G. and Grisebach, H. (1986). *Eur. J. Biochem.* **155**, 311.
Kondo, T., Nakane, Y., Tamuro, H., Goto, T. and Eugster, C. H. (1985). *Tetrahedron Lett.* 5879.
Koster, J., Bussmann, R. and Barz, W. (1984). *Arch. Biochem. Biophys.* **234**, 513.
Kristiansen, K. N. (1986). *Carlsberg Res. Commun.* **51**, 51.
Lowry, J. B. (1976). *Phytochemistry* **15**, 1395.
de Luca, V. and Ibrahim, R. K. (1985). *Arch. Biochem. Biophys.* **238**, 596, 606.
Mabry, T. J. and Markham, K. R. (1975). In: *The Flavonoids* (Eds. Harborne, J. B., Mabry, T. J. and Mabry, H.) pp. 78–126. Chapman & Hall, London.
McClure, J. W. (1986). In: *Plant Flavonoids in Biology and Medicine* (Eds. Cody, V., Middleton, E. and Harborne, J. B.) pp. 77–85. Alan R. Liss, New York.
Markham, K. R. (1981). *Techniques of Flavonoid Identification.* Academic Press, London.
Markham, K. R. and Mabry, T. J. (1975). In: *The Flavonoids* (Eds. Harborne, J. B., Mabry, T. J. and Mabry, H.) pp. 45–77. Chapman & Hall, London.
Markham, K. R. and Porter, L. (1978). *Phytochemistry* **17**, 159.
Markham, K. R., Chari, V. M. and Mabry, T. J. (1982). In: *The Flavonoids: Advances in Research* (Eds. Harborne, J. B. and Mabry, T. J.) pp. 19–134. Chapman & Hall, London.
Matern, V., Heller, W. and Himmelspach, K. (1983). *Eur. J. Biochem.* **113**, 439.
Mentlein, R. and Vowinkel, E. (1984). *Liebig's Ann. Chem.* 1024.
Merlin, J. C., Statona, A. and Brouillard, R. (1985). *Phytochemistry* **24**, 1575.
Mohle, B., Heller, W. and Wellmann, E. (1985). *Phytochemistry* **24**, 465.
Mol, J. N. M., Robbins, M. P., Dixon, R. A. and Veltkamp, E. (1985). *Phytochemistry* **24**, 2267.
Momose, T., Abe, K. and Yoshitama, K. (1977). *Phytochemistry* **16**, 1321.
Nair, A. G. R. and Mohandoss, S. (1985). *Indian J. Chem.* **24B**, 323.
Neumann, D. (1983). *Biochem. Physiol. Pflanzen.* **178**, 405.
Oettmeier, W. and Heupel, A. (1972). *Z. Naturforsch.* **27c**, 177.
Pecket, R. C. and Small, C. J. (1980). *Phytochemistry* **19**, 2571.
Pramono, S., Gleye, J., Dehray, M. and Stanislas, E. (1981). *Plantes Medicinales et Phytother.* **15**, 224.
Preston, N. W. and Timberlake, C. F. (1981). *J. Chromatogr.* **214**, 222.

Proksch, M., Hess, D. and Weissenbock, G. (1981). *Z. Naturforsch.* **36c**, 222.
Reif, H. J., Niesbach, V., Deumling, B. and Saedler, H. (1985). *Mol. Gen. Genet.* **199**, 208.
Reimold, U., Kroger, M., Kreuzaler, F. and Hahlbrock, K. (1983). *EMBO J.* **2**, 1801.
Robinson, G. M. and Robinson, R. (1931). *Biochem. J.* **25**, 1687.
Saito, N. and Harborne, J. B. (1983). *Phytochemistry* **22**, 1735.
Saito, N. and Ueno, K. (1985). *Heterocycles* **23**, 2709.
Saito, N., Osawa, Y. and Hayashi, K. (1971). *Phytochemistry* **10**, 445.
Saito, N., Yokoi, M., Yanaji, M. and Honda, T. (1985). *Phytochemistry* **24**, 2125.
Schram, A. W., Jonsson, L. M. V. and de Vlaming, P. (1983). *Z. Naturforsch.* **38c**, 342.
Schulz, M., Strack, D., Weissenbock, G., Markham, K. R., Dellamonica, G. and Chopin, J. (1985). *Phytochemistry* **24**, 343.
Scogin, R. (1979). *Biochem. System. Ecol.* **7**, 35.
Small, C. J. and Pecket, R. C. (1982). *Planta* **152**, 97.
Sommer, H. and Saedler, H. (1986). *Mol. Gen. Genet.* **202**, 429.
Spribille, R. and Forkmann, G. (1982). *Planta* **155**, 176.
Spribille, R. and Forkmann, G. (1984). *Z. Naturforsch.* **39c**, 714.
Stafford, H. A. and Lester, H. H. (1982). *Plant Physiol.* **70**, 695.
Stafford, H. A. and Lester, H. H. (1985). *Plant Physiol.* **78**, 791.
Stewart, R. N., Asen, S., Massie, D. R. and Norris, K. R. (1977). *Biochem. System Ecol.* **7**, 281.
Stirton, J. and Harborne, J. B. (1980). *Biochem. System. Ecol.* **8**, 285.
Stotz, G. and Forkmann, G. (1981). *Z. Naturforsch.* **36c**, 737.
Stotz, G. and Forkmann, G. (1982). *Z. Naturforsch.* **37c**, 19.
Stotz, G., Spribille, R. and Forkmann, G. (1984). *J. Plant Physiol.* **116**, 173.
Stotz, G., de Vlaming, P., Wiering, H., Schram, A. W. and Forkmann, G. (1985). *Theor. Appl. Genet.* **70**, 300.
Strack, D., Busch, E., Wray, V., Grotjahn, L. and Klein, E. (1986). *Z. Naturforsch.* **41c**, 707.
Sukyok, G. and Raszlo-Benczik, A. (1985). *Phytochemistry* **24**, 1121.
Swain, T. (1976). In: *Chemistry and Biochemistry of Plant Pigments* (Ed. Goodwin, T. W.), 2nd edn., vol. 1, pp. 425–463, vol. 2, pp. 166–206. Academic Press, London.
Sumere, C. F. van and Van de Casteele, K. (1985). *Ann. Proc. Phytochem. Soc. Eur.* **25**, 17.
Takahashi, Y., Miyasaka, S., Tasaka, S. *et al.* (1982). *Tetrahedron Lett.* 5163.
Takeda, K. and Tominaga, S. (1983). *Bot. Mag. Tokyo* **96**, 359.
Takeda, K., Kariuda, M. and Itoi, H. (1985). *Phytochemistry* **24**, 2251.
Takeda, K., Harborne, J. B. and Self, R. (1986). *Phytochemistry* **25**, 1337.
Tamura, H., Kondo, T., Kato, Y. and Goto, T. (1983). *Tetrahedron Lett.* 5749.
Teusch, M. and Forkmann, G. (1987). *Phytochemistry* **26**, 2181.
Teusch, M., Forkmann, G. and Seyffert, W. (1986). *Phytochemistry* **25**, 86.
Timberlake, C. F. (1980). *Food Chem.* **5**, 69.
Timberlake, C. F. and Bridle, P. (1975). In: *The Flavonoids* (Eds. Harborne, J. B., Mabry, T. J. and Mabry, H.) pp. 214–266. Chapman & Hall, London.
Torck, M., Pinkas, M., Jay, M. and Favre-Bonvin, J. (1983). *Pharmacia* **38**, 783.
Varin, L., Barron, D. and Ibrahim, R. K. (1987). *Phytochemistry* **26**, 135.

Weeley, S. van, Bleumer, A., Spruyt, R. and Schram, A. W. (1983). *Planta* **159**, 226.
Weissenbock, G., Schnabl, H., Sachs, G., Elbert, C. and Heller, F. O. (1984). *Physiol. Plant.* **62**, 356.
Weitz, S. and Ikan, R. (1977). *Phytochemistry* **16**, 1108.
Wiermann, R., Wollenweber, E. and Rehse, C. (1981). *Z. Naturforsch.* **36c**, 204.
Williams, C. A., Harborne, J. B. and Mayo, S. J. (1981). *Phytochemistry* **20**, 217.
Wollenweber, E. (1982). *Suppl. Chromatographie*, pp. 50–54. GIT-Verlag, Darmstadt.
Wollenweber, E. and Dietz, V. H. (1981). *Phytochemistry* **20**, 869.
Wollenweber, E., Kohorst, G., Mann, K. and Bell, J. M. (1985). *J. Plant Physiol.* **117**, 423.
Wong, E. (1976). In: *Chemistry and Biochemistry of Plant Pigments* (Ed. Goodwin, T. W.), vol. 1, pp. 464–526. Academic Press, London.
Yasuda, H. and Shinoda, H. (1985). *Cytologia* **50**, 397.
Yazaki, Y. (1976). *Bot. Mag. Tokyo* **89**, 45.
Yoshitama, K. (1977). *Phytochemistry* **16**, 1857.
Yoshitama, K. (1984). *Bot. Mag. Tokyo* **97**, 429.
Yoshitama, K., Hisada, M. and Ishikura, N. (1984). *Bot. Mag. Tokyo* **97**, 31.

Index